Plant Factory

Plant Factory

An Indoor Vertical Farming System for Efficient Quality Food Production

Edited by

Toyoki Kozai
Japan Plant Factory Association
c/o Center for Environment, Health and Field Sciences
Chiba University, Kashiwa, Chiba, Japan

Genhua Niu
Texas AgriLife Research at El Paso
Texas A&M University
El Paso, TX, USA

Michiko Takagaki
Graduate School of Horticulture
Chiba University, Kashiwa, Chiba, Japan

AMSTERDAM • BOSTON • HEIDELBERG • LONDON
NEW YORK • OXFORD • PARIS • SAN DIEGO
SAN FRANCISCO • SINGAPORE • SYDNEY • TOKYO
Academic Press is an imprint of Elsevier

Academic Press is an imprint of Elsevier
125 London Wall, London, EC2Y 5AS, UK
525 B Street, Suite 1800, San Diego, CA 92101–4495, USA
225 Wyman Street, Waltham, MA 02451, USA
The Boulevard, Langford Lane, Kidlington, Oxford OX5 1GB, UK

Notices

Knowledge and best practice in this field are constantly changing. As new research and experience broaden our understanding, changes in research methods, professional practices, or medical treatment may become necessary.

Practitioners and researchers must always rely on their own experience and knowledge in evaluating and using any information, methods, compounds, or experiments described herein. In using such information or methods they should be mindful of their own safety and the safety of others, including parties for whom they have a professional responsibility.

To the fullest extent of the law, neither the publisher nor the authors, contributors, or editors, assume any liability for any injury and/or damage to persons or property as a matter of products liability, negligence or otherwise, or from any use or operation of any methods, products, instructions, or ideas contained in the material herein.

Library of Congress Cataloging-in-Publication Data
A catalog record for this book is available from the Library of Congress

British Library Cataloguing in Publication Data
A catalogue record for this book is available from the British Library

ISBN: 978-0-12-801775-3

For information on all Academic Press publications visit our
website at http://store.elsevier.com/

Publisher: Nikki Levy
Acquisition Editor: Nancy Maragioglio
Editorial Project Manager: Billie Jean Fernandez
Production Project Manager: Caroline Johnson
Designer: Victoria Pearson

Typeset by SPi Global, India
Printed and bound in the United States of America
Transferred to Digital Printing, 2015

Working together
to grow libraries in
developing countries

www.elsevier.com • www.bookaid.org

Contents

Contributors xxi

Preface xxv

PART 1 OVERVIEW AND CONCEPT OF CLOSED PLANT PRODUCTION SYSTEM (CPPS)

CHAPTER 1 Introduction 3

Toyoki Kozai, Genhua Niu

Introduction 3

References 5

CHAPTER 2 Role of the Plant Factory with Artificial Lighting (PFAL) in Urban Areas 7

Toyoki Kozai, Genhua Niu

Introduction 7

Interrelated Global Issues to be Solved Concurrently 7

Resource Inflow and Waste Outflow in Urban Areas............................ 9

Energy and Material Balance in Urban Ecosystems 11

    Photoautotrophs (Plants) and Heterotrophs (Animals and Microorganisms)...... 11

    Waste Produced in Urban Areas as an Essential Resource for Growing Plants.. 12

    Plant Production Systems Integrated with Other Biological Systems 13

    Role of Organic Fertilizers and Microorganisms in the Soil 15

    Stability and Controllability of the Environment in Plant Production Systems 16

    Key Indices for Sustainable Food Production 17

    What is "PFAL"?.. 17

    Plants Suited and Unsuited to PFALs.. 19

Growing Social Needs and Interest in PFALs...................................... 19

Criticisms of PFALs and Responses to Them...................................... 21

    Initial Cost is Too High.. 22

    Production Cost is Too High... 22

    Electricity Cost is Too High, Whereas Solar Light is Free 23

    Labor Cost is Too High.. 25

    PFAL-Grown Vegetables Are Neither Tasty Nor Nutritious............... 25

    Most PFALs Are Not Making a Profit 26

    Land Price is Too High 27

Water Consumption for Irrigation is Too High 27

PFALs Can Only Produce Leafy Greens—Minor
Vegetables—Economically 27

Towards a Sustainable PFAL 29

Requirements for a Sustainable PFAL 29

Factors Affecting the Sustainability of PFALs 30

Similarities Between the Earth, Space Farms, Autonomous Cities,
and PFALs 31

Conclusion 32

References 32

CHAPTER 3   PFAL Business and R&D in the World: Current Status and Perspectives

CHAPTER 3   PFAL Business and R&D in the World: Current Status and Perspectives **35**

Introduction 35

Japan 35

Toyoki Kozai

Brief History and Current Status of PFAL Business 35

Research and Development 37

Public Service 39

Taiwan 39

Wei Fang

Status of PFAL in Taiwan 39

PFAL Expo in Taiwan 41

PFAL Research 42

Business Models of PFALs in Taiwan 47

Conclusions 50

Korea 50

Changhoo Chun

PFAL Industry, a Commitment to the Future 50

Research and Technical Development 51

Private Companies and Farms in the PFAL Business 52

Achievements and Challenges 53

China 53

Qichang Yang, Yuxin Tong, Ruifeng Cheng

Development of PFAL in China 53

Case Study of Typical PFALs 54

Research Projects on Plant Factories in China 58

North America 59

Chieri Kubota

History 59

Contribution of Space Science 60
Current Status and Future Prospective 60
Europe (England, The Netherlands, and Others) 61
Chungui Lu

Background 61
Present Status of Plant Factories in the EU 62
Outlook for Plant Factories in the UK 64
References 66

CHAPTER 4   Plant Factory as a Resource-Efficient Closed Plant Production System 69
Toyoki Kozai, Genhua Niu

Introduction 70
Definition and Principal Components of PFAL 71
Definition of Resource Use Efficiency 72
Water Use Efficiency 73
CO_2 Use Efficiency 75
Light Energy Use Efficiency of Lamps and Plant Community 77
Electrical Energy Use Efficiency of Lighting 78
Electrical Energy Use Efficiency of Heat Pumps for Cooling 79
Inorganic Fertilizer Use Efficiency 79
Representative Values of Resource Use Efficiency 79
Electricity Consumption and Cost 80
Improving Light Energy Use Efficiency 81
Interplant Lighting and Upward Lighting 81
Improving the Ratio of Light Energy Received by Leaves 83
Using LEDs 83
Controlling Environmental Factors Other Than Light 83
Controlling Air Current Speed 83
Increasing the Salable Portion of Plants 84
Increasing Annual Production Capacity and Sales Volume per
    Unit Land Area 84
Estimation of Rates of Photosynthesis, Transpiration, and Water
    and Nutrient Uptake 85
Net Photosynthetic Rate 85
Transpiration Rate 86
Water Uptake Rate by Plants 86
Ion Uptake Rate by Plants 87
Application 87
Coefficient of Performance of Heat Pump 87
References 88

CHAPTER 5   **Micro- and Mini-PFALs for Improving the Quality of Life in Urban Areas** **91**
Michiko Takagaki, Hiromichi Hara, Toyoki Kozai

Introduction 91
Characteristics and Types of m-PFAL............................... 91
Various Applications of m-PFALs 92
　　Homes 92
　　Restaurants and Shopping Centers.............................. 94
　　Schools and Community Centers 95
　　Hospitals 97
　　Offices.. 98
　　Small Shops and Rental m-PFALs............................... 98
Design Concept of m-PFALs 98
m-PFALs Connected to the Internet 99
Advanced Uses of m-PFALs Connecting with a Virtual
　　m-PFAL 102
　　Visualizing the Effects of Energy and Material Balance on
　　　　Plant Growth.. 102
　　Maximizing Productivity and Benefits Using Minimum
　　　　Resources 102
　　Learning the Basics of an Ecosystem 102
　　Challenges.. 102
m-PFAL Connected with Other Biosystems as a Model
　　Ecosystem 103
Light Source and Lighting System Design........................ 104
References 104

CHAPTER 6   **Rooftop Plant Production Systems in Urban Areas** **105**
Nadia Sabeh

Introduction 105
Rooftop Plant Production.. 105
　　Raised-Bed Production 106
　　Continuous Row Farming.. 106
　　Hydroponic Greenhouse Growing................................. 107
Building Integration 107
　　Stormwater Management... 107
　　Energy Use Reductions 109
References 110

PART 2 BASICS OF PHYSICS AND PHYSIOLOGY—ENVIRONMENTS AND THEIR EFFECTS

CHAPTER 7 **Light** **115**
Introduction 115
Physical Properties of Light and its Measurement 115
Genhua Niu
    Physical Properties 115
    Light Measurement 117
Light Sources 118
Kazuhiro Fujiwara
    Classification of Light Sources 119
    Light-Emitting Diodes 119
    Fluorescent Lamps 126
References 127

CHAPTER 8 **Physical Environmental Factors and Their Properties** **129**
Genhua Niu, Toyoki Kozai, Nadia Sabeh
Introduction 129
Temperature, Energy, and Heat 129
    Energy Balance 129
    Radiation 130
    Heat Conduction and Convection 130
    Latent Heat—Transpiration 131
    Measurement of Temperature 131
Water Vapor 132
    Humidity 132
    Vapor Pressure Deficit 132
    Measurement of Humidity 133
Moist Air Properties 133
    Composition of Air 133
    Psychrometric Chart 134
CO_2 Concentration 137
    Nature 137
    Dynamic Changes of CO_2 Concentration in PFAL 137
    Measurement of CO_2 Concentration 137
Air Current Speed 138
    Nature and Definition 138

Measurement.. 138
Number of Air Exchanges per Hour... 139
Nature and Definition... 139
Measurement of Air Exchange... 139
References 139

CHAPTER 9   Photosynthesis and respiration **141**
Wataru Yamori

Introduction 141
Photosynthesis 141
Light Absorption by Photosynthetic Pigments 141
Electron Transport and Bioenergetics.. 143
Carbon Fixation and Metabolism.. 143
C_3, C_4 and CAM Photosynthesis 144
Respiration.. 144
Photorespiration... 146
LAI and Light Penetration 147
Single Leaf and Canopy 148
References 148

**CHAPTER 10   Growth, Development, Transpiration and Translocation as
Affected by Abiotic Environmental Factors**.......................... **151**
Chieri Kubota

Introduction 151
Shoot and Root Growth... 151
Growth: Definition 151
Root Growth 152
Environmental Factors Affecting Plant Growth and Development 153
Temperature and Plant Growth and Development 153
Daily Light Integral 154
Light Quality... 155
Humidity (VPD) 156
CO_2 Concentration.. 156
Air Current Speed.. 157
Nutrient and Root Zone.. 158
Development (Photoperiodism and Temperature Affecting
Flower Development).. 159
Transpiration.. 160
Translocation 162
References 162

CHAPTER 11 Nutrition and Nutrient Uptake in Soilless Culture Systems **165**
Satoru Tsukagoshi, Yutaka Shinohara

Introduction 165
Essential Elements... 165
Beneficial Elements... 169
Nutrient Uptake and Movement.. 169
Nutrient Solution 170
Solution pH and Nutrient Uptake 171
Nitrogen Form 171
New Concept: Quantitative Management............................... 171
References 172

CHAPTER 12 Tipburn **173**
Toru Maruo, Masahumi Johkan

Introduction 173
Cause of Tipburn... 173
    Inhibition of Ca^{2+} Absorption in Root.......................... 173
    Inhibition of Ca^{2+} Transfer from Root to Shoot............... 174
    Competition for Ca^{2+} Distribution 175
Countermeasure... 175
References 176

CHAPTER 13 Functional Components in Leafy Vegetables **177**
Keiko Ohashi-Kaneko

Introduction 177
Low-Potassium Vegetables 177
Low-Nitrate Vegetables 178
    Restriction of Feeding Nitrate Fertilizer to Plants............. 178
    Reduction in Accumulated Nitrate by Assimilation
        of Nitrate.. 179
Improving the Quality of Leafy Vegetables by Controlling
    Light Quality 179
    Leafy Vegetables 180
    Herbs.. 181
Conclusion... 183
References 183

CHAPTER 14 Medicinal Components **187**
Sma Zobayed

Introduction 187
Growing Medicinal Plants Under Controlled Environments: Medicinal
    Components and Environmental Factors 188
    CO_2 Concentration and Photosynthetic Rates 188
    Temperature Stress 188
    Water Stress 189
    Spectral Quality and UV Radiation 189
Conclusion 191
References 191

**CHAPTER 15 Production of Pharmaceuticals in a Specially Designed
Plant Factory** **193**
Eiji Goto

Introduction 193
Candidate Crops for PMPs 194
Construction of GM Plant Factories 195
Optimization of Environment Conditions for Plant Growth 197
    Strawberry 197
    Tomato 198
    Rice 198
Concluding Remarks 200
References 200

PART 3 SYSTEM DESIGN, CONSTRUCTION, CULTIVATION AND MANAGEMENT

**CHAPTER 16 Plant Production Process, Floor Plan, and Layout of
PFAL** **203**
Toyoki Kozai

Introduction 203
Motion Economy and PDCA Cycle 203
    Principles of Motion Economy 203
    PDCA Cycle 204
Plant Production Process 204
Layout 205
    Floor Plan 205
    Operation Room 207
    Culture Room 208

Sanitation Control.. 211
  Biological Cleanness 211
  ISO22000 and HACCP for Food Safety.. 212
References 212

CHAPTER 17 Hydroponic Systems... 213
Jung Eek Son, Hak Jin Kim, Tae In Ahn
Introduction 213
Hydroponic System 213
Sensors and Controllers.. 214
Nutrient Management Systems 214
  Open and Closed Hydroponic Systems... 214
  Changes in Nutrient Balance Under EC-Based Hydroponic Systems 216
Ion-Specific Nutrient Management... 217
Sterilization System.. 220
References 221

CHAPTER 18 Seeding, Seedling Production and Transplanting 223
Osamu Nunomura, Toyoki Kozai, Kimiko Shinozaki, Takahiro Oshio
Introduction 223
Preparation.. 223
Seeding 226
Seedling Production and Transplanting... 229

CHAPTER 19 Transplant Production in Closed Systems 237
Introduction 237
Main Components and Their Functions... 237
Toyoki Kozai

  Main Components... 238
  Light Source, Air Conditioners, and Small Fans.............................. 239
  Electricity Costs... 240
  Nutrient Solution Supply... 242
Ecophysiology of Transplant Production.. 242
Toshio Shibuya

  Introduction.. 243
  Effects of Light Quality on Photosynthetic Performance
    in Transplants 243
  Effects of the Physical Environment on Biotic Stress Resistance
    in Transplants 245

Effects of Plant–Plant Interactions on Gas Exchange Within
    Transplant Canopy.. 246
Effects of Light Quality on Light Competition Between Neighboring
    Plants and Consequent Equality of Plant Growth 249
Conclusions... 250
Photosynthetic Characteristics of Vegetable and Medicinal Transplants
    as Affected by Light Environment 250
Dongxian He
    Introduction.. 251
    Influence of Light Environment on Vegetable Transplant
        Production 251
    Photosynthetic Characteristics of Medicinal
        D. officinale 254
    Blueberry 257
Sma Zobayed
Propagation and Production of Strawberry Transplants........................... 260
Changhoo Chun
    Vegetative Propagation.. 260
    Licensing and Certification 261
    Plug Transplants 261
    Transplant Production in a PFAL 262
    Application of S-PFAL in Korea 265
    References 266

CHAPTER 20 Photoautotrophic Micropropagation 271
Quynh Thi Nguyen, Yulan Xiao, Toyoki Kozai
    Introduction 271
    Development of PAM 271
    Advantages and Disadvantages of PAM for Growth Enhancement of
        In Vitro Plants... 272
    Natural Ventilation System Using Different Types of Small Culture
        Vessels.. 272
    Forced Ventilation System for Large Culture Vessels.......................... 276
    Potential for Secondary Metabolite Production of In Vitro Medicinal Plants
        by Using PAM... 278
    Scaling Up a PAM System to an Aseptic Culture Room—A Closed Plant
        Production System 279
    Conclusion 280
    References 280

CHAPTER 21 Biological Factor Management.. 285
Introduction 285
Controlling Algae... 285
Chieri Kubota

    Hydrogen Peroxide... 285
    Ozonated Water.. 286
    Chlorine.. 286
    Substrates... 286
Microorganism Management 287
Miho Takashima

    Microbiological Testing 287
    Environmental Testing—Airborne Microorganisms 288
    Quality Testing—Testing for Bacteria and Fungi 290
    Examples of Reports of Microbiological Testing in PFALs.................... 290
    Concluding Remarks 293
References 293

CHAPTER 22 Design and Management of PFAL.. 295
Toyoki Kozai, Shunsuke Sakaguchi, Takuji Akiyama, Kosuke Yamada,
Kazutaka Ohshima

Introduction 295
Structure and Function of the PFAL-D&M System...................................... 295
PFAL-D (Design) Subsystem... 297
    Lighting System (LS) 298
PFAL-M Subsystem 298
    Structure of Software... 298
    Logical Structure of Equations... 300
Design of Lighting System 300
    PPFD Distribution 300
    Scheduling the Lighting Cycles to Minimize Electricity Charge 302
Electricity Consumption and its Reduction 303
    Daily Changes in Electricity Consumption... 303
    COP as Affected by the Temperature Difference Between Inside
        and Outside 304
    COP as Affected by the Actual Cooling Load 305
    Monthly Changes in Electricity Consumption.. 306
    Visualization of Power Consumption by Components on the
        Display Screen.. 306
    Rates of Net Photosynthesis, Dark Respiration, and Water
        Uptake by Plants.. 308

Three-Dimensional Distribution of Air Temperature 308
Plant Growth Measurement, Analysis, and Control 309
   Determination of Parameter Values for Plant Growth Curve 309
   Determination of Dates for Transplanting 310
   Determination of the Number of Culture Panels for Different
      Growth Stages 311
Conclusions 311
References 312

**CHAPTER 23 Automated Technology in Plant Factories with
Artificial Lighting** **313**
Hiroshi Shimizu, Kazuhiro Fukuda, Yoshikazu Nishida, Toichi Ogura
Introduction 313
Seeding Device 314
Seedling Selection Robot System 315
Shuttle-Type Transfer Robot 317
Cultivation Panel Washer 318
References 319

CHAPTER 24 Life Cycle Assessment **321**
Yasunori Kikuchi
Standard of LCA 321
   Introduction 321
   Goal and Scope Definition 322
   Life Cycle Inventory Analysis 323
   Life Cycle Impact Assessment 324
   Interpretation 325
Remarks for the Assessment of PFAL 325
   Inventory Data Collection/Impact Assessment 326
   Functional Unit 327
   Interpretation 327
Summary and Outlook 327
References 328

**CHAPTER 25 Education, Training, and Intensive Business Forums
on Plant Factories** **331**
Toshitaka Yamaguchi, Michiko Takagaki, Satoru Tsukagoshi
Introduction 331
Plant Factories in the Center 332

Plant Environment Designing Program 333
Intensive Business Forums on Plant Factories 337
   2010 Business Forums.. 338
   2011 Business Forums.. 340
   2012 Business Forums.. 340
   2013 Business Forums.. 340
   2014 Business Forums.. 342
JPFA's Business Workshops 343
   2009 Business Workshop 344
   2010 Business Workshop 345
   2011 Business Workshop 346
   2012 Business Workshop 346
   2013 Business Workshop 347
   2014 Business Workshop 347
References 348

PART 4 PFALs IN OPERATION AND ITS PERSPECTIVES

CHAPTER 26 Selected Commercial PFALs in Japan and Taiwan.......................... 351
Introduction 351
Representative PFALs in Taiwan 351
Wei Fang
   Cal-Com Bio Corp. of New Kinpo Group.. 352
   Glonacal Green Technology Corp. 352
   Ting-Mao Bio-Technology Corp. 352
   Lee-Pin: A PFAL Building Inside a Greenhouse 353
   Yasai-Lab Corp.: The Largest PF in Taiwan.................................... 353
Spread Co., Ltd. 355
Shinji Inada
   Vision and Mission.. 355
   History and Location 355
   Business Model... 357
   Main Crops 358
   Outline of PFAL 359
   Costs by Component and Sales Price 359
   Markets 360
   Future Plans 361
   Website URL 362
Mirai Co., Ltd. 362

Shigeharu Shimamura

Vision and Mission 362
History and Location 362
Business Model 363
Outline of PFAL 363
Costs by Component and Sales Price 364
Markets 364
Future Plans 364
Website URL 364
Japan Dome House Co., Ltd. 364
Katsuyuki Kitagawa

Background and Geographical Location 364
Business Model 365
Outline of PFAL 365
Sales Price of PFAL 367
Markets 368
Future Plans 368
Website URL 368
Internationally Local & Company (InLoCo) 368
Yasuhito Sasaki

No Agricultural Background 368
The Basis of Profitability 369
How Did InLoCo Achieve Profitability? 369
Business Model 372
Future Plans 374
Website URL 374
Sci Tech Farm Co., Ltd. 375
Kozo Hagiya, Hiroyuki Watanabe

Vision and Mission 375
History and Location 375
Business Model 376
Main Crops 377
Outline of PFAL 377
Cost Breakdown and Sales Price 380
Markets 381
Future Plans 381
Website URL 382
Berg Earth Co., Ltd. 382
Kazuhiko Yamaguchi & Chan-suk Yang

Vision and Mission 382

History and Location 382
Business Model... 383
Main Crops 384
Outline of PFAL.. 384
Costs by Components and Sales Price.. 385
Markets 385
Future Plans 386
Website URL 386

CHAPTER 27 Challenges for the Next-Generation PFAL.............................. **387**
Toyoki Kozai, Genhua Niu

Introduction 387
Lighting System 387
Upward Lighting.. 387
Using Green LEDs 388
Layouts of LEDs.. 389
Breeding and Seed Propagation.. 389
Vegetables Suited to PFAL.. 389
Seed Propagation and Breeding Using PFAL.. 390
Medicinal Plants 390
Cultivation 391
Culture System with Restricted Root Mass 391
Ever-Flowering Berry Production in PFALs 391
PFAL with Solar Cells 391
References 393

CHAPTER 28 Conclusions: Resource-Saving and Resource-Consuming
Characteristics of PFALs **395**
Toyoki Kozai, Genhua Niu

Roles of PFALs in Urban Areas 395
Benefits of Producing Fresh Vegetables Using PFALs in Urban Areas 396
Resource-Saving Characteristics of PFALs.. 396
Possible Reductions in Electricity Consumption and Initial Investment.............. 397
Electricity Consumption.. 397
Initial Resource Investment.. 397
Increasing the Productivity and Quality 397
Dealing with Power Cuts... 398
Challenges 398

Index 401

Contributors

Tae In Ahn
Department of Plant Science, Seoul National University, Seoul, South Korea

Takuji Akiyama
PlantX Corp., Kashiwa, Japan

Ruifeng Cheng
Institute of Environment and Sustainable Development in Agriculture, Chinese Academy of Agricultural Sciences, Beijing, China

Changhoo Chun
Department of Plant Science, Seoul National University, Seoul, South Korea

Wei Fang
Department of Bio-Industrial Mechatronics Engineering, National Taiwan University, Taipei, Taiwan

Kazuhiro Fujiwara
Graduate School of Agricultural and Life Sciences, The University of Tokyo, Tokyo, Japan

Kazuhiro Fukuda
Osaka Prefecture University, Sakai, Osaka, Japan

Eiji Goto
Graduate School of Horticulture, Chiba University, Matsudo, Chiba, Japan

Kozo Hagiya
Sci Tech Farm Co., Ltd., Tokyo, Japan

Hiromichi Hara
Graduate School of Engineering, Chiba University, Chiba, Yayoi, Japan

Dongxian He
College of Water Resources and Civil Engineering, China Agricultural University, Beijing, China

Shinji Inada
Spread Co., Ltd., Kyoto, Japan

Masahumi Johkan
Graduate School of Horticulture, Chiba University, Matsudo, Chiba, Japan

Yasunori Kikuchi
Presidential Endowed Chair for "Platinum Society", The University of Tokyo, Tokyo, Japan

Hak Jin Kim
Department of Biosystems Engineering, Seoul National University, Seoul, South Korea

Katsuyuki Kitagawa
Japan Dome House Co., Ltd., Kaga-shi, Ishikawa, Japan

Toyoki Kozai
Japan Plant Factory Association, c/o Center for Environment, Health and Field Sciences, Chiba University, Kashiwa, Chiba, Japan

Chieri Kubota
School of Plant Sciences, The University of Arizona, Tucson, Arizona, USA

Chungui Lu
Head of Centre for Urban Agriculture, Centre for Urban Agriculture, University of Nottingham, Leicestershire, United Kingdom

Toru Maruo
Graduate School of Horticulture, Chiba University, Matsudo, Chiba, Japan

Quynh Thi Nguyen
Institute of Tropical Biology, Vietnam Academy of Science and Technology, Hochiminh City, Vietnam

Yoshikazu Nishida
Itoh Denki Co., Ltd., Kasai, Hyogo, Japan

Genhua Niu
Texas AgriLife Research at El Paso, Texas A&M University, El Paso, Texas, USA

Osamu Nunomura
Japan Plant Factory Association, c/o Center for Environment, Health and Field Sciences, Chiba University, Kashiwa, Chiba, Japan

Toichi Ogura
Osaka Prefecture University, Sakai, Osaka, Japan

Keiko Ohashi-Kaneko
Research Institute, Tamagawa University, Tokyo, Japan

Kazutaka Ohshima
PlantX Corp., Kashiwa, Japan

Takahiro Oshio
Japan Plant Factory Association, c/o Center for Environment, Health and Field Sciences, Chiba University, Kashiwa, Chiba, Japan

Yasuhito Sasaki
Internationally Local & Company, Itoman-shi, Okinawa, Japan

Nadia Sabeh
Building Performance Engineer, Guttmann & Blaevoet Consulting Engineers, Sacramento, California, USA

Shunsuke Sakaguchi
PlantX Corp., Kashiwa, Japan

Toshio Shibuya
Graduate School of Life and Environmental Sciences, Osaka Prefecture University, Osaka, Japan

Kimiko Shinozaki
Japan Plant Factory Association, c/o Center for Environment, Health and Field Sciences, Chiba University, Kashiwa, Chiba, Japan

Shigeharu Shimamura
Mirai Co., Ltd., Tokyo, Japan

Hiroshi Shimizu
Graduate School of Agriculture, Kyoto University, Kyoto, Japan

Yutaka Shinohara
Japan Greenhouse Horticulture Association, Tokyo, Japan

Jung Eek Son
Department of Plant Science, Seoul National University, Seoul, South Korea

Michiko Takagaki
Center for Environment, Health and Field Sciences, Chiba University, Kashiwa, Japan

Miho Takashima
Japan Research Promotion Society for Cardiovascular Diseases, Sakakibara Heart Institute, Tokyo, Japan

Yuxin Tong
Institute of Environment and Sustainable Development in Agriculture, Chinese Academy of Agricultural Sciences, Beijing, China

Satoru Tsukagoshi
Center for Environment, Health and Field Sciences, Chiba University, Kashiwa, Japan

Hiroyuki Watanabe
Tamagawa University, Tokyo, Machida-shi, Japan

Yulan Xiao
Yangtze Delta Region Institute of Tsinghua University, Jiaxing, China

Chan-suk Yang
Berg Earth Co., Ltd., Uwajima, Ehime, Japan

Kosuke Yamada
PlantX Corp., Kashiwa, Japan

Kazuhiko Yamaguchi
Berg Earth Co., Ltd., Uwajima, Ehim, Japan

Toshitaka Yamaguchi
Center for Environment, Health and Field Sciences, Chiba University, Kashiwa, Japan

Wataru Yamori
Center for Environment, Health and Field Sciences, Chiba University, Kashiwa, and PRESTO, Japan Science and Technology Agency (JST), Kawaguchi, Japan

Qichang Yang
Institute of Environment and Sustainable Development in Agriculture, Chinese Academy of Agricultural Sciences, Beijing, China

Sma Zobayed
SHIRFA Biotech, Pitt Meadows, and JRT Research and Development, Aldergrove, British Columbia, Canada

Preface

This is perhaps the first book in English on plant factories with artificial (or electric) lighting (PFAL, pronounced "P-FAL"). PFALs, also known as indoor plant production systems, are expected to play a vital role in urban development, urban agriculture, and vertical farming in the forthcoming decades.

The tremendous potential of PFAL is attracting much interest among researchers, business people, policymakers, educators, students, community developers, architects, designers, and entrepreneurs. This book explains PFAL's principle, concept, design, operation, social roles, pros and cons, costs and benefits, and the possibilities and challenges for solving local as well as global agricultural, environmental, and social issues.

In the next few decades, urban agriculture and vertical farming will not be just a modified form of existing agriculture and farming practices. This new form of indoor agriculture system will provide entirely new social services in urban areas which are facing ecological, economic, and social constraints. PFALs will be used as a key component of community centers, restaurants, office buildings, hotels, schools, manufacturing factories, supermarkets, convenience stores, resource recycling/reuse systems, and housing in both small and large cities. PFALs provide a means for neighbors, farmers, and growers to share the joys of nature, arts, and science and technology by providing a common platform for "growing life" and "growing with life, together with plants, animals, and microorganisms."

Nevertheless, there are various criticisms of PFALs that must be considered carefully. These include: (1) vegetable production using PFALs could be resource consuming and environmentally unfriendly, (2) PFAL-grown vegetables may not be nutritious or tasty, (3) PFAL-grown vegetables are not organic and will not be accepted by health-conscious citizens, (4) PFAL businesses cannot make a profit because of the high initial and running costs, and (5) PFAL businesses will compete with conventional agriculture and farming, and so may negatively affect conventional agriculture.

Readers of this book will find answers to such criticisms as well as additional challenges to be solved. It is important to remember that recent PFAL R&D and business are merely the start of PFAL technology and new urban agriculture based on the concept and methodology of PFAL. Although there are many issues to be solved, most of them can surely be solved within a decade or so. This book explains the vision, concept, methodology, basic science and technology, current status, perspectives, and challenges of PFAL.

There is no doubt that more and more people are becoming interested in urban agriculture, vertical farming, city farming, and PFALs, as evidenced by the many exhibitions and conferences held in recent years. For example, between September 2012 and May 2015, the following international academic meetings related to vertical farming, urban agriculture, and PFAL were held; such activities will drive PFAL technology and its applications:

(1) Workshop on challenges in vertical farming, September 26, 2012, The University of Maryland, USA (http://www.challengesinverticalfarming.org/)
(2) International conference on vertical farming and urban agriculture 2014, September 9–10, 2014, The University of Nottingham, UK (http://vfua.org/)
(3) International plant factory conference, Kyoto/Osaka, Japan, September 9–11, 2014 (http://www.shita.jp/ICPF2014/)

(4) LED symposium, February 19–20, 2015, The University of Arizona, Tucson, AZ, USA http://leds.hrt.msu.edu/meeting/

(5) Global forum for innovations in agriculture: Edible cities, March 9–11, 2015, Abu Dhabi (http://www.innovationsinagriculture.com/)

(6) The 2015 high-level international forum on protected horticulture (HIFPH 2015) in Shouguang, China (http://www.hifph2015.com/en-index.html)

(7) International congress on controlled environment agriculture 2015 (ICCEA 2015), May 20–22, 2015, Panama City, Panama (http://www.icceapanama.org/)

ACKNOWLEDGMENTS

We would like to thank the following companies and organizations for giving us permission to use their photos in this book: Hokuetsu Industries Co., Ltd., Minoru Industrial Co., Ltd., Mirai Co., Ltd., Mitsubishi Plastics Agri Dream Co., Ltd., Nansei Kobashi Denki Co., Ltd., Nishimatsu Construction Co., Ltd., Paignton Zoo Environmental Park and PlantLab, Sashinami Seisakujo Corporation, and UING Corporation.

We also thank the following publishers for allowing us to republish their figures: International Society for Horticultural Science, Japanese Society of Agricultural, Biological, and Environmental Engineers and Scientists, Johokiko Co., Ltd., and Springer Science+Business Media.

Special thanks are due to Ms. Mami Yoshida for her tireless editorial assistance and dedication. We also appreciate the guidance given by Nancy Maragioglio and Carrie Bolger of Elsevier.

CHAPTER 5, KOZAI

We sincerely thank Mr. J. Kawai of Mitsui Fudosan Co., Ltd., Mr. M. Miyaki of Panasonic Corporation, Mr. S. Shimamura of Mirai Co., Ltd., Dr. M. Nagayama of Sakakibara Memorial Hospital, Mr. T. Maeda of MTI Ltd., Mr. Y. Kojima of Sankyo Frontier Co., Ltd., and Natori City, Miyagi Prefecture for their collaboration.

CHAPTER 19.2, SHIBUYA

We sincerely thank the Japan Society for the Promotion of Science for its financial support through Grants-in-Aid for Young Scientists (A), 15688007; for Challenging Exploratory Research, 20658059 and 25660199; and for Scientific Research (B) (General), 24380140.

CHAPTER 19.3, HE

This work was supported by earmarked funds for Modern Agro-industry Technology Research System (CARS-25), The National Industry of Public Projects (2013018), and National Natural Science Foundation of China (Grant No. 31372089).

Toyoki Kozai, Genhua Niu, and Michiko Takagaki
March 11, 2015

OVERVIEW AND CONCEPT OF CLOSED PLANT PRODUCTION SYSTEM (CPPS)

INTRODUCTION

Toyoki Kozai[1], Genhua Niu[2]

Japan Plant Factory Association, c/o Center for Environment, Health and Field Sciences, Chiba University, Kashiwa, Chiba, Japan[1] Texas AgriLife Research at El Paso, Texas A&M University, El Paso, Texas, USA[2]

INTRODUCTION

Crop production is increasingly threatened by unusual weather, water shortages, and insufficient available land. The world's population is expected to grow from 7 billion in 2011 to 9.3 billion in 2050, and the urban population from 3.6 billion to 6.3 billion, a 72% increase. Due to limited natural resources, 90% of the growth in global crop production is expected from higher yields and increased cropping intensity, with the remaining 10% from expansion of productive land (FAO, 2009). Almost all of the land expansion in developing countries will take place in sub-Saharan Africa and Latin America. The availability of freshwater resources follows a similar trend, i.e., globally more than sufficient but unevenly distributed. In order to feed the world, protect the environment, improve health, and achieve economic growth, a new form of agricultural cultivation is required: indoor vertical farming using a plant factory system with artificial lighting for efficient production of food crops.

The term "plant factory with artificial lighting (PFAL)" refers to a plant production facility with a thermally insulated and nearly airtight warehouse-like structure (Kozai, 2013). Multiple culture shelves with electric lamps on each shelf are vertically stacked inside. Other necessary equipment and devices for a PFAL are air conditioners, air circulation fans, CO_2 and nutrient solution supply units, and an environmental control unit. Stacking more culture shelves vertically increases the efficiency of land use. Fluorescent lamps (FLs) have been mainly used in PFALs due to their compact size, but light-emitting diode (LED) lamps are now attracting great attention in industry and among researchers. LEDs are increasingly being used in recently built PFALs owing to their compact size, low lamp surface temperature, high light use efficiency, and broad light spectra. More information on the light sources and advantages of LEDs is given in Chapter 7.

PFALs are not a replacement for conventional greenhouses or open-field production. Rather, the rapid development of PFALs has created new markets and business opportunities. PFALs are being used in Japan and other Asian countries for commercial production of leafy greens, herbs, and transplants. Indoor vertical farms, which is another term used in North America for concepts similar to PFALs, are also being built in the United States and Canada.

When growing plants in an open field, yield and quality are subject to weather conditions, and so a stable and reliable supply of plant-derived food is always in danger. Greenhouse production is not energy efficient because incident light is not regulated. Solar light intensity is often too low at dawn, sunset, and night, on cloudy and rainy days, and throughout the winter season, while it is too high

Plant Factory. http://dx.doi.org/10.1016/B978-0-12-801775-3.00001-9

around noon on sunny days. The temperature and relative humidity inside a greenhouse are considerably affected by solar light intensity, and thus it is difficult to optimize the environment. In order to lower the temperature, greenhouses are often ventilated, but this allows insects and diseases inside the greenhouse. In addition, CO_2 in a greenhouse with ventilators open cannot be kept higher than outside. Furthermore, light quality and lighting direction are not controllable. Excessive agrochemicals are often used in greenhouse and open-field production and fossil fuels are needed for heating and cooling of greenhouses and for transportation of produce from production site to consumers. Fossil fuels are a nonrenewable energy and excessive use results not only in depletion of resources but also in excessive emission of environmental pollutants including CO_2.

On the other hand, the PFAL is an indoor, advanced, and intensive form of hydroponic production system where the growing environment is optimally controlled. PFAL is one form of "closed plant production system" (CPPS), where all inputs supplied to the PFAL are fixed by plants with minimum emission to the outside environment. If designed and managed properly, the PFAL has the following potential advantages over the conventional production system:

a. It can be built anywhere because neither solar light nor soil is needed;
b. The growing environment is not affected by the outside climate and soil fertility;
c. Production can be year-round and productivity is over 100 times that of field production;
d. Produce quality such as concentrations of phytonutrients can be enhanced through manipulation of the growing environment, especially light quality;
e. Produce is pesticide-free and need not be washed before eating;
f. Produce has a longer shelf life because the bacterial load is generally less than 300 CFU g^{-1}, which is 1/100 to 1/1000 that of field-grown produce;
g. Energy for transportation can be reduced by building PFALs near urban areas; and
h. High resource use efficiency (water, CO_2, fertilizer, etc.) can be achieved with minimum emission of pollutants to the outside environment.

Plants suitable for PFALs are those 30 cm or shorter in height such as leafy greens, transplants, and medicinal plants, because the distance between vertical tiers is typically around 40 cm, the optimum height for maximizing the space usage. Also, plants suitable for PFAL should grow well at relatively low light intensity and thrive at a high planting density. Staple food crops consumed mainly for calories such as wheat, rice, and potatoes are not suitable for PFAL production because their economic value per kilogram of dry mass is generally much lower and they require more time to grow than leafy greens.

In addition to commercial production of leafy greens, small PFALs with a floor area of 15-100 m^2 have been widely used for commercial production of seedlings in Japan because these seedlings can be produced in a short time at a high planting density. Grafted and nongrafted seedlings of tomato, cucumber, eggplant, seedlings of spinach and lettuce for hydroponic culture, and seedlings and cuttings of high-value ornamental plants are produced commercially in these small PFALs in Japan.

There are even smaller PFALs called micro-PFAL or mini-PFAL (m-PFAL), which are described in detail in Chapter 5. These m-PFALs are designed for urban residents who do not have outdoor gardens or for restaurants, cafés, shopping centers, schools, community centers, hospitals, and office buildings. m-PFALs are mainly used for entertainment, as a green interior object, and for the hobby of growing and harvesting plants.

Nevertheless, there are a number of challenges or disadvantages with PFALs that must be addressed. Foremost among these are the high initial and production costs. It is estimated that currently

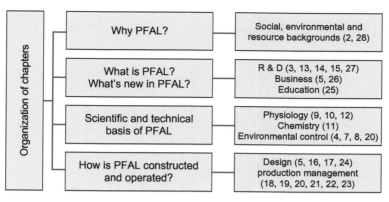

FIGURE 1.1

Organization of this book.

the cost of constructing the outer structure is as high as the cost of installing the PFAL units inside (Kozai, 2013). The initial investment can be reduced significantly through better design. The good news is that production costs are decreasing every year as operational and management experience accumulates. Electricity, labor, and materials (seeds, fertilizers, packing, delivery, etc.) account for similar proportions of the production costs. Lighting accounts for 70-80% of the electricity cost, with air conditioning, pumps, and fans accounting for the rest. There is great potential to reduce the cost of lighting by designing a more efficient lighting system. Other approaches to reducing production costs include increasing the number of vertical tiers, shortening the culture period by optimal environmental control, properly designing the production schedule to assure year-round production with no time loss, increasing planting density, and reducing produce loss. Other PFAL challenges include: culture information and an optimal environmental control strategy for various types of crops, marketing of produce, and breeding of new crops suited to PFALs. Chapter 27 describes the challenges and perspectives of PFALs in detail.

This book was written in response to the increasing interest and demand for information on various aspects of PFAL and indoor vertical farming. The book is organized as shown in Figure 1.1. Historically, there have been several attempts at growing plants indoors using solely artificial light both in Japan and North America. However, these efforts did not lead to successful commercial production mainly because of high equipment and operational costs and fierce competition with conventional production. However, the recent boom in PFALs is different from previous ones and is more realistic. We hope that this book serves as a source of useful information and provides a vision for PFALs, their diverse opportunities, and the influence on future lifestyles.

REFERENCES

FAO, 2009. Global agriculture towards 2050, How to feed the world 2050. <http://www.fao.org/fileadmin/templates/wsfs/docs/expert_paper/How_to_Feed_the_World_in_2050.pdf>.

Kozai, T., 2013. Plant factory in Japan: current situation and perspectives. Chron. Horticult. 53 (2), 8–11.

ROLE OF THE PLANT FACTORY WITH ARTIFICIAL LIGHTING (PFAL) IN URBAN AREAS

Toyoki Kozai[1], Genhua Niu[2]

Japan Plant Factory Association, c/o Center for Environment, Health and Field Sciences, Chiba University, Kashiwa, Chiba, Japan[1] Texas AgriLife Research at El Paso, Texas A&M University, El Paso, Texas, USA[2]

INTRODUCTION

The increasing importance of fresh food production in urban areas is discussed and the methods for achieving efficient production are described. Firstly, resource inflow and waste outflow in urban areas are described. Then, it is pointed out that a large portion of urban waste can be used as an essential resource for growing plants in urban ecosystems, thus significantly reducing resource inflow and waste outflow. Secondly, the role of the plant factory with artificial lighting (PFAL, pronounced "P-Fal") is discussed, and the plants that are suited and unsuited to PFALs are described. Finally, criticisms of PFALs are analyzed, and the methodology toward a sustainable PFAL is discussed.

INTERRELATED GLOBAL ISSUES TO BE SOLVED CONCURRENTLY

The world is facing interrelated issues concerning agriculture, the environment, society, and resources under increasing world population and climate change (Figure 2.1). Since these four issues are closely interrelated, they must be solved concurrently based on a common concept and methodology (Kozai, 2013a). In other words, we must find a concept and methodology for effectively producing high-quality foods, in order to improve social welfare and the quality of life with the minimum consumption of resources and emission of environmental pollutants.

Issues concerning agriculture or food include the dwindling number of farmers due to aging as the urban population increases (Figure 2.2) and loss of arable land area due to urbanization and desertification, salt accumulation on the soil surface, and soil contamination with toxic substances. The global population is forecast to rise to 9 billion people by 2050, 70% of whom will be living in urban areas (UN, 2009, 2011). The urban population in 2050 will require about 70% more food than the 2009 population.

Issues concerning the environment or ecosystem include the decrease in biodiversity and green space, and increase in environmental pollution and abnormal weather (heavy rain/flood, drought, strong wind, etc., likely due to global warming). These problems often make ecosystems less stable.

The shortage of resources including water, fossil fuels, and phytomass (or plant biomass) is becoming increasingly serious. Large areas for producing staple foods are already facing insufficient

Plant Factory. http://dx.doi.org/10.1016/B978-0-12-801775-3.00002-0

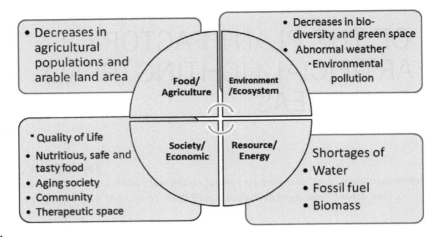

FIGURE 2.1

Four interrelated global issues to be solved concurrently for improved sustainability of the Earth and mankind as well as the quality of our lives.

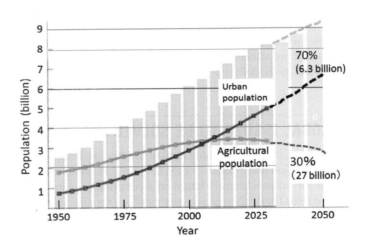

FIGURE 2.2

Trends of world, urban, and agricultural populations (cited and extrapolated from World Bank data).

irrigation water, and drought and/or unstable rainfall, while the demand for domestic water will continue to increase due to expanding urbanization along with the desire for a better quality of life.

Phytomass is becoming an important raw material resource for industrial products as an alternative to fossil fuels, while the reduction in regional phytomass affects the stability of local ecosystems. The shortage of phosphate rock as a raw material for phosphatic fertilizer (PO_4^-, or P), which is an essential fertilizer in agriculture, will become increasingly serious in the coming decades. On the other hand, excess application of P (phosphate) in crop fields is polluting freshwater rivers and lakes, causing eutrophication (Sharpley et al., 2003).

Issues concerning society and economics include an insufficient supply of nutritious, safe and tasty food, social welfare for building mutually supporting communities, therapeutic space and activities, and systems for education, lifelong self-learning, and human resource development.

The so-called "food deserts" is another related issue that has recently emerged, in which people have no access to fresh vegetables and fruits. Instead, they purchase various processed foods laden with sugar and fat at the local convenience store. Fresh foods produced in urban or suburban areas are transported to food processing factories far from the production site, and some of the processed foods are transported back to the production site. Thus, people living in the production area miss the chance to enjoy fresh and healthy vegetables and fruits, while significant resources are used for food processing and transportation.

Despommier (2010) proposed the concept of "vertical farming" to solve the four interrelated global issues, which has had a strong impact on scientists, engineers, policy makers, and architects. Allen (2012) proposed the "good food revolution" for growing healthy food and benefiting people and communities, and started a social business aiming at local production for local consumption of agricultural foods.

Plant Factory: An indoor Vertical Farming System for Efficient Quality Food Production describes the technical, engineering, and scientific aspects of "vertical farming" and "local production for local consumption" with special emphasis on PFAL, an emerging concept and technology.

RESOURCE INFLOW AND WASTE OUTFLOW IN URBAN AREAS

Huge amounts of resources are brought into urban areas, and huge amounts of waste are also produced in urban areas. The waste is processed and produced in both urban and suburban areas (Figure 2.3). The resources include food, water, fossil fuel, electricity, various products, raw materials for industries, and

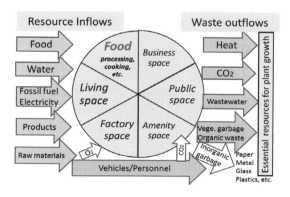

FIGURE 2.3

Resource inflow and waste outflow in urban areas. Most waste produced in urban areas is an essential resource for plant growth by photosynthesis (see also Figure 2.6).

vehicles. Additionally, large amounts of oxygen gas in the atmosphere are absorbed by humans, animals, microorganisms, and chemical reactions in factories. The waste includes heat, CO_2, wastewater, food garbage, grass clippings, pruned plant branches produced in parks, streets, and houses, organic solid waste, and inorganic solid waste (plastics, paper, metal, glass, and so forth).

Food is roughly grouped into plant-derived food, animal meat, fish, and mushrooms. It can also be classified into fresh (or raw), frozen, dried, and processed food. Plant-derived food can be classified into functional food (from specialty crops such as vegetables, edible herbs, fruits, and mushrooms), staple food (from commodity crops such as wheat, potatoes, rice, and corn) and confectionary (such as potato chips, cookies, and pies) (Figure 2.4). Crops that can be economically and commercially grown for sale in urban areas are basically the ones used for fresh functional food.

Fresh food with a water content of 90–95% is heavy and perishable unless cooled or frozen and carefully packaged. In cases where fresh food is transported into urban areas from remote areas by ordinary and small commercial trucks, the CO_2 emission intensity for transportation is, respectively, 0.8 kg and 1.9 kg-CO_2/ton/km (Ohyama et al., 2008). The food in a supermarket in the USA travels about 2000 km on average between its production and consumption sites (Smit and Nasr, 1992). Additional CO_2 is emitted when the food is cooled during transportation. Thus, transporting perishable fresh food over a long distance to urban areas is resource consuming and environment polluting (Figure 2.5). On the other hand, if fresh food is not cooled and/or not carefully packaged during long-distance transportation, much of it is lost. Thus, reduction in food mileage or "local production for local consumption" is particularly important in the case of fresh food.

Food garbage and wastewater produced in urban areas must be processed before it is returned to nature or recycled. Considerable amounts of fossil fuel and/or electricity are consumed for this processing. Thus, reducing the resource inflow reduces the waste outflow and CO_2 emissions in urban areas. In other words, the "throughput" of resources in urban areas needs to be reduced (Smit and Nasr, 1992). This can be achieved by changing the current "consume-dispose open loops" into new "consume-process-/recycle/reuse-produce closed loops" with respect to resource input (Figure 2.6). The same concept can be applied to plant production systems, namely, changing from "open plant production systems" to "closed plant production systems," which are described in Chapter 4.

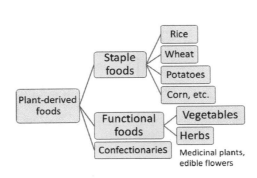

FIGURE 2.4

Classification of plant-derived foods.

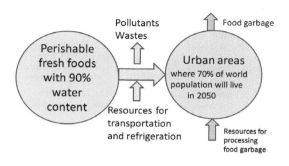

FIGURE 2.5

Transporting perishable fresh foods over a long distance to urban areas is resource consuming and environment polluting. Furthermore, additional resources are consumed to process the food garbage.

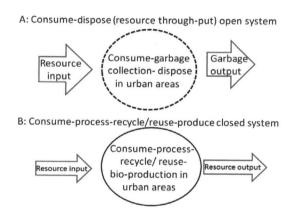

FIGURE 2.6

Open and closed urban areas with respect to resource input.

ENERGY AND MATERIAL BALANCE IN URBAN ECOSYSTEMS

PHOTOAUTOTROPHS (PLANTS) AND HETEROTROPHS (ANIMALS AND MICROORGANISMS)

Plants with green-colored leaves containing chlorophyll require only light, CO_2, water, and inorganic or mineral elements to produce carbohydrates (primary metabolites) for photosynthesis (Figure 2.7), and are called "photoautotrophs." Photoautotrophs synthesize organic compounds such as amino acids, vitamins, proteins, and lipids required for growth and development, using the carbohydrates, minerals, and water absorbed from the roots. Namely, plants do not need organic nutrient elements for their growth, although the plant roots may uptake very small amounts (less than 0.1% of the uptake of inorganic elements) of amino acids, glucose, and others that are present in the soil around the roots.

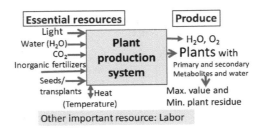

FIGURE 2.7

Essential resources (left) for growing green-colored plants through photosynthesis, and products obtained from plant production system (right).

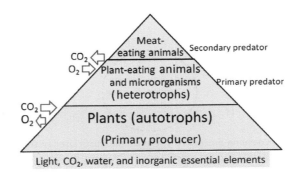

FIGURE 2.8

Animals and microorganisms need plants or plant-derived food for growth. So do humans. On the other hand, plants need light, CO_2, water, and mineral elements only to grow (see also Figure 2.6).

On the other hand, animals and microorganisms require organic elements as essential food for growth, and are called "heterotrophs." Some heterotrophs must take in certain kinds of inorganic elements such as salts, in addition to organic elements. Some heterotrophs (herbivorous) eat plants as food, while others (carnivorous) eat meat. The rest eat both plants and meat, and/or microorganisms. These nutritional relationships among plants, animals, and microorganisms are often presented as the "food chain pyramid" (Figure 2.8). In this pyramid, CO_2 and O_2 have a complementary relationship. Namely, photoautotrophs take in CO_2 and produce O_2, while heterotrophs take in O_2 and produce CO_2 (Figure 2.9).

WASTE PRODUCED IN URBAN AREAS AS AN ESSENTIAL RESOURCE FOR GROWING PLANTS

The waste produced in urban areas and fish/mushroom culture systems can be used as an essential resource, after proper processing, for growing plants in urban areas (Figure 2.10). Figure 2.10 shows an example of the waste recycle/reuse-bioproduction closed system shown in Figure 2.6. Technologies have already been developed for the purification of wastewater, production of compost and fertilizer

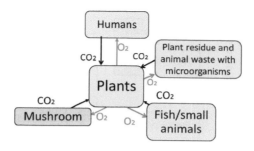

FIGURE 2.9

Complementarity of CO_2 and O_2 flows in urban ecosystems.

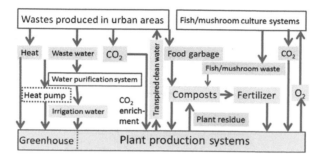

FIGURE 2.10

Waste produced in urban areas and fish/mushroom culture systems can be used as essential resources in plant production systems.

from food garbage and fish/mushroom waste, and collection and distribution of heat energy using heat pumps and/or heat exchangers. It is not difficult to transport CO_2-enriched air produced in offices and factories to nearby plant production systems.

The introduction of various kinds of plant production systems for efficient use of urban waste involves various challenges. By using the waste as an essential resource for plant production, the cost of waste processing and of production and transportation of fresh foods can be reduced. Moreover, recycling water, CO_2, heat, food garbage, and other substances in urban areas will enhance the sustainability of food, the environment, society, and resources.

PLANT PRODUCTION SYSTEMS INTEGRATED WITH OTHER BIOLOGICAL SYSTEMS

Plant production systems for fresh foods can be roughly classified into open fields, greenhouses with or without environmental control units, and indoor systems (Figure 2.11). Some are used for commercial production and others are used for personal, family, and group member activities.

These plant production systems can be integrated with other biological systems (Figure 2.12). An example of the material and energy flow in those biological systems is shown in Figure 2.13. By such

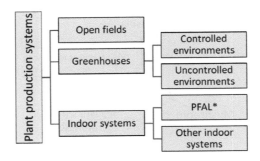

FIGURE 2.11

Types of plant production systems. PFAL denotes "plant factory with artificial lighting," and is precisely defined in Section 2.3.7 of this chapter.

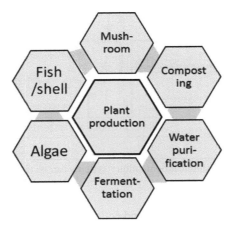

FIGURE 2.12

Various biological systems that can be integrated with different plant production systems for reusing waste as a resource in urban areas.

integration, the amount of waste and environmental pollutants can be further reduced. The amount of resources transported from outside urban areas for producing fresh foods and for processing the waste can also be reduced.

Given that the average daily vegetable consumption is about 300 g per person in many countries (FAOSTAT, http://faostat.fao.org/, 2009), about 1 million tons (=0.3 kg (fresh weight)/day/person × 10 million persons × 365 days) of vegetables are consumed per year in a city with a population of 10 million. It is estimated that around 25% of the vegetables purchased in urban areas goes directly into the garbage.

If a significant percentage of the vegetables were to be produced locally, job opportunities would be created and/or citizens could enjoy growing plants and other living organisms such as fish and mushrooms near their residence. Fresh food production in urban areas for local consumption minimizes the

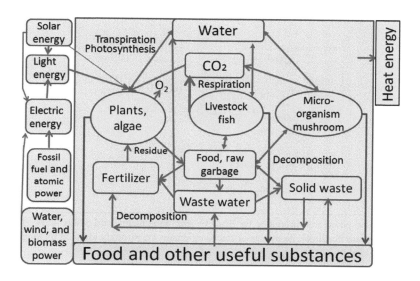

FIGURE 2.13

Material and energy conversion and recycling in urban agriculture.

food mileage and the loss of fresh food during transportation, and enhances communication among residents. Furthermore, people would understand and experience the principles of circulation of materials and energy in ecosystems. In this way, the four global issues concerning food, the environment, society, and resources can be concurrently solved step by step.

ROLE OF ORGANIC FERTILIZERS AND MICROORGANISMS IN THE SOIL

Food garbage, plant residue, fish/animal waste, and other biological waste can be converted to compost or organic fertilizer, or bioenergy (methane gas) for generating electricity. In the future, a system will be developed to extract valuable substances from such waste.

In soil culture in greenhouses and open fields (farmers' fields, community gardens, city farms, and backyard gardens), soil fertility is essential to stabilize the root zone environment and to improve the yield and quality of crops. In general, the application of organic fertilizer improves the soil fertility, which can be expressed and evaluated by the chemical, physical, and biological properties of the soil.

The role of organic fertilizer in plant production is indirect. Organic fertilizer is an essential element for the growth of microorganisms in the soil, and is decomposed into inorganic fertilizer as the microorganisms grow. The rate of decomposition of organic fertilizer into inorganic fertilizer is strongly affected by the soil temperature, water content, and pH (degree of acidity) as well as the characteristics of the microorganism ecosystem. This is why the effects of organic fertilizer on the yield and quality may vary. It can take considerable time for organic fertilizer to become effective in the soil and to make the soil fertile with a high degree of stability. In any case, it should be noted that plant roots uptake the inorganic fertilizer, not the organic fertilizer (Figure 2.14).

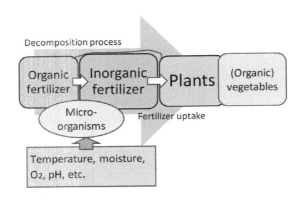

FIGURE 2.14

Organic fertilizer is decomposed into inorganic fertilizer by microorganisms, and plants uptake inorganic fertilizer mostly, not organic fertilizer. Activities of microorganisms are strongly affected by temperature, moisture, oxygen, pH, and other factors.

Aquaponics, a combination of hydroponics and aqua (fish) culture, is becoming popular in the USA. In aquaponics, fish waste is decomposed into inorganic fertilizer with the presence of microorganisms (see Figures 5.11 and 5.12 in Chapter 5). An indoor aquaponics farm with a floor area of 8100 m^2 has been commercially operating since March 2013 by FarmedHere (http://farmedhere.com/) in the Greater Chicago area, USA.

STABILITY AND CONTROLLABILITY OF THE ENVIRONMENT IN PLANT PRODUCTION SYSTEMS

Plant production systems can be characterized by their natural and artificial stability, and natural and artificial controllability (control ability) (Table 2.1). Different plant production systems differing in stability and control ability are needed to maintain the overall sustainability of society under changing climate and social conditions. The plant production system to be chosen depends on its purpose under the given social, environmental, economic, and resource conditions.

In open fields, keeping the soil (root zone) environment naturally stable is of primary importance, because this stability compensates for the low natural and artificial controllability of the aerial environment (weather). The application of organic fertilizer generally enhances the root zone stability with a high buffering function with respect to water content, air porosity, pH, and temperature of the soil, and high population density and diversity of microorganisms (not all kinds of microorganisms are beneficial for plant growth, so it is important to keep the beneficial microorganisms dominant in the soil).

Indoor plant production systems can generally be characterized by their high artificial controllability of both the aerial and root zone environment, or low natural stability and low natural controllability of both the aerial and root zone environment. To keep the artificial stability and controllability at a high level, an intelligent environment controller is necessary to minimize the resource consumption and environmental pollution.

Table 2.1 Classification of Four Types of Plant Production Systems by Their Relative Stability and Controllability, and Other Factors

| Stability and Controllability | Open Fields | Greenhouses | | Indoor Systems[a] |
		Soil Culture	Hydroponics[a]	
Natural stability of aerial zone	Very low	Low	Low	Low
Artificial controllability of aerial zone	Very low	Medium	Medium	Very high
Natural stability of root zone	High	High	Low	Low
Artificial controllability of root zone	Low	Low	High	High
Vulnerability of yield and quality	High	Medium	Relatively low	Low
Initial investment per unit land area	Low	Medium	Relatively high	Extremely high
Yield	Low	Medium	Relatively high	Extremely high

[a]*High/low evaluation is valid only when the manager's skills are fairly high.*

Greenhouses show intermediate stability and controllability between open fields and indoor systems. Soil culture has a higher natural stability than hydroponic culture with respect to the root zone environment. On the other hand, the soil in urban areas is often contaminated with heavy metals, agrochemicals, and other toxic substances (Garnett, 2001). This problem can be avoided by using a hydroponic greenhouse where the culture beds (root zone) are isolated from the soil.

Salt accumulates on the soil surface in the greenhouse due to excessive application of chemical fertilizer and/or livestock manure for many years, reducing crop yield. This reduction in yield can also be avoided by using the hydroponic culture system.

The high cost of greenhouse heating is a major factor in greenhouse crop production in winter. However, in urban areas, waste heat from industrial factories and offices can be used as a heat source for heating (Figures 2.3 and 2.13). In a case where the heat source temperature is too low (10–20°C) but the amount is abundant, heat pumps can be efficiently used to raise the waste heat temperature up to 40–60°C for heating.

KEY INDICES FOR SUSTAINABLE FOOD PRODUCTION

There are three key indices for evaluating the sustainability of fresh food production systems: (1) resource use efficiency (RUE, the ratio of resource amount fixed by produce to resource input); (2) cost performance (CP, ratio of sales amount to production cost); and (3) vulnerability (V) or yearly deviation of yield and unit value (quality) (Figure 2.15 and Chapter 4). Resource input at the system construction stage can be included, if needed, by LCA (life cycle assessment, Chapter 24). In the design and operation of a fresh food production system, its sustainability should be primarily evaluated by these indices.

WHAT IS "PFAL"?

Definition. Indoor plant production systems can be divided into two types in terms of light source: (1) artificial light only and (2) artificial light and (supplemental) solar light. In this book, the former is further divided into two groups. One is PFALs used mostly for commercial production, consisting

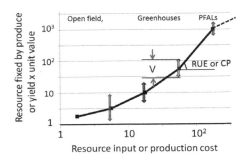

FIGURE 2.15

Scheme showing the (yield × unit value) and its vulnerability as affected by the amount of resource input.
Increases in initial and operation resource input are justified only when the gradient of curve (both RUE and CP) increases
and V decreases. RUE: resource use efficiency, CP: cost performance, and V: vulnerability of (yield × unit value).

of the essential components listed in Table 2.2. The other is the ventilated PFALs and m-PFALs
(mini- and micro-PFAL, Chapter 5) used mostly for other than commercial production.

PFALs aim to grow high-value produce (product of the yield and unit value or quality) with maximum RUE and CP, minimum vulnerability of yield and quality of produce, and minimum emission of
environmental pollutants. However, PFALs are an emerging and thus technically immature production
system, so their commercial application is still very limited.

On the other hand, PFALs have the potential for achieving high RUE, CP, and V and will play an
important role in urban areas in the coming decades. This book primarily focuses on PFAL, except for
the m-PFAL in Chapter 5 and the rooftop plant production system in Chapter 6.

Scientific benefits of PFALs. Growers can gain the experience of field crop cultivation once or twice
a year under changing weather and climate. As for greenhouse cultivation of fruit vegetables, the experience can be acquired one to four times a year in most cases under a controlled environment,

No.	Essential Component of PFAL
	Table 2.2 Essential Components of a PFAL as Defined in this Book
1	Airtight (no. of air exchanges is lower than 0.015 h^{-1})
2	Thermally well-insulated roof and wall (heat transmission coefficient is lower than 0.15 W/m^2/C)
3	Air shower or hot water shower at the culture and operation rooms
4	Multitiers with lighting system and hydroponic culture beds
5	Air conditioners mainly for cooling (dehumidification at the same time) and air circulation
6	CO_2 enrichment to keep its concentration at around 1000 ppm
7	Floor covered with epoxy resin sheet for keeping it clean
8	Collection and reuse system of water condensed at cooling panel (evaporator) of air conditioners
9	Circulation and sterilization system for nutrient solution supply

Other indoor plant production systems with artificial lighting only are called "o-PFAL (open-PFAL)" or "m-PFAL (mini- or micro-PFAL)" in this book.

although the cultivation know-how depends on the season and the presence of diseases often caused by pest insects.

In the case of PFAL cultivation of leafy greens or other herbaceous plants, growers can experience seedling production and cultivation of seedlings until harvesting 10–20 times a year under precisely controlled environments. Thus, experience and know-how about PFAL cultivation can be accumulated much faster than for open field and greenhouse cultivation.

During PFAL cultivation, growers can conduct a simple experiment by changing only one factor—for example, the light source, cultivar, or nutrient composition; other factors remain unchanged. This enables the growers to understand the cause-effect relationship easily (see Chapter 22). Also, the results of PFAL experiments are not affected by the weather, so PFAL experiments are more reproducible than open field and greenhouse experiments. If researchers conduct an experiment using part of a full culture system, there is a high probability that the results can be reproduced in a full-scale culture system. The PFAL manager can also collaborate with researchers at universities or other organizations easily. This is a major benefit of PFALs.

PLANTS SUITED AND UNSUITED TO PFALs

Plants suited to PFALs for commercial production have the following characteristics: (1) short in height (about 30 cm or less) to be adapted to multitier cultivation racks with a vertical distance between the culture beds of 40–50 cm; (2) fast growing (harvestable 10–30 days after transplanting); (3) growing well under low light intensity and at high planting density; (4) high-value product if fresh, clean, tasty, nutritious, and pesticide-free; (5) the product value can be effectively improved by environmental control; (6) about 85% in fresh weight of the plant can be sold as produce (e.g., root weight ratio of leaf lettuce should be lower than 10–15%) (Figure 2.16); and (7) any kind of transplant (Chapter 19).

Plants suited to greenhouses using sunlight rather than PFALs for improved quality and yield include: (1) fruit-vegetables such as tomatoes, green peppers, and cucumbers that contain large amounts of functional components; (2) berries such as strawberries and blueberries; (3) high-end flowers such as *Phalaenopsis*, dwarf loquats; (4) mangoes and grapes, etc. for growing in containers with trickle irrigation; and (5) nonwoody or annual medicinal plants such as *Angelica*, medicinal dwarf *Dendrobium*, Asian ginseng, saffron, and *Swertia japonica*.

Plants that are not suitable to PFAL production are staple crops used primarily as a source of calories (carbohydrates, protein, and fats) for people and livestock, such as rice, wheat, corn, and potatoes, plants such as sugarcane and rapeseed used primarily as a fuel (energy) source, larger fruit trees, and trees used for timber such as cedar and pine, and others including daikon, burdock, and lotus. These plants require large areas for growth and have a harvest cycle of several months to ten or more years, but they have a low ratio of value (price) to mass.

GROWING SOCIAL NEEDS AND INTEREST IN PFALs

While PFALs are currently limited primarily to the production of leafy vegetables, they are expected to help alleviate the following social concerns and demands (Kozai, 2013c):

FIGURE 2.16

Percentage of fresh weight of salable portion to the total plant weight should be 85% or higher to minimize the electricity consumption per kilogram of the salable portion.

(1) Concerns over the safety, security, consistency of supply, and price stability of fresh vegetables, particularly the rising demand for purchasing consistency in the catering (restaurants, etc.) and home-meal replacement industries (prepared food and bento retailers) for the elderly and people living alone;

(2) Demand for highly functional fresh vegetables and medicinal plants arising from concerns about health and improved quality of life;

(3) Demand for consistent all-year-round production of fresh salad vegetables in cold, hot, and arid regions;

(4) Demand for greater local self-sufficiency of fresh vegetables to increase employment opportunities for the aged, disabled, and unemployed (PFALs provide a safe and pleasant work environment and enable year-round employment);

(5) Demand for changes in lifestyle and social population structure in conjunction with convenience stores, supermarkets, restaurants, hospitals, and social welfare facilities, apartment flats, and so forth.;

(6) Demand for new business development in the electrical, information, construction, health care, and food industries;

(7) Demand for efficient use of vacant land, unused storage spaces, shaded areas, rooftops, and basements in urban areas (Hui, 2011);

(8) Demand for high-quality transplants for use in horticulture, agriculture, reforestation, landscaping, and desert rehabilitation; and

(9) Demand for water-saving culture systems in regions with insufficient or saline irrigation water and urban areas (PFALs require only around a fiftieth of the irrigation water required by greenhouses (Kozai, 2013b)).

Recent technical advances have helped to start the PFAL business, which include improvement of electric energy efficiency, cost performance of lighting, air conditioning, and small information control

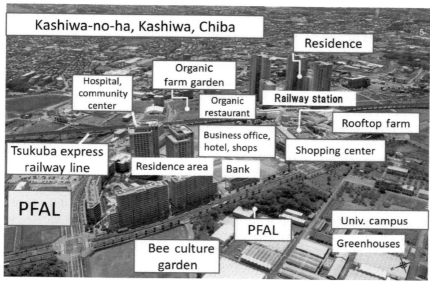

FIGURE 2.17

Example of an ongoing project on urban agriculture development in Kashiwa, Japan, located 30 min by train from central Tokyo.

systems with cloud computing, thermal insulation, electrical energy storage, and generation using natural energy such as solar light, wind, biomass, and geothermal energy (Komiyama, 2014).

Considering the social concerns and demands described here, a social experiment on the introduction of PFALs as part of urban agriculture is underway in accordance with the Smart City Project in Kashiwa-no-ha district, Kashiwa, Chiba, Japan (Figure 2.17). This city has been built since 2003 to be a sustainable city.

CRITICISMS OF PFALs AND RESPONSES TO THEM

Community interest in PFALs has grown in recent years as described previously. On the other hand, PFALs have also attracted considerable criticism and concern. We should accept and analyze the criticism earnestly, study the issues from all sides and search for solutions. When criticisms are based on misundstandings and immature technology, we should clearly explain the true situation based on the facts, and should also improve the technology.

Some typical criticisms of PFALs and urban agriculture (Garnett, 2001; Martin, 2013) are discussed in this section. The numerical data given in the following sections is representative of the PFALs in Japan as of 2014.

INITIAL COST IS TOO HIGH

In 2014, the initial cost of a PFAL building with all the necessary facilities including 15 tiers with a vertical distance between the tiers of 50 cm was about US$4000/m^2 (land area, one US$ = 120 yen) in Japan. This initial cost is about 15 times that of a greenhouse with heaters, ventilators, thermal screens, and other equipment. The wholesale price of leaf lettuce in 2014 was about US$0.7–0.8 per head in Japan, so its maximum annual sales/m^2 is US$2100–2400 (=3000 × 0.7–0.8)/m^2, compared with the initial cost of US$4000/m^2.

On the other hand, annual productivity/m^2 (land area) of an existing PFAL was about 3000 leaf lettuce heads/m^2/year (80–100 g fresh weight per head) compared to that of open fields (32 (=16 × 2 harvests) heads/m^2/year). This productivity of PFALs is about 100 times that of the open field and about 15 times that of the greenhouse (200 heads/m^2/year).

Thus, the initial cost per unit production capacity of the PFAL is more or less the same as that of the greenhouse, although this estimation is rough and varies with many factors. It is difficult to compare the initial cost of the PFAL with that of open fields, which is highly variable. Also, the size and weight of field-grown leaf lettuce heads are often two to three times larger than PFAL-grown ones. On the other hand, the retail prices of field- and PFAL-grown lettuces do not differ significantly in Japan, regardless of the large difference in weight between the two.

The annual PFAL production capacity given above can be calculated as follows: 20 plants/m^2 (culture bed area) × 15 tiers × 20 harvests/year) × 0.9 (ratio of salable plants to transplanted plants) × 0.5 (effective floor ratio of tiers to the total floor area). The other 50% of the floor area is used for the operations room, walkway, seedling production, and equipment. The number of days from transplanting to harvest (80–100 g/aerial part of plant) is 12–15, and from seeding to seedlings ready for transplanting is 20–22.

This annual productivity of PFALs will be improved based on "kaizen" and the PDCA (plan-do-check-action) cycle (Chapter 22) by around 20% within 5–10 years based on the improvements of environment control, cultivar suited to PFAL, layout of tiers, and culture methods. The initial cost of PFALs is expected to continue decreasing by a few percent every year in Japan.

PRODUCTION COST IS TOO HIGH

The component costs for electricity, labor, depreciation, and others in Japan account for, on average, 25–30%, 25–30%, 25–35%, and 20%, respectively. The economic life period for calculating the depreciation differs from country to country. In Japan, it is 15 years for the PFAL building, 10 years for the facilities and 5 years for the LED lamps.

Figure 2.18 shows an example of the production cost by components of a PFAL with daily production capacity of about 7000 leaf lettuce heads (80 g fresh weight/head) (Ohyama, 2015). Electricity, labor, and depreciation are the three major components of the production cost. This PFAL is located in a local city far from any large cities, so the packing and delivery costs account for 12% of the production cost. If the PFAL is located in or near a large city, the packing and delivery cost would be 6–8%. Most routine manual operations in PFALs are conducted by part-time workers whose hourly take-home pay is around US$7–10.

The wholesale price per head was 80–100 yen (US$0.67–0.83, one US$ = 120 yen) and the retail price was 150–200 yen (US$1.25–1.67) in Japan, while it was US$0.39–0.46 and 0.67–3.5, respectively, in Taiwan (Chapter 3, Section 2).

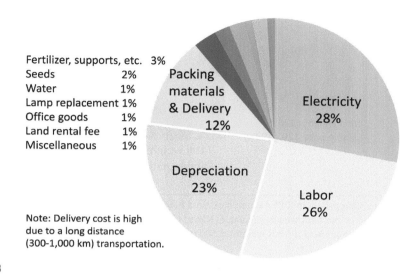

Fertilizer, supports, etc. 3%
Seeds 2%
Water 1%
Lamp replacement 1%
Office goods 1%
Land rental fee 1%
Miscellaneous 1%

Packing materials & Delivery 12%

Electricity 28%

Depreciation 23%

Labor 26%

Note: Delivery cost is high due to a long distance (300-1,000 km) transportation.

FIGURE 2.18

Production costs by components of a PFAL with a daily production capacity of about 7000 leaf lettuce heads (80 g fresh weight/head).

Ohyama (2015).

The wholesale price of leaf lettuce and basil in Japan is, respectively, around US$8 and $23/kg. The price of rocket (*Eruca sativa*), watercress, parsley, and low-potassium leaf lettuce (for patients with kidney problems) is around US$35/kg. The wholesale price is about US$40/kg for coriander and about US$70/kg for peppermint and spearmint.

It should be noted that, in Japan, the monthly basic charge for electricity is US$10–12/kW and the consumption charge is US$0.10–0.12/kWh. Namely, the electricity price in Japan is one of the highest in the world. The costs associated with construction, lighting, and culture tiers are also high in Japan compared with those in many other countries.

In PFAL production cost management, weight percent of the edible or usable part to the total plant weight is an important index for improving the CP (Figure 2.16). In PFAL, electric energy is converted to light energy and further converted to chemical energy in plants. Thus, in the case of leafy greens production, electric energy contained in the plant residue (roots and trimmed leaves) needs to be minimized. This is the opposite from the case of field cultivation—in general, the larger the roots, the better the growth and quality (Figures 2.19 and 2.20).

On the other hand, the roots are indispensable for the intake of water, inorganic fertilizer, and oxygen to support and grow the aerial part. Controlling the root weight to a minimum but not restricting the growth of the aerial part is a key cultivation technique in PFAL production cost management.

ELECTRICITY COST IS TOO HIGH, WHEREAS SOLAR LIGHT IS FREE

Electricity cost accounts for 25–30% of the total production cost (Figure 2.18). The electricity consumption of produce per kilogram can be reduced by 20–30% relatively easily, and by 50–80% theoretically. Figure 2.21 shows the energy conversion process in a culture room of a modern

Total weight : 130 g
Salable part: 120 g (92%)

Total weight: 500 g
Salable part: 200 g (40%)

FIGURE 2.19

Left: Crisp head lettuce plant grown in a PFAL and harvested before head formation at a fresh weight of about 130 g. Ninety-two percent of the total weight is salable. Right: Crisp head lettuce plant grown in the open field and harvested after head formation at a fresh weight of 500 g. Only 40% of the total weight is salable.

Total weight: 250 g
Salable part: 210 g (84%)

Total weight: about 5 kg
Salable part: about 2 kg (40%)

FIGURE 2.20

Left: Chinese cabbage grown in a PFAL and harvested before head formation at a total weight of 250 g for a fire-pot meal. Right: Chinese cabbage grown and harvested in the open field after head formation at a total weight of 5 kg.

energy-efficient PFAL. The electrical energy fixed in the salable part of plants as chemical energy is 1–2% at the highest, and less than 1% in most conventional PFALs. The remaining (98–99%) electric energy is converted into heat energy in the culture room (this is why the heating cost of a thermally well-insulated PFAL even on a very cold winter night ($-40°C$) is 0).

The electricity cost can be reduced by: (1) using advanced LEDs to improve the conversion factor from electric to light energy; (2) improving the lighting system with well-designed reflectors to increase the ratio of light energy emitted by lamps to that absorbed by plant leaves; (3) improving the light quality to enhance growth and the quality of plants; (4) optimally controlling temperature, CO_2 concentration, nutrient solution, humidity, and other factors; and (5) increasing the percentage salable part of plants by the improvement of culture method and selection of cultivars. Once solar light is partially used, all the controllable factors in PFALs become unstable and unpredictable, so this is not a wise choice.

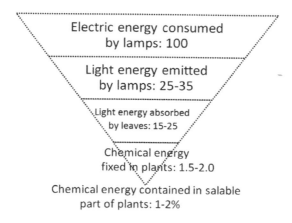

FIGURE 2.21

Energy conversion process in current representative PFALs. The value at each step indicates the percentage of the electrical energy consumed. The electrical energy not converted to chemical energy becomes heat energy. [This is why the heating cost of a thermally well-insulated PFAL is 0 even on very cold winter nights (−40°C)].

LABOR COST IS TOO HIGH

Labor cost accounts for 25–30% of the total production cost in a PFAL, because most PFALs are small-scale, and most handling operations are conducted manually. It is estimated that a 15-tier PFAL with a floor area of 1 ha needs more than 300 full-time employees, if most handling operations are conducted manually. In this sense, even small-scale PFALs create a significant number of job opportunities and promote local industries. Most manual operations are safe and light work performed under comfortable environmental conditions, and are suited to all persons regardless of age, gender, and whether handicapped or not.

On the other hand, in the Netherlands, most handling operations in a greenhouse complex with a floor area of 10 ha or more for production of leafy greens, bedding plants, and transplants are automated, and so need only several employees per ha. Unfortunately, however, those automatic handling systems are too large to be installed in current PFALs.

In recent PFALs with a production capacity of over 10,000 leafy green heads, automation of handling operations is becoming essential to reduce the production cost and to collect and analyze various data automatically. The data includes the environment, plant growth, transplanting, harvesting, resource consumption and the costs, shipping, and sales. PFALs are being automated by using advanced robotic technologies including remote sensing, image processing, intelligent robot hands, cloud computing, big data analysis, and 3-D modeling.

PFAL-GROWN VEGETABLES ARE NEITHER TASTY NOR NUTRITIOUS

Taste and nutrition of fresh vegetables are affected by the genetic properties of plants, time courses of physical, chemical, and biological environments encountered by plants, culture systems and methods, and postharvest treatment. It should be theoretically easier to control the taste and nutrition of PFAL-grown vegetables.

FIGURE 2.22

Factors affecting the quality and economic value of vegetables.

In reality, however, the cause-effect relationship between the environment and taste/nutrition in a PFAL for a particular vegetable is still unclear, so PFAL managers do not know exactly how they should control the environment to improve the taste, nutrition, and growth of plants; the environmental control of PFALs is still at the trial and error stage. This is the main reason why the taste and nutrition of PFAL-grown vegetables are sometimes unstable.

On the other hand, an increasing number of researchers at universities and public institutions as well as private companies are accumulating experimental results and know-how on the cause-effect relationships to improve the taste and nutrition of PFAL-grown vegetables and medicinal herbs by environmental control. Once the cause-effect relationships are revealed, PFALs can reproduce the tasty and nutritious vegetables all year round regardless of the weather. In terms of knowledge and experience of plant production using PFALs, research toward commercialization is still at the initial stage.

One advantage for PFAL researchers is that they can obtain clear cause-effect relationships with less research funds, time, and manpower. It takes time to identify the cause-effect relationships regarding the environment and the taste/nutrition of field-grown vegetables under uncontrollable weather and seasonal changes, even when the cultivar, cultivation system, and postharvest treatment are the same.

The quality and economic value of the vegetables are influenced not only by the taste and nutrition, but also by their safety, which needs to be guaranteed through traceability (Figure 2.22). The traceability of PFAL-grown vegetables is almost 100% and can mostly be done automatically.

Another issue is that taste and preferences are affected by many factors including social and family cultures, personal history, health conditions, and brand image. Thus, it is natural that many people prefer field-grown vegetables to PFAL-grown vegetables.

MOST PFALs ARE NOT MAKING A PROFIT

According to a survey in February 2014 by the Ministry of Agriculture, Forestry, and Fishery of Japan, among 165 PFALs, only 25% made a profit, 50% broke even, and 25% lost money.

Visits to the PFALs that were losing money revealed that the PFAL managers did not know that plants need CO_2 for photosynthesis and they were not aware of the necessity of CO_2 enrichment in an airtight PFAL. This lack of knowledge is not surprising because farmers never supply CO_2 gas to their fields and only a limited number of growers supply CO_2 gas in their greenhouses by opening the ventilators. These findings suggest the importance of human resource development and training for PFAL managers to improve their skills.

In order to make a profit, over 90% of PFAL-grown vegetables must be sold at a reasonable price. In reality, however, the production of vegetables is too poor to do so at the PFALs that are losing money. Also, in order to sell all the produce, a "market-in" strategy is more important than a "product-out" strategy. In other words, there is still considerable room for improving the production cost, productivity, and sales strategy.

LAND PRICE IS TOO HIGH

The price of land is generally very high in urban areas, while PFALs can be built with little disadvantage in shaded, nonfertile soil and idle land areas. They can be built with little difficulty in empty buildings, office rooms, and industrial factories. PFALs need around 1% and 10% of the land area compared with open fields and greenhouses, respectively, for obtaining the same number of leaf lettuce heads and other leafy greens.

WATER CONSUMPTION FOR IRRIGATION IS TOO HIGH

As described in detail in Chapter 4, the net water consumption for irrigation in a PFAL is about 2% of that of a greenhouse, because about 95% of the transpired water vapor from plant leaves is condensed as liquid water at the cooling coil panel (evaporator) of the air conditioners, which is collected and returned to the nutrient solution tank after sterilization.

Drained nutrient solution from the culture beds is also returned to the nutrient solution tank after sterilization. Thus, the amount of water to be added to the tank is almost equal to the amount of water kept or held in harvested plants and the amount that escaped outside as water vapor through air gaps. Similarly, the amount of nutrients to be added is almost equal to the amount of nutrients absorbed by harvested plants. Thus, the efficiency of water and nutrient use is over 0.95 and 0.90, respectively, in most cases.

In addition, there is no need to wash the PFAL-grown vegetables before eating, because they are free of pesticides, insects, and foreign substances, so a considerable amount of water can be saved compared with greenhouse- and field-grown vegetables, which must be washed with tap water, electrolyzed water, and/or hypochlorite-added water. Furthermore, since PFAL-grown vegetables are packed in plastic or paper bags immediately after harvest, the loss of vegetables due to water loss and bacteria growth is minimized. Overall, the total water consumption for irrigation and washing of PFAL-grown vegetables can be reduced by 99% compared with that of greenhouse-grown vegetables.

PFALs CAN ONLY PRODUCE LEAFY GREENS—MINOR VEGETABLES—ECONOMICALLY

Currently, most PFALs only produce leafy greens including herbs (Figure 2.23), because they are easy to grow and there is strong demand for PFAL-grown leafy greens. Approximately 50,000 leaf lettuce heads are produced daily in PFALs (18 million heads/year) in Japan as of 2014, which is only 1% of the field-grown leaf lettuce consumption. Commercial production of spinach in PFALs for fresh salads is currently almost 0 but will increase within several years. Thus, the total production of leafy greens in PFALs is expected to increase sharply as the cost of production falls.

Small rooted vegetables and medicinal plants have recently started trial production. Examples are turnips, carrots, miniradishes, *Panax ginseng*, and *Angelica acutiloba* (Figure 2.24). They are all 30 cm

FIGURE 2.23

Leafy vegetables/herbs commonly produced in PFALs.

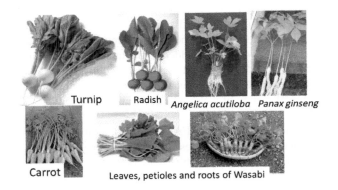

FIGURE 2.24

Small root vegetables and medicinal plants being produced in modern PFALs. Both shoots with leaves and roots are edible, tasty, and nutritious.

in plant height and both the shoots with unfolded leaves and the roots are edible or salable. These plants can be harvested within several weeks after seeding or a few weeks after transplanting. At this growth stage, the shoots are soft and tasty and make a nice side dish to the main course. These kinds of small roots with shoots cannot be produced in open fields, thus eliminating competition with field-grown vegetables.

Tuberous root vegetables generally grow faster than leafy vegetables under optimal environmental conditions, especially at high CO_2 concentration. This is because the carbohydrates (called "source") produced in the leaves by photosynthesis can be efficiently moved through sieve tubes in the stem (translocation) into the tuberous roots (called "sink"). Photosynthesis is enhanced as the amount of

No.	Expected Uses
Table 2.3 Range of Uses of PFAL-Grown Plants in the Coming Decades	
1	(Fresh salads and cut (sliced) vegetables)
2	Frozen vegetables and dried vegetables
3	Paste, sauce, soft drink, drink additives, jam, and juice
4	Pickles, salted vegetables, and kimchi (Korean pickles)
5	Vegetables for fire-pot meal, fried and soup vegetable dish
6	Cosmetics, dyes, aromatics, and spices
7	Medicinal plants for herbal medicine and supplements
8	Functional foods and pharmaceuticals (e.g., vaccine)
9	Transplants of all kinds, mother plants for propagation
10	Strawberry, blueberry, raspberry, and blackberry

carbohydrates in the leaves (source tank) tends to be 0 due to the enhanced translocation to the tuberous roots.

If PFAL-grown vegetables are used commercially not only for fresh salads but also as raw materials as shown in Table 2.3, their market size would increase dramatically, creating "a new branch of urban agriculture" or "a new field of food industries." If the PFAL is well designed and well managed: (1) the production cost per kilogram can be reduced by 20–50%; (2) the economic value per kilogram can be increased by 10–50%; and (3) the initial cost can be easily reduced by 20–40%.

If we could achieve these targets, PFALs could be used at any local and/or rural area to grow high-value crops for processed food materials. The harvested plants would be dried (after boiling), powdered in some cases, and stored there for 1–6 months. Then, they would be taken to a food factory for further processing. In this way, PFALs could be used in any area far from urban areas, where people living in cold, hot, or dry areas could work inside the PFAL under comfortable conditions.

TOWARDS A SUSTAINABLE PFAL
REQUIREMENTS FOR A SUSTAINABLE PFAL

PFALs must be improved to build a sustainable plant production system. A systematic and scientific methodology to realize sustainable PFALs is proposed in this book, although the methodology requires continuous revision and refinement. For PFALs to be sustainable, they must satisfy the following requirements:

(1) They should make simultaneous and parallel contributions to solving global food, environment, resource, and social issues;
(2) The entire plant supply chain from production to consumption should be resource saving and have low CO_2 emissions. In particular, PFALs should dramatically cut the use of water and petrochemicals by greatly reducing the direct use of oil-based products and completely eliminating the use of pesticides and fossil fuels for heating even on very cold winter nights;

(3) They should contribute to protecting the environment by minimizing the release of environmental pollutants;

(4) They should maximize the use efficiency of resource investment (ratio of utility value or rate of conversion of resources to products to the amount of resources committed) and maximize the efficiency of using natural energy;

(5) They should increase the stability of the production system in the face of abnormal weather patterns and the presence of contaminants, and deliver planned high quality and high yield all year round;

(6) The systems should be safe and pleasant for the operators and local residents, and contribute to creating an industry in which environmental health and welfare can coexist;

(7) They should expand employment opportunities and give meaning to life for a broad range of people, including the aged and disabled;

(8) The system as a whole, including its operators, should evolve appropriately with changes in the natural environment and the diverse social environment; and

(9) They should facilitate international technology transfer through the development of standardized systems.

The scientific and engineering aspects of PFALs are described in Chapter 4.

FACTORS AFFECTING THE SUSTAINABILITY OF PFALs

Positive aspects affecting the environmental, resource, social, and economic sustainability of PFALs as well as issues to be solved to improve the sustainability of PFALs are summarized in the following sections.

Positive aspects affecting environmental, resource, social, and economic sustainability

A. Environmental and resource sustainability
   (a) Reductions in: (1) water for irrigation; (2) water for washing products; (3) pesticide application; (4) fertilizer leaching to drains; (5) unusable and/or unsalable parts of plants; and (6) distance for food transportation.
   (b) Promotion of recycling use of: (1) drained water and nutrient solution.
   (c) Improvements in: (1) resource use efficiency (RUE).
B. Social sustainability
   Increase in local employment.
C. Economic sustainability
   (a) Increases in: (1) yield and quality of products and their vulnerability; (2) local fresh food production and sales; (3) local food security and safety; (4) fresh leafy greens edible without washing; and (5) longer shelf life.
   (b) Reductions in: (1) land area per yield (more amenity space for exercise and recreation) (Figure 2.25); (2) loss of produce; and (3) working hours for plant production per yield.

Factors to be solved to improve sustainability

a) Reductions in: (1) electricity consumption for production and product cooling and (2) materials and energy consumption for building construction.

A: 100 ha of open field for leafy vegetable production but it is
bare soil for a half year

B: 1 ha of PFAL for leafy vegetable production and 99 ha of forest
for public green space

FIGURE 2.25

Productivity of leafy greens using a PFAL is 100 times that of open fields which is more sustainable with respect to the environment, resources, society, and economy: A or B?

b) Increases in: (1) recycling use of organic waste including plant residue for composting, biofuels, and other purposes; (2) biodiversity around PFALs in urban areas; (3) amenity space around PFALs for exercise and recreation; and (4) aesthetic value.

c) Promotion of: (1) active community participation; (2) social inclusion: providing fresh food to the poor; (3) education; and (4) software development for design and production management.

SIMILARITIES BETWEEN THE EARTH, SPACE FARMS, AUTONOMOUS CITIES, AND PFALs

In the design and operation of the PFAL with other biological systems toward a sustainable system, the Earth and a virtual residence with space farms on the moon are good natural and artificial models, respectively (Figure 2.26). Both are closed systems with respect to materials, while they are open systems with respect to radiation energy—receiving solar (shortwave) radiation and emitting thermal (long-wave) radiation. All of the systems require plant production systems with high yield and quality in a limited space using minimum resources with minimum emission of environmental pollutants. Controlling the balance of materials and energy flows within the system and between the system and its environment is essential to create a sustainable system. There is an urgent need to minimize the consumption of fossil fuels without reducing the yield and quality of crops both in soil and in PFALs by using natural energy more efficiently.

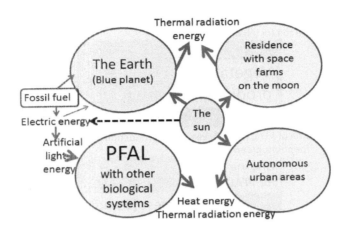

FIGURE 2.26

The Earth and virtual residence with space farms on the moon are both closed systems with respect to materials and are open systems with respect to radiation, receiving solar radiation, and emitting thermal radiation. They are models of PFALs with other biological systems and autonomous (local production for local consumption) urban areas. All the systems require plant production systems with high yield and quality in a limited space using minimum resources and emitting minimum environmental pollutants to be sustainable. There is an urgent need to minimize the consumption of fossil fuels without reducing the yield and quality of crops both in soil and in PFALs by using natural energy more efficiently.

CONCLUSION

This chapter discussed the reasons behind the increasing importance of fresh food production in urban areas, and presented examples of how efficient production could be achieved. Firstly, resource inflow and waste outflow in urban areas were shown. Then, it was pointed out that a large portion of urban waste can be used as an essential resource for growing plants in urban ecosystems, thus significantly reducing resource inflow and waste outflow. The criticisms that urban agriculture is not economically viable and that PFALs make heavy use of electricity, have high production costs, and are not financially viable were also addressed. It was shown that PFALs can produce value-added plants at high yield with minimum resource consumption and environmental pollution, and can contribute to partially solving several important issues occurring in urban areas by producing fresh foods using PFALs in combination with other biological systems.

REFERENCES

Allen, W., 2012. Good Food Revolution: Growing Healthy Food, People and Communities. Gotham Books, USA. pp. 256.

Despommier, D., 2010. The Vertical Farm: Feeding the World in the 21st Century. St. Martin's Press, New York. 305 p.

Garnett, T., 2001. Urban agriculture in London: rethinking our food economy. City Case Study Lindon. 477-500, In: Bakker, N., Dubbeling, M., Guendel, S., Sabel Koshella, U., deZeeuw, H. (Eds.), Growing Cities, Growing Food: Urban Agriculture on the Policy. DSE, Feldafing, Germany.

Hui, C.M.S., 2011. Green roof urban farming for buildings in high-density urban areas. In: Proceedings of Hainan China World Green Roof Conference. March 11-21. China, 1-9.

Komiyama, H., 2014. Beyond the Limits to Growth: New Ideas for Sustainability from Japan. Springer Open, Heidelberg. pp. 100.

Kozai, T., 2013a. Sustainable plant factory: closed plant production system with artificial light for high resource use efficiencies and quality produce. Acta Hortic. 1004, 27–40.

Kozai, T., 2013b. Resource use efficiency of closed plant production system with artificial light: concept, estimation and application to plant factory. Proc. Japan Acad. Ser. B 89, 447–461.

Kozai, T., 2013c. Plant factory in Japan: current situation and perspectives. Chron. Hortic. 53 (2), 8–11.

Martin, G., 2013. Urban agriculture's synergies with ecological and social sustainability: food, nature and community. In: Proceedings of the European Conference on Sustainability, Energy and the Environment, pp. 12.

Ohyama, K., 2015. Actual management conditions on a large-scale plant factory with artificial lighting (written in Japanese: Dai-kibo keiei de no keiei jittai). JGHA Prot. Hortic. (JGHA Shisetsu to Engei) 168, 30–33.

Ohyama, K., Takagaki, M., Kurasaka, H., 2008. Urban horticulture: its significance to environmental conservation. Sustain. Sci. 3, 241–247.

Sharpley, A.N., Daniel, T., Sims, T., Lemunyon, J., Stevens, R., Parry, R., 2003. Agricultural Phosphorous and Eutrophication, second ed. USDA Research Service, Washington, DC. ARS 149, pp. 38.

Smit, J., Nasr, J., 1992. Urban agriculture for sustainable cities: using wastes and idle land and water bodies as resources. Environ. Urban. 4 (2), 141–152.

UN, 2009. Planning Sustainable Cities: Global Report on Human Settlements. UN-Habitat, United Nations, Nairobi.

2011. World Population Prospects. The 2-10 Revision. Department of Economic and Social Affairs, United Nations, New York.

PFAL BUSINESS AND R&D IN THE WORLD: CURRENT STATUS AND PERSPECTIVES

INTRODUCTION

This chapter describes the history, current status, and perspectives of plant factories with artificial lighting (PFAL) in Japan, Taiwan, China, North America, and Europe (England and The Netherlands), including research, development, and business. The governmental subsidy for PFAL research and development (R&D) and business in Japan are introduced. Taiwanese companies have started to export and build turn-key PFALs abroad. The Chinese Academy of Agricultural Sciences started a national project on intelligent plant factory production technology in 2013 supported by the Ministry of Science and Technology, and the project was joined by 15 universities, institutes, and private companies. In Korea, the Ministry of Knowledge Economy (MKE) started a research project named "Development of major components for IT-LED based plant factories" in 2009. The annual domestic market of the PFAL business in Korea is worth nearly US$600 million. In the USA, several large-scale commercial facilities were recently built to produce pharmaceutical protein products (antigens and antibodies). More recently, large commercial PFAL facilities were built close to large cities such as Chicago. In The Netherlands, two relatively large PFALs were built in 2014 and 2015 for research and development by private companies, aiming to commercialize PFALs on a large scale. The perspectives of the PFAL business and R&D in the world are discussed.

JAPAN

Toyoki Kozai

Japan Plant Factory Association, c/o Center for Environment, Health and Field Sciences, Chiba University, Kashiwa, Chiba, Japan

The history and current status of the PFAL business as well as R&D in Japan are briefly described. Then, the governmental subsidy for PFAL R&D and business, which was introduced in 2009, is explained. Finally, recent public service activities for PFAL are outlined.

BRIEF HISTORY AND CURRENT STATUS OF PFAL BUSINESS

In Japan, the first commercial PFAL, Miura Nouen in Shizuoka Prefecture, was established in 1983. This was followed in 1985 by a PFAL at a vegetable sales area in a shopping center in Chiba Prefecture. By the mid-1990s, high-pressure sodium lamps were being used as the light sources. Since the surface temperature of the lamps is over 100 °C, the lamps must be located beyond 1 meter from the plant community.

In the late 1990s, fluorescent lamps became preferred mainly due to their higher PAR (photosynthetically active radiation; wavelength, 400–700 nm) output per watt. Then, PFALs consisting of multi-tiers (4–15 racks) with a vertical separation of about 40 cm between tiers became available. PFALs using LEDs as their light sources began to be commercialized in 2005 in Japan.

As of March 2014, the number of PFALs used for commercial production was 165 and is estimated to exceed 200 by the end of 2014 in Japan, and the number will continue to increase in 2015 and beyond. There were 34 PFALs in March 2009, 64 in March 2011, 106 in March 2012, and 125 in March 2013. In addition, closed plant production systems with artificial lighting (CPPS) units with a floor area of 16.2 m^2 for transplant (seedlings and plantlets from cuttings) production were in commercial use in 2014 at around 300 locations throughout Japan, and some in Australia and China (see Chapter 19, Section 1).

The largest PFAL with fluorescent lamps, which is operated by Spread Co., Ltd. in Kyoto, produces 23,000 leaf lettuce heads daily (see Chapter 26, Section 2). A PFAL with all LEDs producing 10,000 leaf lettuce heads daily was built by Mirai Co., Ltd. in Tagajoh, Miyagi Prefecture in March 2014 (see Chapter 26, Section 3). Figure 3.1 shows the PFAL with 3000 fluorescent lamps (each 1.2 m long) built in June 2011 at Kashiwa-no-ha Campus, Chiba University, operated by Mirai Co., Ltd. On the same campus, a new PFAL was built in September 2014 by Japan Dome House Co., Ltd. (see Chapter 26, Section 4). In the campus of Osaka Prefecture University, a PFAL with all LEDs was built in September 2014, which can produce 5000 leafy greens daily (see Chapter 23).

FIGURE 3.1

PFAL at Chiba University built in 2010, operated by Mirai, Co., Ltd. Total floor area: 406 m^2; Floor area of culture room: 338 m^2, 10 tiers, nine rows. Mainly leaf lettuce and Romaine lettuce. 3000 heads per day (1 million heads/year or 2800 heads/m^2/year).

The size of the current PFAL business market is still very limited, and was probably worth about 12 billion yen (1 US$ = 120 yen) in 2014. According to a survey by the Ministry of Agriculture, Forestry and Fisheries (MAFF) published in March 2014, (1) 75% of PFALs are operated by private companies (the rest are operated mostly by incorporated agricultural organizations); (2) 55% of PFALs have a floor area of less than 1000 m^2 including the operation room and office; (3) 75% of PFALs have annual sales of less than 50 million yen; (4) 75% of PFALs use fluorescent lamps as the light source; and (5) 35% of PFALs received both a subsidy and loan, 30% received neither subsidy nor loan, 20% received a loan, and 15% did not answer. The initial and operation costs are given in Chapter 2, Section 6.

RESEARCH AND DEVELOPMENT

Research in Japan on plant production under artificial light aiming at commercialization was begun in the mid-1970s by Takakura et al. (1974) and Takatsuji (1979). The Japanese Society of High Technology in Agriculture, which was established in 1989 focusing on plant factory research, was merged in 2005 with the Japanese Society of Agricultural, Biological and Environmental Engineers and Scientists. This academic society has been organizing one-day symposiums (the "SHITA Symposium") on plant factories in January every year since 1990. Takatsuji was a group leader of PFAL R&D at the Numazu Campus of Tokai University from 1991 to 2007.

In 2000, a CPPS was built for R&D at Matsudo Campus, Chiba University (Kubota and Chun, 2000; Chun and Kozai, 2001) (Figures 3.2 and 3.3). This PFAL was designed and operated based on the concept of the CPPS (Kozai and Chun, 2002; Kozai et al., 2006). In this PFAL, handling of culture plug trays and operation of the precision irrigation system were automated (Ohyama et al., 2005). Using the PFAL, the production of disease-free sweet potato transplants, tomato seedlings, and medicinal plants were studied (Afreen et al., 2005, 2006; Zobayed et al., 2006, 2007) (Figure 3.4).

FIGURE 3.2

The closed plant production system (CPPS) with seven tiers for disease/pest insect-free transplant production, built in 2000 at the Matsudo Campus of Chiba University. An automatic tray handling/transportation system, automatic precision irrigation system and distributed intelligent control system were installed.

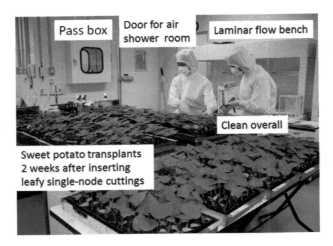

FIGURE 3.3

Operation room next to the culture room. Disease-free, pest insect-free, and pesticide-free sweet potato transplants were propagated. The substrate was autoclaved before use.

FIGURE 3.4

Production of medicinal plants in the CPPS. Left: St. John's wort (*Hypericum perforatum* L.) (Zobayed et al., 2006, 2007). Right: *Glycyrrhiza uralensis* (Afreen et al., 2005, 2006).

Based on this research, CPPS units with fluorescent lamps for transplant (seedling) production were commercialized in 2004 (Kozai, 2007). The same concept was used to develop the sugar-free medium (photoautotrophic) micropropagation system (Kozai et al., 2005).

In 2009, the Ministry of Agriculture, Forestry and Fisheries (MAFF) and Ministry of Economy, Trade and Industry (MITI) started the national project "Plant factory with artificial light and/or solar light" with a total budget for 5 years of 15 billion yen (1 US$ = 120 yen). The former provided subsidies for demonstration, training, extension, and publicity, and the latter provided subsidies for basic research. Private-sector companies could apply for a 50% subsidy from the above budget to build PFALs for commercial use.

Osaka Prefecture University and Chiba University received budgets for PFALs from both MAFF and MITI. Meiji University, Shinshu University, and Shimane University received budgets for PFALs from MITI. The National Agriculture and Food Research Organization (NARO) received a budget for PFALs from MAFF, which covered the construction of the PFAL building and basic infrastructure, while expenses for facilities inside the building were covered by the consortium members (private sector) of each project site.

Apart from these projects, Tamagawa University, Yamaguchi University, and Kyoto University and several other universities have been conducting R&D on PFALs since 2010. Also, many private companies started investigating the PFAL business using their own budgets.

In March 2011, the Great East Japan Earthquake struck the northeastern coast of the main island of Japan, and the ensuing tsunami hit the Pacific coast of Fukushima, Miyagi, and Iwate prefectures (destroying three nuclear power plant units in Fukushima). Since agriculture and horticulture are important industries in those areas, the Japanese government provided subsidies to build new greenhouses and PFALs in 2012 and 2013.

In November 2014, the International Plant Factory Conference was held in Kyoto and Osaka attracting 160 participants including 60 overseas participants (http://www.shita.jp/ICPF2014/).

PUBLIC SERVICE

In June 2010, the Japan Plant Factory Association (JPFA), a nonprofit organization, was established in the Kashiwa-no-ha campus of Chiba University. The number of JPFA corporate members was 60 in 2010 and 98 in 2014. JPFA has two roles: one is collaboration with Chiba University and the other is collaboration with national and international organizations, offering monthly half-day seminars, monthly 2- or 3-day training courses, consulting services, guided tours for visitors to the campus, and collaborative R&D with the corporate member companies (see Chapter 25). In 2012, an association for PFAL managers was newly established, consisting of 58 corporate members and having its office within the JPFA's office.

There are several exhibitions and conferences related to the PFAL business, including: (1) Agro-innovation held at Tokyo Big Sight every year (http://www.jma.or.jp/ai/en/) organized by the Japan Management Organization and (2) Greenhouse Horticulture & Plant Factory Exhibition/Conference held at Tokyo Big Sight every 2 years (http://www.gpec.jp/english/) supported by the Japan Greenhouse Horticulture Association.

TAIWAN

Wei Fang

Department of Bio-Industrial Mechatronics Engineering, National Taiwan University, Taipei, Taiwan

STATUS OF PFAL IN TAIWAN

Whereas the global population will increase from 7 to 9.6 billion in less than 40 years, Taiwan's population is shrinking. It is also estimated that 70% of the global population will live in cities compared with the current 50% (Kozai, 2014), and the trend is the same in Taiwan. As a result, urban agriculture will increase, and so PFALs will play an important role. As mentioned by Glaeser

(2011), cities should develop upwards not outwards, and the same is true for urban agriculture: PFALs are the answer. Many people in Taiwan agree, and consider that PFALs can make us richer, smarter, greener, healthier, and happier.

There are 45 organizations engaged in leafy green production using PFALs in Taiwan as of September 2014, and 56 PFALs of various sizes have been built and operated in the last 4 years. Among these 45 organizations, there are 2 research institutes, 4 universities, and 39 private companies involved, as shown in Figure 3.5. The PFALs built by universities and research institutes are financially supported by their own funds and from the government; the government provides no support to private companies.

Among those 56 PFALs, 73%, 20%, and 7% are located in northern, central, and southern Taiwan, respectively, as shown in Figure 3.6. The scale of PFALs is shown in Figure 3.7: they are categorized into six sizes based on the amount harvested daily, assuming a cropping density of 25 plants per square meter of culture bed. In between the smallest (<100 plants/day) and the largest (>10,000), the size categories are 100–500, 500–1000, 1000–5000 and 5000–10,000. Half of all PFALs are small with daily production of less than 100 plants and only one PFAL with daily production of more than 10,000 plants, which is probably the world's largest PFAL with daily production of 60,000 plants (2.5 tons of leafy greens). Over 90% of PFALs are located in one room on one floor of an office building, typically on an empty floor or basement inside a building in an industrial park in the Taipei area.

Some companies have started to export and build turnkey PFALs abroad, mainly in China as shown in Figure 3.8. To date, there have been 11 such projects, of which two have not been completed (shown by dashed line), one in Beijing and one in Xiamen. These two were suspended due to financial reasons. In 3 out of the 11 projects, the company built the PFAL in its own branch located in China.

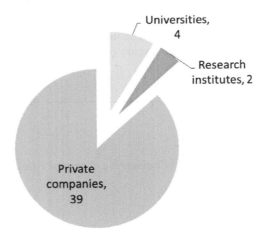

FIGURE 3.5

Distribution of PFALs in Taiwan categorized by organization before September 2014.

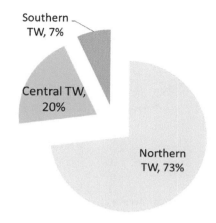

FIGURE 3.6

Geographical distribution of PFALs in Taiwan before September 2014.

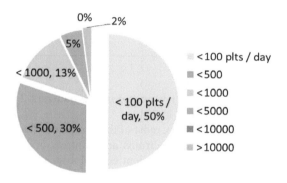

FIGURE 3.7

Number of PFALs in Taiwan categorized by daily production before September 2014.

PFAL EXPO IN TAIWAN

To promote PFALs, various technical books have been translated (Kozai, 2009, 2012; Takatsuji, 2007; Fang, 2011a,c, 2012), and booklets for the general public (Fang, 2011b; Fang and Chen, 2014) have been written. Exhibitions and conferences have been held by the Photonics Industry & Technology Development Association (PIDA) of Taiwan. PIDA has held a photonics festival in Taipei for 23 consecutive years. It is an NPO established by the Taiwan Government to assist the country's optoelectronics industry. Apart from serving as an exhibition organizer, they also provide services such as industry research, consulting, promotion, and communication in the industry and market. 2014 was the third year in which they included PFALs within the festival. Another two NPOs, the Taiwan Plant Factory Industrial Development Association (TPFIDA, founded in 2011) and the Chung-hwa Plant

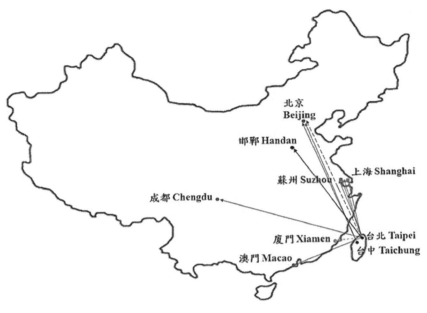

FIGURE 3.8

Export of turnkey PFALs from Taiwan to China.

Factory Association (CPFA, founded in 2012), were major co-organizers. The number of booths related to PFALs increased from 36 in 2012 to 108 in 2014.

Among the PFAL booths of the 2014 Expo, most of the companies demonstrated hardware used in PFALs. Several of them exhibited various spectrums and controls of LED tubes and panels, while others showed locally developed or imported nutrient control systems. One booth was that of Mirai Co., Ltd., Japan, which showed its PFAL turnkey capabilities. Also, several local companies promoted their ability to set up PFALs abroad. At least five companies demonstrated home appliance-style plant growth desktop devices and three showed growth benches for shops, restaurants, and supermarkets with or without environmental control capability. One company showed an LED illuminated green wall that cleans the air, another exhibited an aquaponics system, and another showed various by-products from PFAL-grown vegetable ingredients.

PFAL RESEARCH

Cost comparison of PFALs

Crops grown in PFALs can be classified into four types: RTC (ready to cook), RTE (ready to eat), CAW (cook after wash), and EAW (eat after wash). The retail price of RTE lettuce and CAW Pak-Choi varies widely, from NT$500 to 2000 and from NT$200 to 300 per kg (1 USD = 31.46 TWD), respectively. Table 3.1 shows the average retail price and cost of lettuce produced in PFALs in Japan and Taiwan with the same daily production of 1000 plants.

Table 3.1 Comparison of Retail Price and Cost of Lettuce Produced in PFALs in Japan and Taiwan

Lettuce	Japan	Taiwan
Retail price[a]	¥150~200	¥81~420[b]
Cost[a]	¥80~100	¥47~56[b]

[a]Japanese yen per 70 g fresh mass produced.
[b]Exchange rate at 1 NT$=3 Japanese yen.

There are some fundamental reasons for this dramatic difference in production cost. In particular, the high cost of construction and equipment, especially LEDs, leads to high depreciation cost, while the high cost of labor and electricity leads to high operating cost.

Spectra of LEDs used in PFALs

Figure 3.9 shows spectra of artificial light used in PFALs in Taiwan. Assuming the same size of culture bed (1.8×1.2 m), the light efficiency of various lights is compared in Table 3.2. The row with the shaded background shows that LED panels are less efficient than LED tubes with reflective film between the tubes. Also, the longer the tube, the higher the overall quantitative efficiency measured in micro-mole per joule.

Wireless sensor networks in PFALs

A wireless sensor network (WSN) is used to evaluate the uniformity of air temperature, humidity, and light intensity horizontally in a layer and vertically within layers of a PFAL (Chang et al., 2011; Juo et al., 2012). Each wireless sensor module is equipped with temperature, relative humidity, and light sensors, hanging from the tops of crops in each layer of culture beds to measure the uniformity of distribution of light and air. As shown in Figure 3.10, the temperature distribution is clearly related to the distribution of the fresh weight harvested. This means that greater uniformity of temperature reduces the variation in final fresh weight, and so a smart fan system has been developed to increase the uniformity (Lee et al., 2013).

Ion-selective sensors for nutrient detection

Traditional ion-selective sensors are expensive and have a short usable lifespan. New ion-selective sensors for detecting macro-elements in nutrient solutions have been developed. Figure 3.11 shows the sensing responses of screen-printed ion-selective electrodes (ISEs) for Ca^{2+}, K^+, Mg^{2+}, NH^{4+}, and NO^{3-}.

Nondestructive plant growth measurement system

A measurement system with cameras attached to a sliding rail on each layer of the culture bed accompanied by weighing devices for each plant was developed for continuous and automatic measurement of plant growth. The system takes images at preset time intervals and stitches all images across the culture bed to form a panoramic image of the entire bed using a computer with image-processing capability. In the recording process, the cameras move across the whole culture bed and capture images. Temperature and humidity sensors are also integrated with the imaging system to acquire

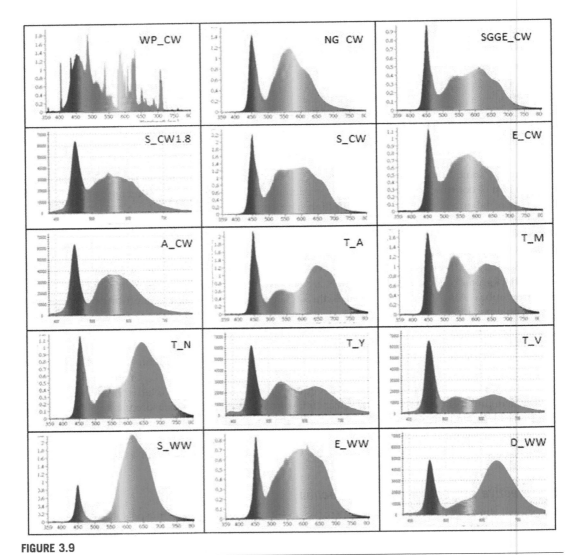

FIGURE 3.9

Spectra of various artificial lights used in PFALs (Fang, 2014).

spatial-temporal environmental information during the plant growth period. The image-processing algorithms, which calculate geometric features such as projected leaf area, plant height, volume and diameter, have been developed and incorporated into the automated measurement system (Yeh et al., 2014). The accompanying automatic weighing system using load cells was also developed to record the fresh weight of individual plants throughout the growth period. The weighing system

Table 3.2 Comparison of Efficiency of Various Light Sources Used in PFALs (Fang, 2014)

Comp_Spec of Light Sources	Reflectors on Top	No. of Tubes or Panels	PPF[a] μmol m^{-2} s^{-1}	PPF $\times$ Area[b] μmol s^{-1}	Power Consumption, W	Efficiency, μmol J^{-1}
S_CW1.8	Y	6 Tubes	334.9 $\pm$ 86.6	723.4	172.97	4.2
T_V	Y	9 Tubes	281.6 $\pm$ 59.4	608.4	195	3.1
H_CW	Y	9 Tubes	273.9 $\pm$ 79.8	73.9	212	3.1
T_A	Y	9 Tubes	225.8 $\pm$ 49.1	487.8	189	2.6
E_CW	Y	9 Tubes	263.8 $\pm$ 64.2	569.8	228	2.5
T_A	No need	12 panels	411.5 $\pm$ 103.2	888.9	432	2.1
T_N	Y	9 Tubes	186.7 $\pm$ 39.3	403.2	196	2.1
E_WW	Y	9 Tubes	210.8 $\pm$ 51.0	455.3	229	2.0
T_N	No need	12 panels	378.4 $\pm$ 95.4	809.6	428	1.9
T_M	Y	9 Tubes	168.0 $\pm$ 39.1	326.9	193	1.9
T_M	No need	12 panels	387.3 $\pm$ 99.4	836.5	442	1.9
T_Y	Y	9 Tubes	168.3 $\pm$ 34.7	363.5	188	1.9
S_R	No need	12 panels	291.0 $\pm$ 71.3	628.6	334	1.9
T5FL_CW	Y	9 Tubes	250.0 $\pm$ 57.5	540	283	1.9

[a]Measured at a distance of 10 cm under light except the T5FL treatment (at 20 cm).
[b]Area of culture bed on bench layer = 1.8 m $\times$ 1.2 m = 2.16 m^2.

FIGURE 3.10

Wireless sensing nodes in PFAL.

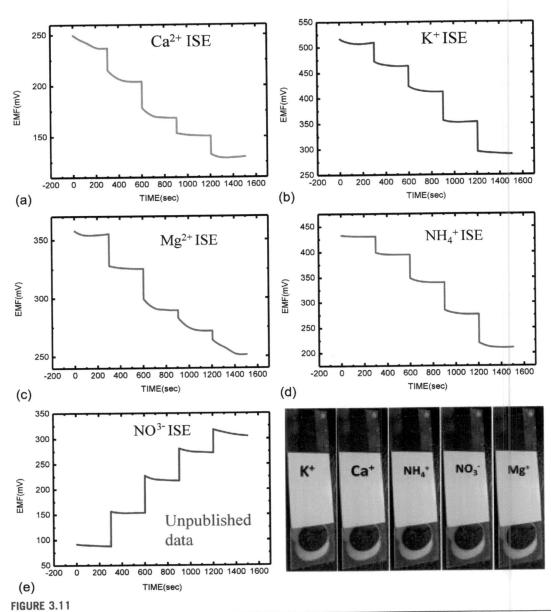

FIGURE 3.11

Sensing responses of screen-printed ion-selective electrodes (ISEs) for Ca^{2+}, K^+, Mg^{2+}, NH^{4+}, and NO^{3-}.

can also be used to measure plant growth as an independent system. Figure 3.12 shows a schematic diagram of the plant growth measurement system. For the weighing system, the load cell signals are calibrated, acquired and displayed in real time. The data are analyzed in correspondence with the plant geometric features obtained from the imaging system, enabling a plant growth model to be developed for various controlled environmental conditions. This plant growth measurement system provides a nondestructive, real-time processing approach over the traditional measuring methods. Furthermore, because it is automated, the system can gather a large number of plant measurements easily. Hence, the system is an efficient, practical tool for optimizing the parameters of the growing environment and plant factory environment.

BUSINESS MODELS OF PFALs IN TAIWAN

PFALs are eye-catching and new to the general public, and are attractive to consumers who are concerned about the environment and health. However, without a proper business model and careful planning, a PFAL business may fail. There are various business models being tested in Taiwan.

(a)

FIGURE 3.12

The nondestructive plant growth measurement system. (a) Schematic diagram of the system,

(Continued)

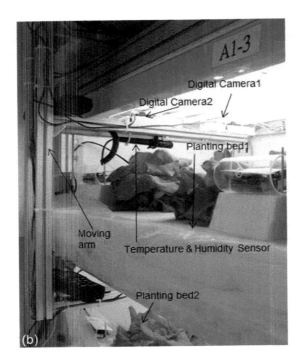

FIGURE 3.12—CONT'D

(b) the imaging system integrated with temperature and humidity sensors, (c) the plant weight measurement devices.

The product can be the plant itself such as whole plant, loose leaf, or baby leaf. The way to present the product is very important, such as using a sealed soft plastic bag, a soft plastic bag with tiny holes, or a sealed hard plastic box for the packaging. Different packaging methods express different concepts to the customer. Is it a product that does not need to be washed before eating, or like products grown in a greenhouse? It is important to emphasize that the product is locally made, not imported. Some products are provided with salad dressing, so the taste of the dressing is also important. One common matter is that the packaging bags and boxes be well designed to give a better appearance than traditional agricultural products.

The sales channels can be membership based, through a web site, within the company, or within the local community. It is important to limit the amount sold through third parties. The products can be sold through supermarket chains (owned by others) only temporarily, since the shelf charge is normally too high. Thus, for PFAL products, B2C (business to consumer) is much more favorable than B2B (business to business). C2B is even better and will be the direction to go.

If the company cannot sell all its products, other product lines may be considered. One company in Taiwan has developed more than 10 kinds of processed products such as ice cream, egg roll, bread, noodle, face mask, and skincare soap. The products can also be used as a nutritional additive in various forms such as juice, powder, and tablet. Different kinds of vegetable additives have different prices for the same product. For example, noodles with butter-lettuce and with ice-plant differ in price.

One construction company combined the PFAL concept in its community construction plan. Each family will have a home appliance style device for growing vegetables at home, and the community will have a service division that provides seeds, seedlings, stock nutrient solutions, and other needed supplies to community residents, thus promoting a green lifestyle.

Shops with a PFAL in the back or to one side, and with a restaurant or stand selling organic products in front, are a popular business model in Taiwan. Such shops are normally chain stores and are located throughout a city.

Several companies focus on the development and sale of home appliance style plant production units and indoor green walls for home use. One company produces aquaponic units for hobby growers and home owners.

Some companies are capable of constructing PFALs for others, and most of them have a demonstration site that potential customers can visit for further consideration. Some successful companies will have a demonstration site with a more convincing scale, with a daily production of no less than 100 plants, and need to be operated smoothly for more than several months. An established marketing channel is a great advantage. Unfortunately, few companies meet these requirements, and so are less convincing.

Many PFAL-related hardware providers such as LED providers, clean room constructors, air conditioning system providers, hydroponics system providers, power supply providers, and thermal insulation providers have started to build PFAL demonstration rooms and are learning how to grow plants. Their common goal is to become a turnkey provider of PFALs.

In short, there are several distinct business models in Taiwan:

1. The PFAL produces leafy greens for their own usage. For example, restaurant owners and corporations with more than 1000 employees.
2. The PFAL produces leafy greens for internet customers and members. Some companies are quite flexible, and even exchange membership with other health-related organizations such as yoga clubs.
3. The PFAL produces leafy greens and processed products as vegetable additives.
4. The PFAL produces leafy greens as an eye-catching sales point, but profit is made by other means such as the construction of chain stores selling organic dry foods.
5. Home appliance style PFAL module providers with a demonstration room.
6. Home appliance style PFAL module providers having alliances with the construction industry.
7. PFAL related hardware providers which build PFAL demonstration rooms to sell their products and which plan to become turnkey providers.
8. PFAL turnkey builders and consultants with or without a PFAL demonstration room.

As described above, PFAL companies in Taiwan are still small in scale but are flexible and willing to try various business models. Some look promising, some have failed. Even with the same business model, some companies succeed while others fail, as in any emerging industry.

CONCLUSIONS

The PFAL business is booming in Taiwan. The number of booths attending the PFAL Expo has tripled in 3 years. Without financial and policy support from the government, private companies are entering this new industry with great zeal. PFAL-related NPO organizations have been established, enabling the horizontal and vertical connection and integration of companies.

At present, there is no private agricultural organization involved in the PFAL business in Taiwan. Several farmers' associations have considered converting unused warehouses to PFALs, but finally abandoned the idea. The high initial cost is the first concern, followed by the difficulty of finding skilled workers and managers to run the PFAL. At present, there are not enough skilled, qualified managers and workers in the PFAL industry in Taiwan. Besides academic training in undergraduate and graduate schools of Taiwan, our team also runs 30-h workshops twice a year and has trained more than 400 people so far, but less than 10% of them have gone into the business afterward. To capture worldwide business opportunities, it is crucial to train skilled managers and workers for the PFAL industry.

Many companies have become involved with a view to the business opportunities of turnkey projects. However, some failed to prove that their system can grow quality plants efficiently. Unfortunately, some companies saw PFALs as a quick way to make money, resulting in law suits and public distrust. Of the 11 international turnkey PFALs built by Taiwanese companies so far, are all in China.

Some consumers question the use of artificial lights and hydroponics, and there were complaints about the non-natural use of chemicals 3 or 4 years ago. Public awareness, food safety problems, environmental problems, and frequent media reports about PFALs help consumers to learn about the technologies, appreciate them and be willing to pay extra to buy PFAL products. Nevertheless, it is necessary to reduce the cost, increase the value, and increase the varieties that can be grown in PFALs. PFALs will coexist with organic agriculture and traditional agriculture, and will also play a key role in urban agriculture in the smart cities of the future.

KOREA

Changhoo Chun

Department of Plant Science, Seoul National University, Seoul, South Korea

PFAL INDUSTRY, A COMMITMENT TO THE FUTURE

The convergence of information and communication technology (ICT) in the farming business has been considered to be one of the measures for mitigating climate change, heralding a paradigm shift in agricultural production, and creating new growth in Korea. In 2013, the Korean government announced its "Plan for promotion of agri-food and ICT convergence" in order to apply ICT technologies to the production, distribution, and consumption of agricultural products. One of its major projects which is already being implemented is the supply of smart greenhouse systems that support the monitoring and control of plant cultivation environments via smart phones, by providing high-tech sensing, monitoring, and controlling equipment to farms. PFAL, the most advanced type of farming system, has gained general acceptance as a highly successful application of ICT-converged smart farming systems.

In 2009, the Ministry of Knowledge Economy (MKE) started a research project named "Development of major components for IT-LED based plant factories." Several national policy support projects such as "Construction of business ecosystems based on plant factories" (2012, MKE) and "Demonstration support project for promoting the plant factory business" (2013, Ministry of Agriculture, Food and Rural Affairs) were also implemented along with some support projects funded by local governments. These policy support projects have been created because PFALs are recognized as a promising export to foreign countries even though this high-tech equipment industry was seldom economically feasible in the domestic market at the time.

RESEARCH AND TECHNICAL DEVELOPMENT

Korea has developed various technologies for vegetable breeding, hydroponics, greenhouse structure, as well as hardware and software for environmental control in greenhouses, which can easily be diverted to vegetable production in PFALs. Even though many element technologies for PFALs have been developed and related research results have been published, no commercialized PFAL was introduced until 2009, whereas vegetable production in greenhouses has become crucial to Korea's agriculture sector. Universities and national and prefectural research agencies were the main agents for research and technical development (RTD) in the area of controlled environment agriculture (CEA) that enables the grower to manipulate the cultivation environment to the desired conditions and that is useful for isolating specific environmental variables for more precise studies on plant responses to modified sets of environment. There is obvious commonality in research areas between PFAL and CEA with artificial lighting.

Prior to 2009, most practical research studies on PFAL (or CPPS) were performed using fluorescent lamps as the artificial light source. Some representative samples of related RTD in those days were CPPS for producing plug transplants of seed-propagated vegetables for the nursery industry; CPPS for producing vegetatively propagated strawberry transplants for consolidating the national proliferation program; CPPS for producing a monocotyledonous vegetable as a raw material for an instant ramen company; and CPPS for producing a salad vegetable for a restaurant chain, a fast-food chain, and a vegetable processing and distribution company. Total solutions were developed for each system and delivered to contracted enterprises, which included the cropping system, cultivar selection, seed sterilization, seed germination, raising transplants, transplanting, planting density, temperature settings for air and nutrient solution, setting of photo and dark periods, PPF setting, harvest scheduling, and so on.

The biggest change in RTD between the time before and after the year 2009 was the use of LEDs as the sole light source of PFALs. Of course, the physiological and morphological responses of plants grown under red, blue, green, or even far-red monochromic light sources and under combinations of those monochromic sources were studied scientifically from the late 1990s. Due to the low luminance and high cost of LEDs, however, they were not widely used as a light source for PFALs until 2009 in Korea.

Research and development of PFALs with LEDs dramatically increased from 2009. The trigger was the adoption of a "Low-carbon, green growth campaign" as a national policy and "Development of major components for IT-LED based plant factories" was selected as one of the smart projects in that year. Four other major research projects funded by the government relating to PFALs with LED lighting were: "Development of LED-IT based plant production techniques for environment-friendly horticultural products," "Development of lighting techniques using LEDs that have a particular spectra promoting plant growth," "Commercialization for exports of plant factories," and "Development of urban-type PFAL technology."

As various types of LED chips with different wavelengths and wattages and LED lighting fixtures with different shapes, combinations of spectra, and luminance have become available in the market, research on PFALs with LEDs has sharply increased not only at universities and public research agencies but also private companies. A large number of research results on the photosynthesis, growth, and morphology of various vegetables in response to different PPF ratios of blue and red LEDs have been published in international and domestic journals including *Horticulture, Environment, and Biotechnology*, the *Korean Journal of Horticultural Science & Technology*, and *Protected Horticulture and Plant Factory*. And as the PFAL business has expanded, private companies have started accumulating their own confidential research data on lighting equipment such as white LEDs having different spectra, productivity, cropping systems, cultivation methods, specialized quality, and even profitability.

PFALs for research by national research agencies have been installed at the Antarctic King Sejong Station (55 m^2) of the Korea Institute of Ocean Science; the National Academy of Agricultural Science (446 m^2 in Suwon-si, and 1506 m^2 in Wanju-gun) of the Rural Development Administration (RDA); Protected Horticulture Research Station (142 m^2 in Haman-gun) of RDA; National Institute of Horticultural and Herbal Science (55 m^2 in Umseong-gun) of RDA; and Korea Institute of Science and Technology (33 m^2 in Gangneung-si). Meanwhile, PFALs of prefectural research agencies have been installed in the Agricultural Research & Extension Services of Gyeonggi-do (115 m^2 in Whaseong-si); Chungcheongbuk-do (413 m^2 in Cheongju-si); Gyeongsangbuk-do (132 m^2 in Daegu-si); and Gyeongsangnam-do (198 m^2 in Jinju-si). PFALs for research by universities have been installed at Seoul National University in Seoul-si and Suwon-si; Chungbuk National University in Cheongju-si; Gongju National University in Yesan-gun; Gyengsang National University in Jinju-si; and Jeonbuk National University in Jeonju-si and Iksan-si.

PRIVATE COMPANIES AND FARMS IN THE PFAL BUSINESS

As mentioned above, several private companies have entered the PFAL business since 2009 and currently about 30 companies are operating PFALs for demonstration and/or production. They have also installed PFALs for their customers in Korea and other countries including Japan, China, Mongolia, and Qatar, the customers being local cities and boroughs, agricultural research centers of local government bodies, universities and schools, cafes and restaurants, hospitals, mega marts and department stores, community centers in apartments, farms, and others.

Most of the companies which have installed PFALs for research, demonstration, and/or production and are currently operating them can mainly be classified as small and medium-sized enterprises. These include Paru (20 m^2 in Suncheon-si), Insung Tec (165 m^2 in Yongin-si), Taeyoun Eco & Agro-Industry (165 m^2 in Seoul-si), KAST Agricultural System & Technology (132 m^2 in Gumi-si), Korea Refrigerated Foods (50 m^2 in Gimhae-si), Yuyang DNU (9 m^2 in Hwanseong-si), Cham Farm (50 m^2 in Goseong-gun), Jinwon Farm (330 m^2 in Gwangju-si), Wise Control (33 m^2 in Yongin-si), Refresh Hamyang (1694 m^2 in Hamyang-gun), Vegetechs (661 m^2 in Goyang-si), Future Green (50 m^2 in Suwon-si), Miraewon Farm (604 m^2 in Pyeongtaek-si), Happy Enjoy Farm (200 m^2 in Gyeongsan-si), Maxfor (115 m^2 in Yongin-si), and Eum Farm (226 m^2 in Gimpo-si). Some of the companies classified as large-scale enterprises such as Dongbu Lightec (9 m^2 in Bucheon-si), Nongshim Engineering (230 m^2 in Anyang-si), Lotte Mart (9 m^2 in Seoul-si), and Lotte R&D Center (17 m^2 in Seoul-si) are operating PFALs mainly for research.

ACHIEVEMENTS AND CHALLENGES

The PFAL and related industries in Korea have grown dramatically in just 5 years. A recent report from RDA (Lee, 2014) estimated that the market size of the domestic PFAL business is about US$577 million per year; with US$77 and 500 million as added values from the PFAL itself and industries with forward/backward linkages, respectively.

All the element technologies for PFALs have been successfully developed from a series of systematically designed joint researches among multidisciplinary experts. In practice, the lighting technology for PFALs in Korea skipped the stage of fluorescent lamps and LED technology was directly applied, causing serious problems such as high initial investment and difficulties in cultivation due to lack of information. Nevertheless, the Korean PFAL industry has had a great opportunity to confidently compete against other PFAL-advanced countries (Chun, 2014).

Policy support projects prepared by the government were also efficient for the early stage of development of PFALs in Korea. Further investment from the private and public sectors is needed to achieve stable growth of this industry in terms of networking of minor PFALs for ensuring successful marketing of their products, and the development of high value-added products by meeting the various demands of today's aging, health-aware society. Unlike in Japan, PFALs in Korea are not subsidized by national or provincial governments. The PFAL industry would develop further if a subsidy program were to be introduced as PFAL-related technologies approach successful deployment on a commercial scale.

CHINA

Qichang Yang

Institute of Environment and Sustainable Development in Agriculture, Chinese Academy of Agricultural Sciences, Beijing, China

Yuxin Tong

Institute of Environment and Sustainable Development in Agriculture, Chinese Academy of Agricultural Sciences, Beijing, China

Ruifeng Cheng

Institute of Environment and Sustainable Development in Agriculture, Chinese Academy of Agricultural Sciences, Beijing, China

DEVELOPMENT OF PFAL IN CHINA

In China, studies on PFAL technologies began in 2002 and mainly focused on hydroponic technologies and their control systems, supported by the Ministry of Science and Technology of China. Since then, R&D on PFAL in China has advanced rapidly. By 2013, about 35 plant factories had been built, as shown in Figure 3.13 (Table 3.3), distributed in nine cities or provinces, including Beijing, Shandong, Shanghai, Nanjing, Changchun, and Guangdong. The PFALs in China are mostly located in research institutes and parks and are used for research and demonstration.

FIGURE 3.13

Regional distribution of PFALs in China (2013).

Table 3.3 Supplementary Table of Figure 3.13		
No.	**City/Province**	**Floor Area (m²)**
1	Harbin, Heilongjiang	12
2	Jilin	200
3	Beijing	3069
4	Inner Mongolia	100
5	Shouguang, Shandong	200
6	Gaoqing, Shandong	616
7	Ningyang, Shandong	72.9
8	Xian, Shanxi	200
9	Nanjing, Jiangsu	300
10	Changxing, Zhejiang	800
11	Zhuhai, Guangdong	30
12	Jianjiang, Guangdong	12

CASE STUDY OF TYPICAL PFALs

PFALs in the Chinese Academy of Agricultural Sciences

Since 2002, researchers in the Institute of Environment and Sustainable Development in Agriculture (IESDA) at the Chinese Academy of Agricultural Sciences (CAAS) have been studying PFAL technologies. In 2005, a PFAL with fluorescent lamps, a PFAL half with LED lamps, and a PFAL with

FIGURE 3.14

PFAL labs built in CAAS in 2005 (Left: fluorescent lamps, 20 m^2); in 2009 (Middle: half with LED lamps, 100 m^2); and in 2013 (Right: LED lamps, 100 m^2).

all LED lamps were built in turn at CAAS (Figure 3.14). The research conducted at CAAS focused on energy-saving technologies, LED energy-saving technologies, nutrient solution management, and vegetable quality control technologies for PFALs.

In 2012, a PFAL with a floor area of 80 m^2 was built in a greenhouse complex in the demonstration center of CAAS, which has a total floor area of 40,000 m^2 (Figure 3.15). The demonstration center is operated by Beijing IESDA Protected Horticulture Co., Ltd., which manages more than 10 PFALs in China (Table 3.4). The demonstration center is divided into seven function halls for exhibiting new PFAL technologies, new cultivation methods, and new ideas on urban, household and/or office horticulture. Recently, a three-layer vertical farming model was built in the demonstration center, which has an underground layer for mushroom production, a middle layer for leaf vegetable production with artificial light, and a top layer for fruit vegetable production with sunlight.

PFAL of Beijing Kingpeng International Hi-Tech Corporation

In 2010, Beijing Kingpeng International Hi-Tech Corporation, belonging to the Beijing Agriculture Machinery Institute, built a PFAL with a total area of 1300 m^2 (Figure 3.16). The PFAL includes a tissue culture room, seedling room, artificial light culture room with solar light generation technologies, vegetable storage room, etc.

Plant factory of Zhejiang University

The plant factory of Zhejiang University was built in 2013 in Changxing Agricultural Station. The total area of the plant factory is 1600 m^2 (area of PFAL: 800 m^2), and it has 10 layers of movable cultivation beds to increase the efficiency of land use (Figure 3.17). Many research teams of Zhejiang University, including control science and engineering, light engineering, computer science and technology, agricultural engineering, biological engineering, and horticulture and plant nutrition, are working together to improve PFAL technologies.

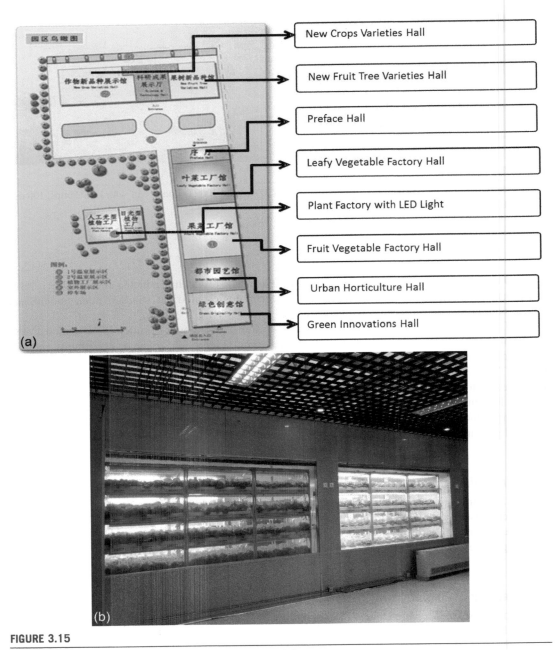

New Crops Varieties Hall

New Fruit Tree Varieties Hall

Preface Hall

Leafy Vegetable Factory Hall

Plant Factory with LED Light

Fruit Vegetable Factory Hall

Urban Horticulture Hall

Green Innovations Hall

FIGURE 3.15

(a, b) Demonstration center of PFAL built in 2012 in CAAS.

Table 3.4 PFALs Built by Beijing IEDA Protected Horticulture Co., Ltd			
Name of PFAL	Area (m²)	Year	City or Province
Changchun	200	2009	Jilin
Zhuhai	30	2010	Guangdong
Nanjing Tangshan Cuigu	300	2010	Jiangsu
National Agricultural Technology Demonstration Park	60	2011	Beijing
Kanjiangislets	12	2011	Guangdong
Ningyang	73	2012	Shandong
Qinlan	9	2012	Shanxi
Bayannur	73	2012	Inner Mongolia
Beijing	76	2012	Beijing
Haerbin	12	2012	Heilongjiang
Fengdong	40	2013	Shanxi

FIGURE 3.16

PFAL built by Beijing Kingpeng International Hi-Tech Corporation in 2010. Left: external appearance of the PFAL, 1300 m²; Right: internal appearance of the PFAL.

FIGURE 3.17

PFAL built in Zhejiang University in 2013. Left: Bird's eye view of the PFAL, 1600 m²; Middle and Right: Movable cultivation beds with LED lighting.

PFAL with LED in Shouguang

Shouguang city plays a very important role for developing protected horticulture in China, as it hosts the China International Vegetable Fair every May, attracting more than 2 million visitors. New technologies for protected horticulture are exhibited at the fair. In 2009, a PFAL with LEDs (40 m^2) was first exhibited at the 10th China International Vegetable Fair, and this new plant production system attracted the attention of millions of visitors. The area of the PFAL was enlarged to 200 m^2 at the 11th Fair in 2010. New technologies for LED energy saving, hydroponics, multicultivation of fruit vegetables and leaf vegetables, intelligent control systems, and so forth were demonstrated at the PFALs (Figure 3.18).

RESEARCH PROJECTS ON PLANT FACTORIES IN CHINA

In 2013, the National High Science & Technology Project on intelligent plant factory production technology (2013–2017; 46 million yuan (1US$ = 6.27 yuan)), organized by CAAS, was supported by the Ministry of Science and Technology of China. Fifteen universities, institutes, and companies joined the project (Table 3.5).

FIGURE 3.18

PFAL in Shouguang city. Left: PFAL with LED lighting, 200 m^2; Middle: multi-cultivation of fruit vegetables; Right: multi-cultivation of leaf vegetables.

Table 3.5 National High Science & Technology Project on Intelligent PFAL Production Technology

Item	Key Technologies	Participating Universities, Institutes or Companies
1	Energy-saving LED light source and intelligent light environment control	CAAS
		Institute of Semiconductors, Chinese Academy of Sciences
		Zhejiang University
		Beijing IEDA Protected Horticulture Co., Ltd.
2	Equipment of multilayer cultivation system	Beijing Agriculture Machinery Institute
		China Agriculture University
		Beijing Kingpeng International Hi-Tech Corporation
		Lhasa National Agricultural Science and Technology Demonstration Park
3	Energy-saving environmental control based on light and temperature coupling	Tech. Top Photoelectric Technology Company
		Nanjing Agricultural University
4	Nutrient solution management and vegetable quality control	CAAS
		Northwest A&F University
		China Agriculture University
		Beijing University of Aeronautics and Astronautics
5	Intelligent control system based on the network management	The National Engineering Research Center for Information Technology in Agriculture
		Vegetable Research Center of Beijing Academy of agriculture and Forestry
		China Agriculture University
		Jilin University
6	Integrated technology demonstration of solar light plant factory	Dushi Green Engineering Company
		Shanghai Jiao Tong University
		Tongji University
7	Integrated technology demonstration of PFAL	Zhejiang University
		IEDA, CAAS

NORTH AMERICA

Chieri Kubota

School of Plant Sciences, The University of Arizona, Tucson, Arizona, USA

HISTORY

The United States introduced PFAL for crop production in the 1980s. Those earlier facilities were producing leafy crops using deep flow technique (DFT) hydroponic systems under high-intensity discharge (HID) lamps inside a building. Among them, a commercial lettuce production facility located in Dekalb (west of Chicago) in Illinois operated for many years before their closure in the early 1990s. In the early 2000s, production of plant-made pharmaceuticals (AKA molecular farming) became a viable application of PFAL. In fact several large-scale commercial facilities were built in the US and Canada to produce pharmaceutical protein products (antigens and antibodies) using genetic modification or transient expression of plants. Then more recently, coinciding with the increasing interest of local food production, several large commercial PFAL facilities were built within close proximity to mega cities (such as Chicago). In addition, as a unique application, a small PFAL was developed to produce fresh vegetables in the US South Pole Station. This 22 m^2 footprint facility produced over 30 different types of crops (several at one time) including tomatoes, lettuces, and herbs for the crew year-round (Patterson et al., 2012). The facility also has an adjacent room (9 m^2) separated by a glass wall for research station personnel to experience the bright plant growing environment. This human psychological support by providing green plants was critical in such an isolated environment as the South Pole.

CONTRIBUTION OF SPACE SCIENCE

It should be noted that many key technologies employed in plant-growing systems in PFAL originated from the US. The most significant technological contributions are hydroponics/soilless culture technique and use of light emitting diodes (LEDs) for plant production. Hydroponic systems were invented as a research tool of plant nutrition and later implemented as a leafy crop production system to provide fresh vegetables for US troops stationed on several islands in the western Pacific during World War II (Jones, 2000). Around 1980, the US National Aeronautics and Space Administration (NASA) initiated its Controlled Ecological Life Support System (CELSS) Program (Wheeler, 2004), where plants were grown hydroponically under artificial (electric) lighting. The production efficiency was a key focus area in NASA's life support research. An efficient lighting system based on LEDs was first developed for NASA's life support applications by a group of scientists and engineers in Wisconsin (Bula et al., 1991). Much of our knowledge regarding light quality requirements was the outcome of many research groups in US land grant institutions funded by NASA. For example, the necessity of adding a small amount of blue light to red light was first reported by Bula et al. (1991) followed by others, establishing the general understanding of light qualities required to grow plants using monochromatic light sources.

CURRENT STATUS AND FUTURE PROSPECTIVE

As of 2014, there were a small number of commercial PFALs for food production in North America. For transplant production, there is at least one company producing grafted tomato seedlings using PFAL. More PFALs could be reported as being operated if statistics included medicinal crop production (such as medicinal *Cannabis spp.* or plant-made pharmaceutical production). However, finding the actual number and production capacity is challenging as there are no statistics generated by reliable organizations. Typical crop species grown in North American PFALs include leaf lettuces, basil, and micro/baby greens. Baby greens and microgreens definitions are often vague, but generally microgreens are at the beginning of first true leaf expansion and baby greens include a few true leaves. Because of the limited shelf life of these small greens, PFALs are well suited for such applications. Typically many different leafy crops are produced in the same growing facility. Some facilities are certified organic producers and/or implement aquaponics.

Regarding facilities, many of these utilize a multitiered production system with LED, induction or fluorescent lamps installed in each tier. The distance between tiers is relatively large ($\sim$1 m) compared with what is commonly seen in Japan and other Asian countries, presumably primarily considering logistics and accessibility to the plants instead of space use or energy use efficiency. CO_2 enrichment is not a common practice as the production systems are often located inside the voluminous space of warehouse buildings. Such buildings often have air handling systems with minimum ventilation to assure human health and the growing space is often shared with CO_2 emitting workers who are engaged in activities such as transplanting, harvesting, and mixing nutrient solution. As a result the CO_2 concentration inside the growing facility is, anecdotally, not at problematic low levels such as we may experience in a truly contained environment. Use of automation is limited. Logistical improvement of workers moving plants from seed to harvest may be a critical area for future R&D.

It is hard to generalize the business model of North American PFALs. However, one prominent trend among the commercial PFALs in North America may be the strong support from high-end retailers (grocery stores) that promote local food production and organic produce. Another trend is collaborations with resort hotels and restaurants. For example, there seem to be a few projects planned to build PFALs in the Las Vegas area to provide the fresh produce consumed by visitors and tourists. Given that the traditional supply chain in North America is driven by a small number of large-scale produce industries with a limited number of production regions (open field) selected based on the production scale and costs (mainly in California), local production of fresh produce is expected to continuously attract American consumers. For example, California (spring to fall) and Arizona (winter) together supply more than 90% of lettuces and other leafy crops consumed in US and Canada. PFAL will not replace this traditional open-field production currently going in California/Arizona but could be a sustainable alternative of specialty crops in the future. It is also critical to have academic research capacity to support this emerging industry sector. Forming industry associations may help but may not work as it does in Asian countries due to the competitive nature of venture capital in North America. The potential immediate role of academia may be providing opportunities for information exchange as well as needed education and training for those PFAL practitioners and planners. Increasing the number of online course opportunities in the area of controlled environment agriculture may help meet such needs.

EUROPE (ENGLAND, THE NETHERLANDS, AND OTHERS)

Chungui Lu

Head of Centre for Urban Agriculture, Centre for Urban Agriculture, University of Nottingham, Leicestershire, United Kingdom

BACKGROUND

Urban agriculture refers to agricultural practices in urban areas and their surrounding regions (peri-urban), and is a centralized operation involving horticulture, animal husbandry, aquaculture, and other practices for producing fresh food or other agricultural products. There are many different approaches to urban agriculture, including ground-level farming, rooftop farming, hydroponics, greenhouses and other new technologies. Urban agriculture has the potential to produce food for local consumption, especially perishables and high-value horticultural crops. Also, there is increasing interest in commercial-scale cultivation of nonfood crops in urban areas, such as flowers, green walls, and the like. Urban agriculture plays a key role in food security and is found in smart cities, which are a phenomenon closely related to urban economies, culture, science, and technology; urban agriculture indicates that a city's economic development has reached a higher level. Compared with other agricultural practices, urban agriculture makes intensive use of capital, facilities, technology, and labor. It is also an industrialized, market-oriented agriculture, and can take advantage of the developed markets, information and transportation networks of international cities to boost agricultural production and interregional trade.

Greenhouses were introduced several decades ago to protect plants from weather conditions. Initially they were used by farmers in farming areas as alternative ways to protect production. More recently, greenhouses including plant factories and roof gardens have been built in cities. An increasing number of companies and researchers have become involved in urban farming, successfully producing fresh food and other products in a sustainable way. Urban agriculture is being seen as an emerging business opportunity in urban areas.

PRESENT STATUS OF PLANT FACTORIES IN THE EU

Agriculture in the European Union faces some serious challenges in the 21st century that include: changes in climate, significant increase in urbanization, competition for water and vital resources, rising costs, decreasing growth in agricultural productivity, competition for international markets, and uncertainty about the effectiveness of current European policies (EGTOP/6/13). Plant factories (protected cultivation) have the potential to address some of the upcoming challenges.

The EU is one of the main global producers by glasshouse, particularly in countries such as the Netherlands, Spain and Italy. Global production in greenhouses is growing, currently with an estimated area of 800,000 ha, of which 20% (160,000 ha) is situated in Europe (EuroStat, 2011). This production system is characterized by the ability to change climatic conditions using various technologies and practices for greenhouse vegetable production.

Plant factory technologies have developed remarkably in the past decade in the EU, including computer-integrated systems for optimizing the growing conditions and efficiently using water, nutrients and energy (Morimoto et al., 1995; Vänninen et al., 2010). In a controlled environment, precise control over light quality, light intensity, photoperiod, humidity, carbon dioxide concentration, nutrient solution, pH, and temperature can all be achieved (Stutte, 2006; Kozai, 2013) to improve productivity and quality. In terms of lighting applications, new LED technology has made it possible to build plant factories.

Philips' Horticulture LED Solutions Group in the Netherlands has been developing LED lighting solutions for horticultural applications for more than 7 years, has proven the commercial feasibility of LED lighting for horticulture, and has recently started working on plant factories with artificial lighting (PFAL). Working in farming, which is defined as multilevel horticulture in a controlled environment, with an open innovation structure with both academic and commercial partners, Philips has developed commercial-scale solutions for city farming that are currently in operation and is working on several more. Examples include Green Sense Farms in the USA, Osaka University in Japan, and National Urban Park in the CAAS in China. The company also has a high-tech research facility at Stockbridge

Technology Centre in the UK, which conducts trials on the optimum light recipes to save energy and make production more profitable for future upscaling.

PlantLab, a privately owned Dutch company that specializes in controlled environment agriculture, was founded in the Netherlands in 2010. PlantLab's new international research center was officially opened by King Willem-Alexander on 30th September 2014 (Figure 3.19a). The company built its first commercial plant production units (PPU) in March 2010, and is currently investing over $22 million in a 200,000 square-foot headquarters and research facility, which will allow optimization of crop production and further improvements in its technology and expertise. PlantLab developed a patented technology to grow crops in plant production units, which are completely closed climate chambers without daylight. The opening of the new facility marks the start of a large-scale roll-out of the PlantLab

(a)

(b)

FIGURE 3.19

PlantLab new headquarters and R&D center (a) and plants growing under LED light (b).

concept with international partners in various industries. PlantLab launched an important first partnership with the Swiss breeding company Syngenta last year. PlantLab's method can be applied in many industries, ranging from the processing and production of vegetables, fruit and flowers, to the production of ingredients for children's food, medicines, flavors, fragrances and cosmetics.

By utilizing the advanced technology of PPUs and proprietary mathematical models, which have been developed over the past 25 years by the founders of PlantLab, as well as state-of-the-art LED systems (Figure 3.19b), air control technologies and optimum water control solutions, PlantLab's innovations remove the typical variables that hinder crop growth and delivery. This means it is now possible to accurately control crop yields, growing times, nutritional content, and other factors such as taste and appearance. In 2015 PlantLab will build a Dutch-based PPU for the production of fresh herbs and lettuces as a showcase for urban farming.

The Netherlands is the most advanced country regarding plant factories and protected glasshouse cultivation in the EU. In other parts of Europe, some small-scale vertical farms have been built, while construction has begun on some large-scale ones. In 2012, the Swiss company UrbanFarmers, a spin-off of the Zurich University of Applied Sciences (ZHAW), built a 260 m^2 greenhouse farm on an industrial rooftop in Basel (Graber et al., 2014). Together with the research team at ZHAW, this aquaponic farm was planned and built, and is now operated and monitored. In Sweden, Plantagon's Greenhouse, which will be a 17-story vertical farm that integrates sustainable growing systems, energy solutions and recycling, is approaching the start of construction (HortiDaily, 2014). Meanwhile, rooftop greenhouses (RTG) with an area of 250 m^2 are being built on rooftops for food production in Barcelona, Spain. Two small-scale land-based cyprinid fish farms that divert recirculating water into a closed-loop system with hydroponic beds have been developed in Slovenia. Such closed-loop systems could contribute to urban food production and higher sustainability by reducing the usage of water and chemicals while minimizing nutrient losses.

OUTLOOK FOR PLANT FACTORIES IN THE UK

Alongside the many challenges, the UK is becoming a world leader in agricultural technology, innovation and sustainability. The vision of the government's new UK strategy for agricultural technologies (www.gov.uk/government/publications/uk-agricultural-technologies-strategy) is to exploit opportunities to develop and adopt new and existing technologies, products and services to increase productivity, and thereby contribute to global food security and international development. To achieve this, the government is investing £70 m in a new Agri-Tech Catalyst. The government will invest £90 million over 5 years to establish a small number of Centers for Agricultural Innovation to support advances in sustainable intensification.

The University of Nottingham (UoN) is a leading academic institution in the agrifood industry; its School of Biosciences is the largest group of plant and crop scientists in any UK university. In order to improve food security, a Center for Urban Agriculture (CUA) has been established. Interestingly, a multidisciplinary group of staff from biosciences, engineering, environmental technology, and economics has been formed, with great potential to deliver research results; the group led by Dr. Chungui Lu has successfully organized two international conferences on urban agriculture and vertical farming. Two research projects on smart LED lighting systems (£1.5 m) and smart glasshouse LEDs (£0.5 m) at UoN have recently received funding under the Innovation UK Agri-Tech Strategy. Four PhD research projects in this area funded by EPSRC and CFFRC have begun. Through the projects, CUA has established a strong multidisciplinary team on sustainable agriculture. The Smart LED Lighting System uses sensors in precision farming to optimize horticultural crop yield, quality, and resource use efficiency in real time, providing light energy tailored to the demands of the plants. The Smart Glasshouse project provides a sustainable solution to the inherent problems of the greenhouse industry by the introduction of a low-cost energy saving and climate control system. Innovative technologies including effective heat-insulating solar glass, vacuum insulation panels, windcatchers, and LED lights will be used to reduce the heating and cooling load of commercial greenhouses, thereby reducing carbon emissions and improving the cost effectiveness of greenhouse production.

CambridgeHOK is a leading designer and constructor of glasshouses and control systems for academic and commercial growers with an annual turnover of £8–12 m. The company is already providing turnkey conventional glasshouse projects around the world, working at the forefront of production technologies. Its recent projects include those for the University of Nottingham, Stockbridge, and Cornerway's Nursery. In 2012 CambridgeHOK won the BCIA award for a control system at The Eden Project.

Stockbridge Technology Centre has built a new multitier LED4CROPS facility with LED lighting for studying the production of a wide range of crops from all sectors of the horticulture industry including ornamentals. The energy and cost price model varies widely according to each situation, so it is important to identify key "first to market" products and countries.

Paignton Zoo is currently experimenting with a high-density vertical growing system with a VertiCrop technology. VertiCrop is an experimental vertical irrigation system (Figure 3.20) designed to grow food for the animals of the zoo. In the summer of 2009, 11,900 plants were simultaneously grown on a set of rotating structures with 8 levels, 3 m high, covering a total ground area of 70 m^2. The system gives five times the yield of a typical field (Bayley and Yu, 2010), and can still be used during the winter months, demonstrating the adaptability of the system with minimal energy inputs.

In the UK, plant factories are still at the developing stage, and hence the scientific community is highly involved. While many small city farms already exist and are being developed, we believe that a truly commercial-scale alternative to conventional growing is yet to be developed. There are many

FIGURE 3.20

The VertiCrop growing production system in Paignton Zoo, UK.

companies in the UK developing products for plant factories. One such pioneer in the UK's hydroponics industry is HydroGarden, which is developing the latest technologies in hydroponic systems and LED lighting, and is moving products to market.

The development of plant factories not only involves "close to market" urban farm solutions but also creates networks of researchers to discuss current ideas, technologies, potential businesses and research. Recently, an International Vertical Farming and Urban Agriculture Conference (VFUA) was held in September 2014 in Nottingham, UK (http://VFUA.org). The conference successfully evaluated the benefits, opportunities, risks and challenges of urban farming and provided a forum for establishing research collaboration and networking among academic researchers and companies. The conference kicked off with Dickson Despommier, who is a pioneer of the VF movement. Jack Ng from Skygreens talked about his commercial vertical farm, which was the world's first. Other keynote speakers from China (Qichang Yang), Japan (Toyoki Kozai) and South Korea (Jung-Eek Son) presented their excellent works on urban agriculture and plant factories. Many companies from the UK, EU, USA, and other countries, including HydroGarden, Phillips, Illumates, PlantLab, and The FarmHere, showcased their products for plant factories and the new technologies they have used.

REFERENCES

Afreen, F., Zobayed, S.M.A., Kozai, T., 2005. Spectral quality and UB-V stress stimulate glycyrrhizin concentration of *Glycyrrhizia uralensis* in hydroponic and pot system. Plant Physiol. Biochem. 43, 1074–1081.

Afreen, F., Zobayed, S.M.A., Kozai, T., 2006. Melatonin in *Glycyrrhiza urarensis*: response of plant roots to spectral quality of light and UB-V radiation. J. Pineal Res. 41 (2), 108–115.

Bayley, J.E., Yu, M., 2010. VertiCrop™ Yield and Environmental Data. Valcent EU Ltd, Launceston, Cornwall. Central Government issue 20.

Bula, R.J., Morrow, R.C., Tibbitts, T.W., Barta, D.J., Ignatius, R.W., Martin, T.S., 1991. Light-emitting diodes as a radiation source for plants. HortSci. 26, 203–205.

Chang, Y.W., Lin, T.S., Wang, J.C., Chou, J.J., Liao, K.C., Jiang, J.A., 2011. The effect of temperature distribution on the vertical cultivation in plant factories with a WSN-based environmental monitoring system. In: 2011 International Conference on Agricultural and Natural Resources Engineering (ANRE-2011) paper ID: 146.

Chun, C., 2014. Selection of crops and optimized cultivation techniques in PFAL. In: Proc. KIEI Seminar 2004-42 (IoT-based agro-systems and new business models). December 16, 2014, Seoul, Korea.

Chun, C., Kozai, T., 2001. A closed transplant production system, A hybrid of scaled-up micropropagation system and plant factory. J. Plant Biotechnol. 3 (2), 59–66.

Eurostat, 2011. Sustainability and Quality of Agriculture and Rural Development.

Fang, W., 2011a. Plant Factory with Solar Light. Harvest farm magazine, Taiwan (in traditional Chinese).

Fang, W., 2011. Some remarks regarding plant factory. Agriculture extension booklet number 67. College of Bioresource and agriculture. National Taiwan University.

Fang, W., 2011c. Totally Controlled Plant Factory. Harvest farm magazine, Taiwan (in traditional Chinese).

Fang, W., 2012. Plant Factory with Artificial Light. Harvest farm magazine, Taiwan (in traditional Chinese).

Fang, W., 2014. Industrialization of plant factory in Taiwan. In: Proceedings of Invited lecture in. Greenhouse Horticulture & Plant Factory Exhibition/Conference (GPEC). Japan Protected Horticulture Association, pp. 131–181 (in Japanese).

Fang, W., Chen, G.S., 2014. Plant Factory: A New Thought for the Future. Grand Times Publisher, Taiwan (in traditional Chinese).

Glaeser, E.L., 2011. Triumph of the City: How Our Greatest Invention Makes Us Richer, Smarter, Greener, Healthier, and Happier. Penguin Press, USA.

Graber, A., Durno, M., Gaus, R., Mathis, A., Junge, R., 2014. UF001 LokDepot, Basel: the first commercial rooftop aquaponic farm in Switzerland. In: International Conference on Vertical Farming and Urban Agriculture A16, p. P24.

HortiDaily, 2014. http://www.hortidaily.com/article/13878/Sweden-Plantagon-to-build-unique-vertical-greenhouse-for-urban-agriculture.

Jones, B., 2000. Hydroponics. A practical Guide for the Soilless Grower. St. Lucie Press, Boca Raton, FL 230 p.

Juo, K.T., Lin, T.S., Chang, Y.W., Wang, J.C., Chou, J.J., Liao, K.C., Shieh, J.C., Jiang, J.A., 2012. The effect of temperature variation in the plant factory using a vertical cultivation system. In: Proceeding of the 6th international symposium on machinery and mechatronics for agriculture and biosystems engineering (ISMAB2012), pp. 963–968.

Kozai, T., 2007. Propagation, grafting and transplant production in closed systems with artificial lighting for commercialization in Japan. Prop. Ornam. Plants 7 (3), 145–149.

Kozai, T., 2009. Plant Factory with Solar Light (written in Japanese: Taiyoko-gata shokubutsu kojo). Ohmsha Ltd., Japan.

Kozai, T., 2012. Plant Factory with Artificial Light (written in Japanese: Jinkoko-gata shokubutsu kojo). Ohmsha Ltd., Japan.

Kozai, T., 2013. Resource use efficiency of closed plant production system with artificial light: Concept, estimation and application to plant factory. Proc. Japan Acad. Ser. B Phys. Biol. Sci. 89, 447–461.

Kozai, T., 2014. Topic and future perspectives of plant factory. In: Proceedings of Invited lecture in. Greenhouse Horticulture & Plant Factory Exhibition/Conference (GPEC). Protected Horticulture Association, pp. 63–96 (in Japanese).

Kozai, T., Chun, C., 2002. Closed systems with artificial lighting for production of high quality transplants using minimum resource and environmental pollution. Acta Hortic. 578, 27–33.

Kozai, T., Afreen, F., Zobayed, S.M.A. (Eds.), 2005. Photoautotrophic (Sugar-Free Medium) Micropropagation as a New Micropropagation and Transplant Production System. Springer, Dordrecht, p. 316.

Kozai, T., Ohyama, K., Chun, C., 2006. Commercialized closed systems with artificial lighting for plant production. Acta Hortic. 711, 61–70 (Proc. Vth IS on Artificial Lighting).

Kubota, C., Chun, C. (Eds.), 2000. Transplant Production in the 21st Century. Kluwer Academic Publishers, Dordrecht, p. 290.

Lee, G.I., 2014. Current status and development plans for PFAL in Korea and foreign countries. In: Proc. KIEI Seminar 2004-42 (IoT-based agro-systems and new business models). December 16, 2014, Seoul, Korea.

Lee, C.Y., Huang, Y.K., Lin, T.S., Shieh, J.C., Chou, J.J., Lee, C.Y., Jiang, J.A., 2013. A smart fan system for temperature control in plant factory. EFITA/WCCA/CIGR 2013, paper ID: C0154.

Morimoto, T., Torii, T., Hashimoto, Y., 1995. Optimal control of physiological processes of plants in a green plant factory. Control. Eng. Pract. 3, 505–511.

Ohyama, K., Murase, H., Yokoi, S., Hasegawa, T., Kozai, T., 2005. A precise irrigation system with an array of nozzles for plug transplant production. Trans. ASAE 48 (1), 211–215.

Patterson, R.L., Giacomelli, G.A., Kacira, M., Sadler, P.D., Wheeler, R.M., 2012. Description, operation and production of the South Pole food growth chamber. Acta Horticult. 952, 589–596.

Stutte, G.W., 2006. Process and product: recirculating hydroponics and bioactive compounds in a controlled environment. HortSci. 41, 526–530.

Takakura, T., Kozai, T., Tachibana, K., Jordan, K.A., 1974. Direct digital control of plant growth -I. Design and operation of the system. Trans. ASAE 17 (6), 1150–1154.

Takatsuji, M., 1979. Plant factory with artificial lighting (written in Japanese: Shokubutsu kojo), Koudan-sha (Blue backs), p. 232.

Takatsuji, M., 2007. Totally Controlled Plant Factory (written in Japanese: Kanzen seigyo-gata shokubutsu kojo). Ohmsha Ltd., Japan.

Vänninen, I., Pinto, D.M., Nissinen, A.I., Johansen, N.S., Shipp, L., 2010. In the light of new greenhouse technologies: 1. Plant-mediated effects of artificial lighting on arthropods and tritrophic interactions. Ann. Appl. Biol. 157, 393–414.

Wheeler, R.M., 2004. Horticulture for Mars. Acta Horticult. 642, 201–215.

Yeh, Y.H.F., Lai, T.C., Liu, T.Y., Liu, C.C., Chung, W.C., Lin, T.T., 2014. An automated growth measurement system for leafy vegetables. Biosyst. Eng. 117, 43–50.

Zobayed, S.M.A., Afreen, F., Kozai, T., 2006. Plant-Environment interactions: Accumulation of Hypericin in dark glands of *Hypericum perforatum*. Ann. Bot. 98, 793–804.

Zobayed, S.M.A., Afreen, F., Kozai, T., 2007. Phytochemical and physiological changes in the leaves of St. John's wort plants under a water stress condition. Environ. Exp. Bot. 59, 109–116.

PLANT FACTORY AS A RESOURCE-EFFICIENT CLOSED PLANT PRODUCTION SYSTEM

Toyoki Kozai[1], Genhua Niu[2]

Japan Plant Factory Association, c/o Center for Environment, Health and Field Sciences, Chiba University, Kashiwa, Chiba, Japan[1] Texas AgriLife Research at El Paso, Texas A&M University, El Paso, Texas, USA[2]

LIST OF ABBREVIATIONS

COP   coefficient of performance for cooling
CPPS   closed plant production system
CTPS   closed transplant production system
EC   electrical conductivity of nutrient solution
FL   fluorescent lamp
LAI   leaf area index
LCA   life cycle assessment
LED   light emitting diode
PAR   photosynthetically active radiation
PFAL   plant factory with artificial lighting
RUE   resource use efficiency
SPPS   sustainable plant production system
VPD   water vapor partial pressure deficit

LIST OF SYMBOLS, VARIABLE AND COEFFICIENT NAMES, UNITS AND EQUATION NUMBERS

A_A (MJ m^{-2} h^{-1})   Electricity consumption of air conditioners (heat pumps) (Equations 4.9, 4.11, 4.12)
A_L (MJ m^{-2} h^{-1})   Electricity consumption of lamps (Equations 4.8, 4.11, 4.12)
A_M (MJ m^{-2} h^{-1})   Electricity consumption of water pumps, air fans, etc. (Equations 4.11 and 4.12)
A_T (MJ m^{-2} h^{-1})   Total electricity consumption ($A_A + A_L + A_M$) (Equation 4.11)
CUE CO_2 use efficiency (Equation 4.3)
COP   Coefficient of performance of heat pumps for cooling (Equations 4.9 and 4.12)
C_{in} (µmol mol^{-1}) CO_2 concentration of room air (Equations 4.4 and 4.13)
C_{out} (µmol mol^{-1}) CO_2 concentration of outside air (Equations 4.4 and 4.13)
C_L (µmol m^{-2} h^{-1}) CO_2 loss to the outside (Equations 4.3 and 4.4)
C_P (µmol m^{-2} h^{-1}) CO_2 fixed by plants (Equations 4.3 and 4.13)

Plant Factory. http://dx.doi.org/10.1016/B978-0-12-801775-3.00004-4

C_R (μmol m^{-2} h^{-1})	CO_2 released in room air by human respiration (Equations 4.3 and 4.13)
C_S (μmol m^{-2} h^{-1})	CO_2 supplied to room air from CO_2 cylinder (Equations 4.3, 4.7, 4.13)
D (μmol m^{-2} h^{-1})	Dry mass increase rate of plants (Equations 4.5 and 4.6)
EUE_L	Electric energy use efficiency (Equation 4.7)
F (m^2)	Floor area of culture room (Equations 4.13, 4.14, 4.16, 4.17)
f (MJ kg^{-1})	Conversion factor from plant dry mass to chemical energy, 20 MJ kg^{-1} (Equations 4.5–4.7)
FUE_I	Inorganic fertilizer use efficiency (Equation 4.10)
h	Conversion factor from electric energy to PAR_L (Equation 4.7 and 4.8)
H_h (MJ m^{-2} h^{-1})	Heat energy removed from culture room by heat pumps (Equation 4.9 and 4.12)
H_V (MJ m^{-2} h^{-1})	Heat energy exchange by air infiltration and penetration through walls (Equation 4.12)
I_{in} (mol mol^{-1})	Ion concentration of 'I' in nutrient solution at the inlet of culture beds (Equation 4.17)
I_{out} (mol mol^{-1})	Ion concentration of 'I' in nutrient solution at the outlet of culture beds (Equation 4.17)
I_S (mol m^{-2} h^{-1})	Supply rate of inorganic fertilizer ion element "I" supplied to the PFAL (Equation 4.10)
I_U (mol m^{-2} h^{-1})	Absorption rate of inorganic fertilizer ion element "I" by plants (Equations 4.10 and 4.17)
k_C (kg m^{-3})	Conversion factor from volume to mass of CO_2 (1.80 kg m^{-3} at 25 °C and 101.3 kPa) (Equations 4.4 and 4.13)
k_{WV} (kg m^{-3})	Conversion factor from volume to mass of water (0.736 kg m^{-3} at 25 °C and 101.3 kPa) (Equation 4.2)
k_{LW} (kg m^{-3})	Conversion factor from volume to mass of liquid water (997 kg m^{-3} at 25 °C and 101.3 kPa) (Equation 4.16)
LUE_L	Light energy use efficiency with respect to PAR_L (Equations 4.5 and 4.7)
LUE_P	Light energy use efficiency with respect to PAR_P (Equation 4.6)
N (h^{-1})	Number of air exchanges (Equations 4.2, 4.4, 4.13, 4.14)
PAR_L (MJ m^{-2} h^{-1})	Photosynthetically active radiation emitted from lamps (Equations 4.5, 4.7, 4.8)
PAR_P (MJ m^{-2} h^{-1})	Photosynthetically active radiation received at plant community surface (Equation 4.6)
V_A (m^3)	Volume of room air (Equations 4.2, 4.4, 4.13, 4.14)
V_{LW} (m^3)	Volume of nutrient solution in culture beds (Equations 4.16 and 4.17)
X_{in} (kg m^{-3})	Water vapor density of room air (Equations 4.2 and 4.14)
X_{out} (kg m^{-3})	Water vapor density of outside air (Equations 4.2 and 4.14)
W_C (kg m^{-2} h^{-1})	Liquid water collected for recycling use in the PFAL (Equations 4.1 and 4.14)
W_L (kg m^{-2} h^{-1})	Water vapor loss rate from the PFAL to the outside (Equations 4.1 and 4.2)
W_P (kg m^{-2} h^{-1})	Water held in plants in the PFAL (Equation 4.1)
W_S (kg m^{-2} h^{-1})	Liquid water supply rate into the PFAL (Equations 4.1 and 4.15)
W_T (kg m^{-2} h^{-1})	Transpiration rate of plants in the PFAL (Equations 4.14 and 4.15)
W_U (kg m^{-2} h^{-1})	Water uptake rate of plants in culture beds in the PFAL (Equations 4.15 and 4.16)
W_{in} (kg m^{-2} h^{-1})	Water inflow rate to hydroponic culture beds in the PFAL (Equations 4.16 and 4.17)
W_{out} (kg m^{-2} h^{-1})	Water outflow rate from hydroponic culture beds in the PFAL (Equations 4.16 and 4.17)
WUE	Water use efficiency (Equation 4.1)

INTRODUCTION

This chapter describes the characteristics and principal components of a plant factory with artificial lighting (PFAL), which is one type of closed plant production system (CPPS). Then, the concept and definition of resource use efficiency (RUE) are described for each resource component.

The characteristics of the PFAL are compared with those of a greenhouse, mainly from the viewpoint of RUE. It is shown that the use efficiencies of water, CO_2 and light energy are considerably higher in the PFAL than in a greenhouse. On the other hand, there is much room for improvement in the light and electric energy use efficiencies of the PFAL. Challenging issues for the PFAL and RUE are also discussed. This chapter is an extended and revised version of Kozai (2013c).

DEFINITION AND PRINCIPAL COMPONENTS OF PFAL

A plant factory with artificial lighting (PFAL), a type of closed plant production system (CPPS), is defined as a warehouse-like structure covered with opaque thermal insulators, in which ventilation is kept at a minimum, and artificial light is used as the sole light source for plant growth (Kozai, 1995, 2005). In a PFAL, the environment for plant growth can be controlled as precisely as desired, regardless of the weather. In addition to the recirculating nutrient solution in a hydroponic system, the water transpired by plants can be condensed and collected at the cooling panel of air conditioners and then recycled for irrigation. Closed transplant production systems and closed micropropagation systems are two kinds of PFAL (Kozai et al., 2000; Kozai et al., 2005).

A PFAL consists of six principal structural elements (Figure 4.1): (1) a thermally well-insulated and nearly airtight warehouse-like structure covered with opaque walls; (2) a multitier system (mostly 4–16 tiers or layers; about 40 cm vertically between tiers) equipped with lighting devices such as fluorescent lamps (FLs) and/or light-emitting diodes (LEDs) over the culture beds; (3) air conditioners (also known as heat pumps), principally used for cooling and dehumidification to eliminate heat generated by lamps and water vapor transpired by plants in the culture room, and fans for circulating air to enhance photosynthesis and transpiration and to achieve a uniform spatial air distribution; (4) a CO_2 delivery unit to maintain CO_2 concentration in the room at around 1000 $\mu mol\ mol^{-1}$ (or ppm) during the photoperiod for enhancing plant photosynthesis; (5) a nutrient solution delivery unit; and (6) an environmental control unit including electrical conductivity (EC) and pH controllers for the nutrient solution (Kozai, 2007; Kozai et al., 2006).

A PFAL must be designed and operated to achieve the following goals: (1) maximize the amount of usable or salable parts of plants using the minimum amount of resources; (2) maintain the highest RUE;

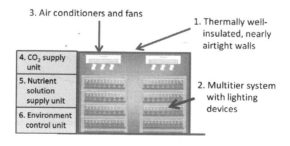

FIGURE 4.1

Configuration of plant factory as a closed plant production system (CPPS) consisting of six principal components. Electricity is needed for operating Nos. 2 to 6.

(3) minimize pollutants released into the environment; (4) minimize costs while achieving the previous three goals (Kozai, 2007; Kozai et al., 2006).

Among the resources, a considerable amount of electrical energy is consumed in a PFAL mainly for lighting and air conditioning. Thus, electrical energy and light energy are the two most important resources for improving the use efficiencies in a PFAL.

DEFINITION OF RESOURCE USE EFFICIENCY

The concept of resource use efficiency (RUE) is schematically shown in Figure 4.2. With respect to essential resources for growing plants in a PFAL, namely, water, CO_2, light energy, electrical energy, and inorganic fertilizer, the use efficiencies per unit time interval ("hour" is used for unit time in this chapter) are defined as in Equations (4.1)–(4.10) and are illustrated schematically in Figure 4.3 (Li et al., 2012a,b; Ohyama et al., 2000; Yokoi et al., 2003, 2005; Yoshinaga et al., 2000). These are water use efficiency (WUE), CO_2 use efficiency (CUE), light energy use efficiency with respect to PAR_L (LUE_L), light energy use efficiency with respect to PAR_P (LUE_P), electrical energy use efficiency (EUE_L), and inorganic fertilizer use efficiency (FUE_I). PAR_L and PAR_P are photosynthetically active radiation emitted from lamps and received at the plant community surface, respectively. The unit for all variables on the right-hand side of Equations (4.1) and (4.3) is kg m^{-2} (floor area) h^{-1}. These use efficiencies are defined with respect to the PFAL containing plants. It should be noted that in the fields of plant ecology and agronomy, WUE is defined with respect to plants or plant community (Salisbury and Ross, 1991). CUE in Equation (4.3) is defined only when CO_2 is supplied or enriched in the PFAL. Methods for estimating the values of variables given on the right-hand side of Equations (4.1)–(4.7) are described in Li et al. (2012a,b,c).

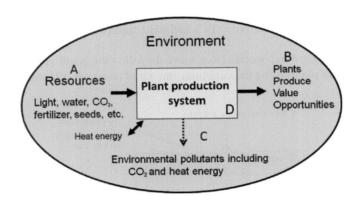

FIGURE 4.2

Scheme showing the concept of resource use efficiency (RUE) of the plant production system. RUE is defined as the ratio of B to A where A = B+C+D. RUE is estimated for each resource component. The plant production system is called a "closed plant production system" (CPPS) when each of the resources is converted into produce at its maximum level, so that the resource consumption and emission of environmental pollutants are minimized, resulting in the maximum RUE and lowest cost for resources and pollution processing. A PFAL is one type of CPPS.

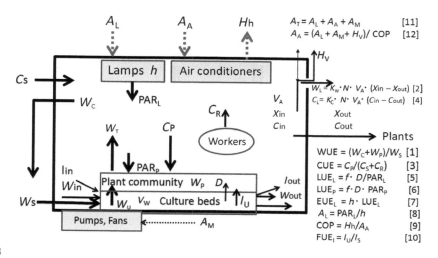

FIGURE 4.3

Schematic diagram showing the rate and state variables in the PFAL. The solid line represents the flow of materials, and the dotted line represents the flow of energy. Numbers in brackets represent equation numbers in the text. For the meanings of symbols, see the list of symbols.

WATER USE EFFICIENCY

Water use efficiency (WUE) can be estimated as follows:

$$\text{WUE} = \frac{W_c + W_P}{W_S} = \frac{W_S - W_L}{W_S} \tag{4.1}$$

$$W_L = V_A N \frac{X_{in} - X_{out}}{F} \tag{4.2}$$

where W_c is the mass (or weight) of water collected at the cooling panel of the air conditioner for recycling $(\text{kg m}^{-2}\,\text{h}^{-1})$; W_P is the change in the mass of water held in plants and substrates $(\text{kg m}^{-2}\,\text{h}^{-1})$; and W_S is the mass of water irrigated (or supplied) to the PFAL $(\text{kg m}^{-2}\,\text{h}^{-1})$; W_L is the mass of water vapor lost to the outside by air permeation through small gaps in the entrance/exit and walls $(\text{kg m}^{-2}\,\text{h}^{-1})$. In Equation (4.2), N is the number of air exchanges per hour in the culture room (h^{-1}); V_A is the air volume in the culture room (m^3); F is the floor area of the culture room (m^2); and X_{in} and X_{out} are, respectively, the mass of water vapor per unit volume of air containing water vapor inside and outside the culture room (kg m^{-3}). Generally, no liquid wastewater is drained from the culture room to the outside. If there is any, it must be added as a variable on the right-hand side of Equation (4.1).

It is assumed in Equation (4.2) that X_{in} at time t is the same as at time $(t + \delta t)$ where δ is the estimation time interval. If not, the term $\frac{X_{in}(t) - X_{in}(t + \delta)V_A}{\delta t}$ is added on the right-hand side of Equation (4.2), the value of which is negligibly small in most cases. In a case where X_{out} changes with time during the estimation time interval, the average value is used in Equation (4.2).

WUE is 0.93–0.98 in the PFAL (Figure 4.4, Table 4.1) and 0.02–0.03 in a greenhouse, meaning that WUE of the PFAL is approximately 30–50 (nearly equal to 0.95/0.03–0.98/0.02) times greater than that

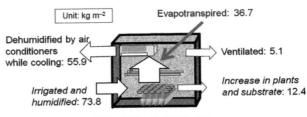

$$WUE = (55.9 + 12.4)/73.8 = 0.93$$

FIGURE 4.4

An experimental result of water use efficiency (WUE) for 14 days in the CPPS. In the case of a greenhouse, the evapotranspired water cannot be reused. Relative humidity was kept at about 80% and air temperature at 30 °C (Ohyama et al., 2000).

Table 4.1 Estimated Representative Values of Use Efficiencies of Water (WUE), CO_2 (CUE), Light Energy (LUEL), and Electrical Energy (EUEL) for a PFAL (Plant Factory with Artificial Lighting), and WUE, CUE, and LUEP for a Greenhouse with Ventilators Closed and/or Open

Use Efficiency	PFAL	Greenhouse with Ventilators Closed and Enriched CO_2	Greenhouse with Ventilators Open	Theoretical Maximum Value for PFAL
WUE (water)	0.95–0.98	N/A	0.02–0.03	1.00
CUE (CO_2)	0.87–0.89	0.4–0.6	N/A	1.00
LUE_L (lamps, PAR_L)	0.027	–	–	About 0.10
LUE_P (plant community)	0.032–0.043 0.05	N/A	0.017 0.003–0.032	About 0.10
EUE_L (electricity)	0.007	–	–	About 0.40

LUE_L and EUE_L are defined only for a PFAL using artificial lamps. N/A (not available) means that the efficiencies of WUE, CUE, and LUE_P can be obtained experimentally but the data were not found in the literature.
The maximum value for a PFAL based on a theoretical consideration of each use efficiency is also given in the column "Theoretical maximum value." For the definitions of WUE, CUE, LUEL, EUEL, and EUEP, see Equations (4.1)–(4.7) in the text.
The numerical data in this table is cited from or based on the literature Bugbee and Salisbury (1988), Chiapale et al. (1984), Li et al. (2012a,b,c), Mitchell et al. (2012), Ohyama et al. (2000), Sager et al. (2011), Shibuya and Kozai (2001), Yokoi et al. (2003, 2005), Wheeler (2006), Wheeler et al. (2006).

of a greenhouse (Ohyama et al., 2000; Kozai, 2013a). Namely, the PFAL is a highly water-saving plant production system compared with a greenhouse. It is noted that WUE increases with increasing leaf area index (LAI) and decreasing N (Figure 4.5). Thus, airtightness of the CPPS is essential to achieve a high WUE value.

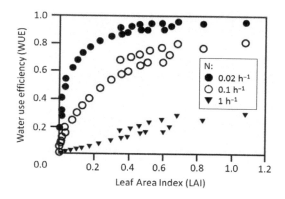

FIGURE 4.5

WUE (water use efficiency) of the CPPS as affected by LAI and N (number of air exchanges: 0.02, 0.1, and 1 h^{-1}). Water vapor density inside and outside the closed system was assumed to be 16 and 6 g m^{-3}, respectively. All values are simulated values (Yokoi et al., 2005).

A PFAL with thermally insulated walls and a high level of airtightness (N is 0.01–0.02 h^{-1}) must be cooled by air conditioning while the lamps are turned on, even during the cold winter nights in order to maintain a suitable internal temperature by eliminating the heat generated from the lamps. With cooling, a large portion of the evaporated water, W_C in Equation (4.1), can be collected as condensation on the cooling panels of the heat pump, and recycled for irrigation. Only a small percentage of W_s is lost to the outside, thanks to the high airtightness of the PFAL. The N value must be lower than about 0.02 h^{-1} to minimize CO$_2$ loss to the outside and prevent entry of insects, pathogens, and dust into the culture room.

On the other hand, water vapor evapotranspired in a greenhouse cannot be collected for recycling, because most of it is lost to the outside through ventilation, while the rest of the water vapor mostly condenses on the inner surfaces of the greenhouse walls and so cannot be collected. The amount of water vapor lost to the outside, W_L in Equation (4.2), increases with increasing $(X_{in} - X_{out})$ and N of a greenhouse with ventilators open. N varies in the range of 0.5–100 h^{-1} depending on the number of ventilators and degree of opening as well as the wind velocity outside (Chiapale et al., 1984).

If all the lamps are turned off, the relative humidity of the room air in a PFAL approaches 100%, resulting in little transpiration, which may cause physiological disorders of the plants. To avoid this, the tiers in the PFAL are often divided into two or three groups and the lamps in each group are turned on for 12–16 h per day in rotation to generate heat from the lamps at any time of the day so that the heat pumps are running 24 h for dehumidification and cooling of room air.

CO$_2$ USE EFFICIENCY

CO$_2$ use efficiency (CUE) is defined as:

$$CUE = \frac{C_P}{C_S + C_R} = \frac{C_S - C_L}{C_R + C_S} \tag{4.3}$$

$$C_L = k_C N \frac{V_A (C_{in} - C_{out})}{F} \qquad (4.4)$$

where C_P is the net photosynthetic rate ($\mu mol\ m^{-2}\ h^{-1}$); C_S is the CO_2 supply rate ($\mu mol\ m^{-2}\ h^{-1}$); C_R is the respiration rate of workers in the culture room ($\mu mol\ m^{-2}\ h^{-1}$), if any; and C_L is the rate of loss of CO_2 to the outside due to air infiltration ($\mu mol\ m^{-2}\ h^{-1}$). C_R can be estimated using data on the number of workers, working hours per person, and hourly amount of CO_2 released by human respiration per person (about 0.05 kg h^{-1}) (Li et al., 2012b). In Equation (4.4), N, F, and V_A are the same as in Equation (4.2), and C_{in} and C_{out} are, respectively, CO_2 concentration inside and outside the culture room ($\mu mol\ mol^{-1}$); k_C is the conversion factor from mol to volume (0.0245 $m^3\ mol^{-1}$ at 25 °C). It should be noted that CO_2 is often purchased by weight. The conversion factor from volume to mass (weight) is 1.80 kg m^{-3} at 25 °C.

It is assumed in Equation (4.4) that C_{in} at time t is the same as at time $(t+\delta)$ where δ is the estimation time interval. If not, the term $\frac{C_{in}(t) - C_{in}(t+\delta) V_A}{\delta t}$ is added on the right-hand side of Equation (4.4), the value of which is negligibly small in most cases. In a case where C_{out} changes with time during the estimation time interval, the average value is used in Equation (4.4).

CUE is about 0.87–0.89 in a PFAL with N of 0.01–0.02 h^{-1} and CO_2 concentration of 1000 $\mu mol\ mol^{-1}$ (Figure 4.6, Table 4.1), whereas CUE is around 0.5 in a greenhouse with ventilators closed having N of about 0.1 h^{-1} and enriched CO_2 concentration of 700 $\mu mol\ mol^{-1}$ (Yoshinaga et al., 2000; Ohyama et al., 2000). Thus, CUE is roughly 1.8(=0.88/0.50) times higher in a PFAL than in a greenhouse with all ventilators closed and CO_2 enrichment (Yokoi et al., 2005). This is because the amount of CO_2 released to the outside, C_L in Equation (4.4), increases with increasing N and $(C_{in} - C_{out})$ (Figure 4.6). It is thus natural that the set point of CO_2 concentration for CO_2 enrichment is generally higher (1000–2000 $\mu mol\ mol^{-1}$) in a PFAL than in a greenhouse (700–1000 $\mu mol\ mol^{-1}$).

When N and $(C_{in} - C_{out})$ of the PFAL are constant with time, CUE increases with increasing LAI (leaf area index or ratio of leaf area to cultivation area) in the range of 0–3 (Yoshinaga et al., 2000). This is because, in Equation (4.3), C_p increases with increasing LAI but C_L is not affected by LAI. CUE also increases with increasing net photosynthetic rate per unit planting area (Figure 4.7). In order to maintain a high CUE regardless of LAI, the set point of the CO_2 concentration for CO_2 enrichment should be increased with increasing LAI and/or the net photosynthetic rate.

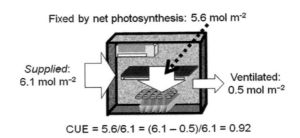

Fixed by net photosynthesis: 5.6 mol m⁻²

Supplied:
6.1 mol m⁻²

Ventilated:
0.5 mol m⁻²

CUE = 5.6/6.1 = (6.1 − 0.5)/6.1 = 0.92

FIGURE 4.6

An experimental result of CUE (CO_2 use efficiency) during 15 days in the CPPS at $N = 0.01\ h^{-1}$. CO_2 concentration during the photoperiod was kept at about 1000 ppm (Yoshinaga et al., 2000). One mole of CO_2 is equal to 44 g.

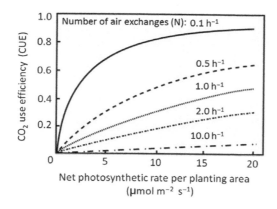

FIGURE 4.7

CUE (CO_2 use efficiency) as affected by net photosynthetic rate and N (number of air exchanges). Assumptions for calculation: floor area: 1000 m^2, room air volume: 3000 m^3; N: 0.1, 0.5, 2.0, and 10 h^{-1}; CO_2 conc. inside and outside: 1000 and 350 μmol mol^{-1}; air temperature inside and outside: 27 °C; planting area is covered with plants; soil respiration is negligibly small (Yoshinaga et al., 2000).

LIGHT ENERGY USE EFFICIENCY OF LAMPS AND PLANT COMMUNITY

The light energy use efficiencies of lamps and plant community (LUE_L and LUE_P) are defined as:

$$LUE_L = \frac{fD}{PAR_L} \qquad (4.5)$$

$$LUE_P = \frac{fD}{PAR_P} \qquad (4.6)$$

where f is the conversion factor from dry mass to chemical energy fixed in dry mass (about 20 MJ kg^{-1}); D is the dry mass increase rate of whole plants or salable parts of the plants in the PFAL (kg m^{-2} h^{-1}); and PAR_L and PAR_P are, respectively, photosynthetically active radiation (PAR) emitted from the lamps and received at the plant community surface in the PFAL (MJ m^{-2} h^{-1}).

LUE_L and LUE_P can also be defined, respectively, as $b \times C_p/PAR_L$ and $b \times C_p/PAR_P$ where b is the minimum PAR energy to fix one mole of CO_2 in plants (0.475 MJ mol^{-1}) and C_p is the net photosynthetic rate of plants (μmol m^{-2} h^{-1}). The ratio of PAR_P to PAR_L is often referred to as the "utilization factor" in illumination engineering.

The average LUE_P for tomato seedling production in a greenhouse was 0.017 (Shibuya and Kozai, 2001). On the other hand, the average LUE_L for tomato seedling production in a PFAL was 0.027 (Yokoi et al., 2003). It should be noted, however, that LUE_L of the PFAL in Equation (4.5) is estimated based on PAR_L emitted from the lamps (Takatsuji and Mori, 2011; Yokoi et al., 2003), whereas LUE_P in Equation (4.6) of a greenhouse is estimated based on PAR_P received at the community surface. The ratio PAR_P/PAR_L is estimated to be 0.63–0.71 (Yokoi et al., 2003). Thus, LUE_P in the PFAL would be approximately in the range between 0.038(=0.027/0.71) and 0.043(=0.027/0.63), which is 1.9 (=0.032/0.017) to 2.5(=0.043/0.017) times higher than that in a greenhouse.

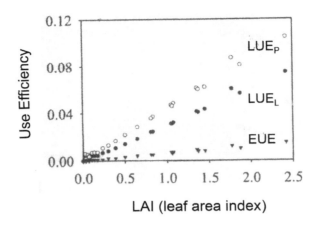

FIGURE 4.8

LUE_P (LUE with respect to PAR at the plant community level), LUE_L (LUE with respect to PAR emitted from lamps) and EUE (electric energy use efficiency) as affected by LAI (leaf area index) (Yokoi et al., 2003). PAR: photosynthetically active radiation.

The maximum LUE_P, which is closely related to the inverse of "quantum yield," is estimated to be in the order of 0.1 (Salisbury and Ross, 1991; Bugbee and Salisbury, 1988). The LUE_P of various crops in space research by NASA in the US is around 0.05 (Massa et al., 2008; Sager et al., 2011; Wheeler, 2006; Wheeler et al., 2006). It should be noted that the LUE_P and LUP_L values shown in Figure 4.8 are significantly higher than those found by NASA; further careful experiments are necessary to verify the accuracy of measurements. Regardless, there should be room to improve the LUE_L and LUE_P of the PFAL. Approaches to doing so will be described later in this chapter.

ELECTRICAL ENERGY USE EFFICIENCY OF LIGHTING

The electrical energy use efficiency of lighting (EUE_L) is defined as:

$$EUE_L = hLUE_L = \frac{fhD}{PAR_L} \tag{4.7}$$

where h is the conversion coefficient from electrical energy to PAR_L energy, which is around 0.25 for white fluorescent lamps, and 0.3–0.4 for recently developed LEDs (Mitchell et al., 2012; Takatsuji and Mori, 2011). The electrical energy consumed by lamps, A_L, can be expressed as:

$$A_L = \frac{PAR_L}{h} \tag{4.8}$$

EUE_L, LUE_L and LUE_P in a PFAL increase almost linearly with increasing LAI at LAI of 0–2.5 (Yokoi et al., 2003) (Figure 4.8).

ELECTRICAL ENERGY USE EFFICIENCY OF HEAT PUMPS FOR COOLING

The electrical energy use efficiency of heat pumps for cooling (COP) is defined as:

$$COP = \frac{H_h}{A_A} \tag{4.9}$$

where H_h is the heat energy removed from the culture room by heat pumps (air conditioners) (MJ m^{-2} h^{-1}) and A_A is the electricity consumption of the heat pumps (MJ m^{-2} h^{-1}). This efficiency is often referred to as the coefficient of performance (COP) for cooling of heat pumps; more information regarding COP is described later in this chapter.

INORGANIC FERTILIZER USE EFFICIENCY

Inorganic fertilizer use efficiency (FUE$_I$) is defined as:

$$FUE_I = \frac{I_U}{I_S} \tag{4.10}$$

where I_U is the absorption rate of ion element I of inorganic fertilizer by plants, and I_S is the supply rate of I to the PFAL. I includes nitrogen (NO_3^-, NO_4^+), phosphorus (PO_4^-), potassium (K^+), etc. N is dissolved in water as NH_4^+ or NO_3^-, but is counted as N in Equation (4.10). Thus, N use efficiency, P use efficiency, K use efficiency, and so forth, can be defined for each nutrient element.

Culture beds in the PFAL are isolated from soil, and the nutrient solution drained from the culture beds is returned to the nutrient solution tank for recirculation. Nutrient solution is rarely discharged to the outside, usually about once or twice a year when certain ions such as Na^+ and Cl^-, which are not well absorbed by plants, have excessively accumulated in the nutrient solution or when the culture beds have become accidentally contaminated by certain pathogens. In these cases, the supply of fertilizer is stopped for several days before removing the plants from the culture beds so that most nutrient elements in the culture beds and nutrient tank are absorbed by the plants. As a result, the water discharged from the PFAL to the outside is relatively clean. Therefore, FUE$_I$ of the PFAL should be fairly high, although no literature regarding this subject could be found.

On the other hand, FUE$_I$ of a greenhouse and an open field is relatively low, which occasionally causes salt accumulation on the soil surface (Nishio, 2005). In an open field, excess supply of nitrates and phosphates occasionally causes nutrient run-off and leaching, resulting in eutrophication of rivers and lakes (Nishio, 2005; Sharpley et al., 2003).

REPRESENTATIVE VALUES OF RESOURCE USE EFFICIENCY

WUE, CUE, LUE and EUE for a PFAL found in the literature are summarized in Table 4.1 and compared with those of a greenhouse with ventilators closed and/or open, together with their theoretical maximum values. In the table, LUE$_L$ and EUE$_L$ are defined only for a PFAL using lamps; N/A (not available) means that the efficiencies of WUE, CUE and LUE$_P$ can be obtained experimentally but data were not found in the literature.

The table shows that WUE and CUE of the PFAL are considerably higher than those of a greenhouse with ventilators open. LUE_P of PFAL ranges from 0.032 to 0.043, while LUE_P of a greenhouse with ventilators open varies from 0.003 to 0.032. As for PFAL, LUE_L of 0.027 is 63–84% of LUE_P of 0.032–0.043, which means that PAR_P was 63–84% of PARL.

LUE_L and EUE_L in Table 4.1 are, respectively, 0.027 and 0.007, which means that 2.7% of PAR energy and 0.7% of electrical energy were converted into chemical energy contained in dry matter of plants. The chemical energy contained in dry matter is 20.0 MJ kg^{-1} (Equations 4.5 and 4.6). Thus, the electricity required to produce 1 kg of dry matter is 2857($=20.0/0.007$) MJ/kg or 794($=2857/3.6$) kWh. Similarly, the PAR energy required to produce one kg of dry matter is 740($=20.0/0.027$) MJ/kg or 205 ($=740/3.6$) kWh.

The WUE and CUE in a PFAL for tomato seedling production were determined by Yokoi et al. (2005). WUE and CUE increased as LAI increased and the number of air exchanges decreased. The maximum WUE (0.95–0.98) and CUE (0.87–0.89) were obtained at the maximum LAI of 1.2 and minimum number of air exchanges of 0.02 h^{-1}.

ELECTRICITY CONSUMPTION AND COST

Commercial production from a PFAL is limited to value-added plants because the electricity consumption for lighting to increase the dry mass of plants is significant: the electricity cost typically accounts for 25% of the total production cost (Kozai, 2012). Accordingly, the plants that are suitable for production in a PFAL are those that can be grown at high planting density (50–1000 plants/m^2) to a harvestable stage in a short cultivation period (30–60 days) under relatively low light intensity (100–300 µmol m^{-2} s^{-1}). Also, mature plants need to be shorter than around 30 cm because the multitier system must be used to increase annual productivity per unit floor area.

The electricity consumption in the culture room of a PFAL per unit floor area (MJ m^{-2} h^{-1}), A_T, and its components are given by:

$$A_T = A_L + A_A + A_M \tag{4.11}$$

$$A_A = \frac{A_L + A_M + H_V}{COP} = \frac{H_h}{COP} \tag{4.12}$$

where A_L, which is expressed as PAR$_L/h$ in Equation (4.8), is the electricity consumption for lighting; A_A is that for air conditioning; and A_M is that for other equipment such as air circulation fans and nutrient solution pumps. H_V is the cooling load due to air infiltration and heat penetration through the walls, both of which account for only a small percentage of A_A. H_h is the heat energy removed from the PFAL using heat pumps. COP is the coefficient of performance of heat pumps as defined by Equation (4.9), which increases with decreasing outside temperature at a given room air temperature.

On the annual average, A_L, A_A, and A_M of a PFAL account for, respectively, 80%, 16%, and 4% of A_T in Tokyo (Ohyama et al., 2002b) (Table 4.2). The annual average COP for cooling in a PFAL with room air temperature of 25 °C was 5–6 in Tokyo where the annual average temperature is about 15 °C; COP was approximately 4 in summer and 10 in winter (Ohyama et al., 2002a) (Figure 4.9). Thus, in order to reduce the total electricity consumption, it is most efficient to reduce A_L by improving LUE_L, followed by improving COP.

Table 4.2 Percentages of Annual Electricity Consumption by Components (Ohyama and Kozai, 2004)

Purpose	Percentage	Equipment
Lighting	80%	Fluorescent lamps 40 W
Cooling	16%	Heat pumps (air conditioners)
Others	4%	Water pumps, fans, etc.

The COP of the heat pump is estimated to be ca. 5.25(=(80+4)/16=5.25).

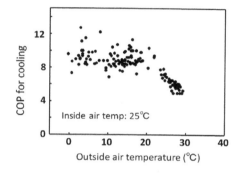

FIGURE 4.9

COP for cooling of room air conditioner (or heat pump) as affected by outside air temperature (Ohyama et al., 2002a).

IMPROVING LIGHT ENERGY USE EFFICIENCY

There are various methods to improve LUE_L in Equation (4.5) and thus EUE_P in Equation (4.6) (Kozai, 2011). However, as there has been little research on improving LUE_L, some of the methods described in the following paragraphs are not based on experimental data, but rather on a theoretical consideration and practical experience in commercial plant production using PFALs.

Figure 4.10 shows the process of converting electrical energy into chemical energy for the salable portion of plants with current representative values. At present, less than one percent of electrical energy is converted into the salable portion of plants; the remaining ca. 99% is converted into heat energy, which is removed to the outside by air conditioners. The research described in the following paragraphs shows that this conversion factor can be improved to 3% or higher (Figure 4.11).

INTERPLANT LIGHTING AND UPWARD LIGHTING

Traditional overhead lighting provides an uneven light distribution and causes mutual shading in a dense canopy and senescence of lower leaves. Interplant or intracanopy lighting can provide more light to the lower leaves, improve the light distribution and thus enhance photosynthesis of the whole canopy. This is because the net photosynthetic rate of the lower leaves, which is often negative or nearly zero, will be made positive by interplant lighting (Dueck et al., 2006).

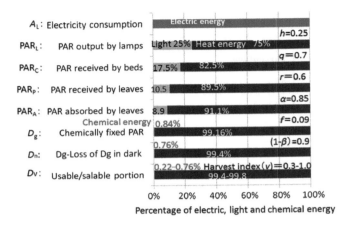

FIGURE 4.10

Process of converting electrical energy for lighting in the PFAL into chemical energy contained in the salable portion of plants.

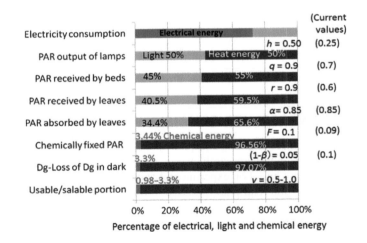

FIGURE 4.11

Maximum possible and existing processes of converting electrical energy for lighting in the PFAL into chemical energy contained in the salable portion of plants.

The benefits of interplant lighting have already been reported in greenhouse plant production and under controlled environments. Adding additional light within a cowpea canopy significantly delayed senescence of the interior leaves in the canopy (Frantz et al., 2000). Greenhouse cucumber yields were increased through intracanopy supplemental lighting in Canada and Europe (Hao et al., 2012; Pettersen et al., 2010). LEDs are a feasible light source for interplant lighting because of their small volume and lower surface temperature compared to other light sources.

IMPROVING THE RATIO OF LIGHT ENERGY RECEIVED BY LEAVES

The LUE_L increases with increasing ratio of the light energy received by leaves to the light energy emitted by lamps. This ratio can be improved by well-designed light reflectors, reduction in vertical distance between lamps and plants, increase in distance between plants (or planting density) as plants grow, and other such measures (Massa et al., 2008). The ratio increases linearly from zero up to unity with increasing LAI from 0 to 3 (Yokoi et al., 2003). Thus, LUE_L can be significantly improved by maintaining LAI at around 3 throughout the culture period by automatic or manual spacing of plants.

USING LEDs

A direct method of improving EUE_L is to use a light source with a high h value in Equation (4.7). The h value for recently developed LEDs is around 0.4 (Liu et al., 2012), whereas that of fluorescent lamps (FL) is around 0.25 (Kozai, 2011). Although the price of LEDs with h of about 0.4 is several times higher than that of FL as of 2013, the price has decreased considerably every year and this trend will continue. The spectral distribution or light quality of LED lamps affects plant growth and development and consequently LUE_L (Jokinen et al., 2012; Liu et al., 2012; Massa et al., 2008; Mitchell et al., 2012; Morrow, 2008).

CONTROLLING ENVIRONMENTAL FACTORS OTHER THAN LIGHT

LUE_L is largely affected by the plant environment and ecophysiological status of plants, as well as by the genetic characteristics of plants. An optimal combination of temperature, CO_2 concentration, air current speed, water vapor pressure deficit, and the composition, pH and electrical conductivity (EC) of the nutrient solution must be found in order to improve LUE_L for each crop (Bugbee and Salisbury, 1988; Evans and Poorter, 2001; Dueck et al., 2006; Kozai, 2011; Liu et al., 2012; Morrow, 2008; Nederhoff and Vegter, 1994; Sager et al., 2011; Thongbai et al., 2011; Wheeler, 2006; Wheeler et al., 2006; Yabuki, 2004).

The optimal light level for plant growth and development varies with species and growth stage. For example, high light intensity and high CO_2 concentration may have a limited effect on increasing biomass when plants are too small. Similarly, no further increase in light intensity is needed when the plant response to light reaches a plateau. Therefore, it is important to find the most economic minimum light levels in a PFAL for crops in order to increase LUE_L.

CONTROLLING AIR CURRENT SPEED

A horizontal air current speed of 0.3–0.5 m s^{-1} within the plant community enhances the diffusion of CO_2 and water vapor from room air into the stomatal cavity in leaves, and consequently photosynthesis, transpiration and plant growth, if the vapor pressure deficit (VPD) is controlled at an optimum level (Yabuki, 2004). The control of air current speed in accordance with plant growth stage by changing the rotation speed of fans is an advantage of the PFAL for improving plant growth. Yokoi et al. (2007) showed that the growth and uniformity of tomato seedlings in a CPPS are enhanced at a higher air current speed (0.7 m s^{-1}) than at a lower speed (0.3 m s^{-1}). They also reported suppression of growth and less uniformity at higher planting density at higher air current speed than at low air current speed.

INCREASING THE SALABLE PORTION OF PLANTS

Electricity is consumed to produce whole plants consisting of leaves with petioles, roots, stems, and sometimes flower stalks and buds. Thus, in order to improve EUE_L and LUE_L of the salable parts of plants, it is important to minimize the dry mass of the nonsalable portion. Leaf vegetables such as lettuce plants should be grown with a minimum percentage of root mass, typically less than 10% of the total mass. This is not so difficult because water stress of plants can be minimized by controlling VPD of the room air and water potential of the nutrient solution in the culture beds of a PFAL. In the case of root crops such as turnip plants, the salable portion can be significantly increased by harvesting them earlier than usual so that the aerial part will be edible. This is often practiced in commercial PFALs (Figure 4.12).

INCREASING ANNUAL PRODUCTION CAPACITY AND SALES VOLUME PER UNIT LAND AREA

The relative annual production capacity per unit land area of an existing PFAL with 10 tiers is more than 100 and 150 times, respectively, that of an open field and can be 200 and 250 times, respectively, by improving the following factors: (1) shortening the culture period from transplanting to harvest through optimal environmental control; (2) increasing the ratio of cultivation area of each tier to floor area; and (3) increasing percentages of salable plants and salable portion of plants. The relative annual sales price of PFAL-grown lettuce is generally 1.2–1.5 times that of field-grown plants due to improved quality, resulting in an increase in relative annual sales volume.

FIGURE 4.12

Salable and unsalable portions (roots and trimmed leaves) of lettuce plants grown in the PFAL. Trimmed leaves are yellowed, small outer leaves and/or damaged. Fresh weight of roots and trimmed leaves account for less than 10% and ca. 5% of total fresh weight, respectively.

ESTIMATION OF RATES OF PHOTOSYNTHESIS, TRANSPIRATION, AND WATER AND NUTRIENT UPTAKE

Interactions between plants and their environments in a PFAL equipped with a hydroponic culture system are fairly simple compared with those in a greenhouse and open field. This is because the PFAL is thermally well-insulated and nearly airtight, so the environment is not affected by the weather, especially the fluctuations in solar radiation with time. Therefore, the energy and mass balance of the PFAL and related rate variables (namely, flows that include the time dimension in units) such as those given in Equations (4.1) and (4.2) can be measured, evaluated and controlled relatively easily (Li et al., 2012a,b,c).

The quality of environmental control will be significantly improved by monitoring, visualizing, understanding and controlling the RUE [WUE, CUE, LUE_L, EUE_L and FUE] in Equations (4.1)–(4.10) and rate variables. The rate variables include the rates of net photosynthesis (gross photosynthesis minus dark and photorespirations), dark respiration, transpiration, water uptake, and nutrient uptake, which represent the ecophysiological status of the plant community. Another group of rate variables includes the supply rates of CO_2, irrigation water, fertilizer constituents, and electricity consumption by its components in the PFAL.

The set points of environmental factors can be determined, for example, for maximizing LUE_L, with the minimum resource consumption or at the lowest cost (Kozai, 2011). In order to maximize the cost performance (ratio of economic benefit to production cost) instead of maximizing LUE_L, the concept and method of integrative environmental control must be introduced (Dayan et al., 2004; Kozai et al., 2011a). Continuous monitoring, visualization and control of the rate variables and RUE will become important research subjects in integrated environmental control for plant production using the PFAL (Dayan et al., 2004; Kozai et al., 2011a).

NET PHOTOSYNTHETIC RATE

The net photosynthetic rate (C_p) at time t during the time interval of δt can be estimated by modifying Equations (4.3) and (4.4) as follows:

$$C_P = C_S + C_R - \left(k_C N \frac{V_A}{F}\right)(C_{in} - C_{out}) + \frac{\left(\frac{k_C V_a}{F}\right)\left(C_{in}\left(t + \frac{\delta t}{2}\right) - C_{in}\left(t - \frac{\delta t}{2}\right)\right)}{\delta t} \qquad (4.13)$$

where the fourth term expresses the change in CO_2 mass in the air of the culture room during the time interval of δt between $t - \frac{\delta t}{2}$ and $t + \frac{\delta t}{2}$. In practical applications, δt would be 0.5–1.0 h. C_P can be estimated fairly accurately using Equation (4.13), since (1) k_c and V_A are constant; (2) C_S, C_{in}, and C_{out} can be accurately measured; and (3) N can be estimated using Equation (4.13) as described later. Dark respiration rate can be estimated using the same Equation (4.13) during the dark period. Under steady-state conditions, Equation (4.13) is simplified and estimation of C_P and CUE is easy (Figure 4.13).

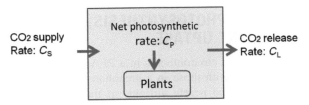

$$C_L = k N V (C_{in} - C_{out}), \text{ CUE} = C_P/C_S = (1-C_L)/C_S$$
For k, N, V, C_{in}, and C_{out}, see List of symbols.

FIGURE 4.13

Scheme showing the method for estimating hourly net photosynthetic rate (C_S) and CUE based on the CO_2 balance of the PFAL under steady-state conditions. For unsteady conditions, see Equation (4.13). The set point of C_{in} is determined by considering CUE, costs for C_S and C_L, and benefit by C_P.

TRANSPIRATION RATE

In a PFAL, the transpiration rate of plants grown in the hydroponic culture beds in the PFAL at time t during the time interval of δt, W_T, can be estimated by:

$$W_T = W_C + \left(N \frac{V_A}{F}\right)(X_{in} - X_{out}) + \frac{\left(\frac{V_A}{F}\right)\left(X_{in}\left(t + \frac{\delta t}{2}\right) - X_{in}\left(t - \frac{\delta t}{2}\right)\right)}{\delta t} \tag{4.14}$$

where the third term expresses the change in water vapor mass in the air of the culture room during the time interval of δt between $t - \frac{\delta t}{2}$ and $t + \frac{\delta t}{2}$. Since V_A and F are constants, and W_C, X_{in} and X_{out} can be measured relatively accurately, N can be estimated for the time interval of $t - \frac{\delta t}{2}$ and $t + \frac{\delta t}{2}$ using Equation (4.14') (Dayan et al., 2004; Kozai et al., 2011a,b; Li et al., 2012a):

$$N = \frac{W_T - W_C - \frac{\left(\frac{V_A}{F}\right)\left(X_{in}\left(t + \frac{\delta t}{2}\right) - X_{in}\left(t - \frac{\delta t}{2}\right)\right)}{\delta t}}{\left(\frac{V_A}{F}\right)(X_{in} - X_{out})} \tag{4.14'}$$

In the PFAL, no evaporation from the substrate in the culture can be assumed, because the culture beds are covered with plastic panels with small holes for growing plants. Furthermore, neither evaporation nor condensation from/to the walls and floor can be assumed, because the walls and floor are thermally well-insulated.

WATER UPTAKE RATE BY PLANTS

Water uptake rate (W_U) can be expressed by modifying Equation (4.14) as:

$$W_U = W_P + W_t \tag{4.15}$$

where W_P is the change in water mass of plants during time interval δt which is difficult to estimate accurately. On the other hand, W_U can be estimated relatively easily by the equation:

$$W_U = W_{in} - W_{out} - \frac{\left(\frac{k_{LW}}{F}\right)\left(V_{LW}\left(t + \frac{\delta t}{2}\right) - V_{LW}\left(t - \frac{\delta t}{2}\right)\right)}{\delta 2} \tag{4.16}$$

where W_{in} and W_{out} are, respectively, the flow rate of nutrient solution into and from the culture beds. k_{LW} is a conversion factor from volume to mass of liquid water (997 kg m^{-3} at 25 °C and 101.3 kPa). $V_{LW}\left(t+\frac{\delta t}{2}\right) - V_{LW}\left(t-\frac{\delta t}{2}\right)$ is the change in nutrient solution volume in the culture beds and/or the nutrient solution tank during the time interval of δt. W_{U} in Equation (4.16) can be estimated for each culture bed or for all the culture beds, by changing the measuring points of W_{in} and W_{out}.

ION UPTAKE RATE BY PLANTS

The ion uptake rate by plants in the culture bed, I_{U}, can be estimated by:

$$I_{U} = I_{in}W_{in} - I_{out}W_{out} + \frac{\left(\frac{I_{in} + I_{out}}{2}}{F}\right)\left(V_{LW}\left(t+\frac{\delta t}{2}\right) - V_{LW}\left(t-\frac{\delta t}{2}\right)\right)}{\delta t} \qquad (4.17)$$

where I_{in} and I_{out} are, respectively, the concentration of ion I at the inlet and outlet of the culture bed. W_{in} and W_{out} are, respectively, the flow rate of nutrient solution at the inlet and outlet of the culture bed. The third term shows the change in amount of ion I in the culture bed during the time interval of δt. I_{U} can be measured for each culture bed, more than one culture bed or all the culture beds.

APPLICATION

The methods of estimating rate variables given previously can be applied to small PFAL-type plant growth chambers used for research, education, self-learning, and hobby purposes (Kozai, 2013b), as well as PFALs for commercial plant production. Since all the major rates and state variables in the PFAL can be measured and graphically displayed, the quantitative relationships among these variables can be analyzed and understood relatively easily. This benefit could be increased by displaying the relationships together with camera images of the plants. Major rates and state variables of the PFAL are schematically shown in Figure 4.3.

COEFFICIENT OF PERFORMANCE OF HEAT PUMP

A heat pump is a device that provides heat energy from a source of heat to a destination called a "heat sink." Heat pumps are designed to move heat energy in the direction opposite to that of spontaneous heat flow by absorbing heat from a cold space and releasing it to a warmer one, and vice-versa. A heat pump uses external power to accomplish the work of transferring energy from the heat source to the heat sink. Electrically driven heat pumps are commonly used for heating, ventilation, and air conditioning.

The term *coefficient of performance* (COP), as defined by Equation (4.9), is the ratio of heating or cooling provided to electrical energy consumed. Higher COPs equate to lower operating costs. COP is highly dependent on operating conditions, especially the difference in temperature between the inner and outer units (or the evaporator and condenser) of the heat pump and the ratio of actual cooling/heating load. The COPs for heating and cooling are different because the heat reservoir of interest is different. For cooling, the COP is the ratio of the heat removed from the cold reservoir to the input work.

However, for heating, the COP is the ratio of the heat removed from the cold reservoir plus the heat added to the hot reservoir to the input work.

In a plant factory under artificial light, heat pumps are principally working in the cooling mode even in the winter months in cold regions due to the heat produced by lights. To prevent high humidity, the lights are turned on alternately at different tiers in the PFAL so that some lights always remain on to keep the heat pumps running. In this case, the COP for cooling ranges from 8 to 12 with an inside air temperature of 25 °C and outside temperatures ranging from 0 to 15 °C (Kozai, 2012). The electricity consumption for cooling is usually 1/10 to 1/15 that of lighting because 80% of the cooling load is required to remove the heat produced by the lights.

Thanks to technological progress, COP values have improved considerably in recent years. For example, at an outside temperature of 34 °C and an inside temperature of 27 °C, the COP of household heat pumps (air conditioners) for cooling was improved from 2.5–3.0 in 1994 to 5–6 in 2009, doubling the efficiency in 15 years (Kozai, 2012). Research has shown the potential of using heat pumps for controlling the environment in greenhouses. Tong et al. (2010) reported an average hourly COP of 4.0 when heat pumps were used for heating a greenhouse to maintain the inside temperature at 16°C with outside temperatures ranging from −5 to 6 °C. In the heating mode, COP decreases as the outside temperature decreases. For cooling a greenhouse to maintain the inside temperature at 18 °C with outside air temperatures ranging from 21 to 28 °C, the average hourly COP was 9.3 (Tong et al., 2013). Similarly, in the cooling mode, COP decreases as outside temperature increases.

REFERENCES

Bugbee, B.G., Salisbury, F.B., 1988. Exploring the limits of crop productivity I. Photosynthetic efficiency of wheat in high irradiance environments. Plant Physiol. 88, 869–878.

Chiapale, J.P., de Villele, O., Kittas, C., 1984. Estimation of ventilation requirements of a plastic greenhouse. Acta Horticult. 154, 257–264 (in French with English abstract).

Dayan, E., Presnov, E., Dayan, J., Shavit, A., 2004. A system for measurement of transpiration, air movement and photosynthesis in the greenhouse. Acta Horticult. 654, 123–130.

Dueck, T.A., Grashoff, C., Broekhuijsen, G., Marcelis, L.F.M., 2006. Efficiency of light energy used by leaves situated in different levels of a sweet pepper canopy. Acta Horticult. 711, 201–205.

Evans, J.R., Poorter, H., 2001. Photosynthetic acclimation of plants to growth irradiance: the relative importance of specific leaf area and nitrogen partitioning in maximizing carbon gain. Plant Cell Environ. 24, 755–767.

Frantz, J.M., Joly, R.J., Mitchell, C.A., 2000. Intracanopy lighting influences radiation capture, productivity, and leaf senescence in cowpea canopies. J. Am. Soc. Hortic. Sci. 125 (6), 694–701.

Hao, X., Zheng, J., Little, C., Khosla, S., 2012. LED inter-lighting for year-around greenhouse mini cucumber production. Acta Horticult. 956, 335–340.

Jokinen, K., Sakka, L.E., Nakkila, J., 2012. Improving sweet pepper productivity by LED interlighting. Acta Horticult. 956, 59–66.

Kozai, T., 1995. Development and Application of Closed Transplant Production Systems. Yoken-Do, Tokyo. 191 p. (in Japanese).

Kozai, T., 2005. Advanced Transplant Production Technology—Commercial Use of Closed Transplant Production System. Noh-Den Kyokai, Tokyo. 150 p. (in Japanese).

Kozai, T., 2007. Propagation, grafting, and transplant production in closed systems with artificial lighting for commercialization in Japan. J. Ornam. Plants 7 (3), 145–149.

Kozai, T., 2011. Improving light energy utilization efficiency for a sustainable plant factory with artificial light. In: Proceedings of Green Lighting Shanghai Forum 2011; pp. 375–383.

Kozai, T., 2012. Plant Factory with Artificial Light (written in Japanese: Jinkoko-gata shokubutsu kojo). Ohm Publishing Company, Tokyo.

Kozai, T., 2013a. Sustainable plant factory: closed plant production systems with artificial light for high resource use efficiency and quality produce. Acta Horticult. 1004, 27–40.

Kozai, T., 2013b. Plant factory in Japan: current situation and perspectives. Chron. Hortic. 53 (2), 8–11.

Kozai, T., 2013c. Resource use efficiency of closed plant production system with artificial light: concept, estimation and application to plant factory. Proc. Japan Acad. Ser. B 89, 447–467.

Kozai, T., Kubota, C., Chun, C., Afreen, F., Ohyama, K., 2000. Necessity and concept of the closed transplant production system. In: Kubota, C., Chun, C. (Eds.), Transplant Production in the 21st Century. Kluwer Academic Publishers, The Netherlands, pp. 3–19.

Kozai, T., Afreen, F., Zobayed, S.M.A. (Eds.), 2005. Photoautotrophic (Sugar-Free Medium) Micropropagation as a New Micropropagation and Transplant Production System. Springer, Dordrecht, The Netherlands.

Kozai, T., Ohyama, K., Chun, C., 2006. Commercialized closed systems with artificial lighting for plant production. Acta Horticult. 711, 61–70.

Kozai, T., Ohyama, K., Tong, Y., Tongbai, P., Nishioka, N., 2011a. Integrative environmental control using heat pumps for reductions in energy consumption and CO_2 gas emission, humidity control and air circulation. Acta Horticult. 893, 445–449.

Kozai, T., Li, M., Tong, Y., 2011b. Plant environment control by integrating resource use efficiencies and rate variables with state variables of plants. In: Proceedings of Osaka Forum 2011. Japanese Society of Agricultural, Biological and Environmental Engineers and Scientists, Kagawa University, Japan, pp. 37–54 (in Japanese).

Li, M., Kozai, T., Niu, G., Takagaki, M., 2012a. Estimating the air exchange rate using water vapour as a tracer gas in a semi-closed growth chamber. Biosyst. Eng. 113, 94–101.

Li, M., Kozai, T., Ohyama, K., Shimamura, S., Gonda, K., Sekiyama, S., 2012b. Estimation of hourly CO_2 assimilation rate of lettuce plants in a closed system with artificial lighting for commercial production. Eco-engineering 24 (3), 77–83.

Li, M., Kozai, T., Ohyama, K., Shimamura, D., Gonda, K., Sekiyama, T., 2012c. CO_2 balance of a commercial closed system with artificial lighting for producing lettuce plants. HortSci. 47 (9), 1257–1260.

Liu, W.K., Yang, Q., Wei, L.L., 2012. Light Emitting Diodes (LEDs) and Their Applications in Protected Horticulture as a Light Source. China Agric. Sci. Tech. Pub, Beijing (in Chinese).

Massa, G.D., Kim, H.-H., Wheeler, R.M., 2008. Plant productivity in response to LED lighting. HortSci. 43 (7), 1951–1956.

Mitchell, C.A., Both, A.J., Bourget, C.M., Burr, J.F., Kubota, C., Lopez, R.G., Morrow, R.C., Runkle, E.S., 2012. LEDs: the future of greenhouse lighting!. Chron. Hortic. 52 (1), 6–11.

Morrow, R.C., 2008. LED lighting in horticulture. HortSci. 43 (7), 1947–1950.

Nederhoff, E.M., Vegter, J.G., 1994. Photosynthesis of stands of tomato, cucumber and sweet pepper measured in greenhouses under various CO_2-concentrations. Ann. Bot. 73, 353–361.

Nishio, M., 2005. Agriculture and environmental pollution: soil environment policy and technology in Japan and the world (written in Japanese: Nogyo to kankyo osen: Nihon to sekai no dojo kankyo seisaku to gijutsu). Nohbunkyo.

Ohyama, K., Yoshinaga, K., Kozai, T., 2000. Energy and mass balance of a closed-type transplant production system (Part 2): water balance. J. SHITA 12 (4), 217–224 (in Japanese with English abstract and captions).

Ohyama, K., Kozai, T., Kubota, C., Chun, C., 2002a. Coefficient of performance for cooling of a home-use air conditioner installed in a closed-type transplant production system. J. SHITA 14, 141–146 (in Japanese with English abstract and captions).

Ohyama, K., Kozai, T., Kubota, C., Chun, C., Hasegawa, T., Yokoi, S., Nishimura, M., 2002b. Coefficient of performance for cooling of a home-use air conditioner installed in a closed-type transplant production system. J. Soc. High Technol. Agric. 14 (3), 141–146.

Pettersen, R.I., Torre, S., Gislerød, H.R., 2010. Effects of intracanopy lighting on photosynthetic characteristics in cucumber. Sci. Hortic. 125, 77–81.

Sager, J.C., Edwards, J.L., Klein, W.H., 2011. Light energy utilization efficiency for photosynthesis. Trans. ASAE 25 (6), 1737–1746.

Salisbury, F.B., Ross, C.W., 1991. Plant Physiology. Wadsworth Publishing Company, USA. p. 609.

Sharpley, A.N.T., Daniel, T., Sims, J., Lemunyon, R., Stevens, R., Parry, R., 2003. Agricultural phosphorous and eutrophication. Second ed. US Department of Agriculture, Agricultural Research Service, ARS-149.

Shibuya, T., Kozai, T., 2001. Light-use and water-use efficiencies of tomato plug sheets in the greenhouse. Environ. Control Biol. 39, 35–42 (in Japanese with English abstract and captions).

Takatsuji, M., Mori, Y., 2011. LED Plant Factory (written in Japanese: LED shokubutsu kojo). Nikkan Kogyo Co., Tokyo. p. 4.

Thongbai, P., Kozai, T., Ohyama, K., 2011. Promoting net photosynthesis and CO_2 utilization efficiency by moderately increased CO_2 concentration and air current speed in a growth chamber and a ventilated greenhouse. J. ISSAAS 17, 121–134.

Tong, Y., Kozai, T., Nishioka, N., Ohyama, K., 2010. Greenhouse heating using heat pumps with a high coefficient of performance (COP). Biosyst. Eng. 106, 405–411.

Tong, Y., Kozai, T., Ohyama, K., 2013. Performance of household heat pumps for nighttime cooling of a tomato greenhouse during the summer. Appl. Eng. Agric. 29 (3), 414–421.

Wheeler, R.M., 2006. Potato and human exploration of space: some observations from NASA-sponsored controlled environment studies. Potato Res. 49, 67–90.

Wheeler, R.M., Mackowiac, C.L., Stutte, G.W., Yorio, N.C., Rufffe, L.M., Sager, J.C., Prince, R.P., Knott, W.M., 2006. Crop productivities and radiation use efficiencies for bioregenerative life support. Adv. Space Res. 41, 706–713.

Yabuki, K., 2004. Photosynthetic Rate and Dynamic Environment. Kluwer Academic Publishers, Dordrecht. 126 p.

Yokoi, S., Kozai, T., Ohyama, K., Hasegwa, T., Chun, C., Kubota, C., 2003. Effects of leaf area index of tomato seedling population on energy utilization efficiencies in a closed transplant production system. J. SHITA 15, 231–238 (in Japanese with English abstract and captions).

Yokoi, S., Kozai, T., Hasegawa, T., Chun, C., Kubota, C., 2005. CO_2 and water utilization efficiencies of a closed transplant production system as affected by leaf area index of tomato seedling populations and the number of air exchanges. J. SHITA 18, 182–186 (in Japanese with English abstract and captions).

Yokoi, S., Goto, E., Kozai, T., Nishimura, M., Taguchi, K., Ishigami, Y., 2007. Effects of planting density and air current speed on the growth and that uniformity of tomato plug seedlings in a closed transplant production system. J. SHITA 19 (4), 159–166.

Yoshinaga, K., Ohyama, K., Kozai, T., 2000. Energy and mass balance of a closed-type transplant production system (Part 3): carbon dioxide balance. J. SHITA 13, 225–231 (in Japanese with English abstract and captions).

MICRO- AND MINI-PFALs FOR IMPROVING THE QUALITY OF LIFE IN URBAN AREAS

5

Michiko Takagaki[1], Hiromichi Hara[2], Toyoki Kozai[3]

Center for Environment, Health and Field Sciences, Chiba University, Kashiwa, Japan[1] Graduate School of Engineering, Chiba University, Chiba, Yayoi, Japan[2] Japan Plant Factory Association, c/o Center for Environment, Health and Field Sciences, Chiba University, Kashiwa, Chiba, Japan[3]

INTRODUCTION

Today, approximately 3.6 billion people worldwide live in urban areas. By 2030, the population in agricultural areas is predicted to decrease to 3 billion due to the aging of farmers, while the population in urban areas will exceed 5 billion. Therefore, the current system of producing almost all food including fresh food in agricultural areas and transporting it to urban areas needs to be changed in the coming decades (Despommier, 2010).

In order to enhance local production for local consumption of fresh and clean vegetables, urban agriculture, especially plant factories with artificial lighting (PFAL), have become increasingly important in Japan and other countries having densely populated urban areas, as described in Chapters 2 and 3.

In parallel with the commercial production of leafy vegetables using PFALs, some residents living in urban areas with little chance to grow plants outdoors have recently started enjoying indoor farming using a household PFAL or micro-PFAL in Japan, Taiwan, and China and a few other Asian countries (Takagaki et al., 2014). In addition, mini-PFALs have recently been set up for various purposes at restaurants, cafés, shopping centers, schools, community centers, hospitals, etc. Such micro- and mini-PFALs are both called "m-PFALs" in this chapter. This chapter explains how m-PFALs and their networks could bring new lifestyles to people living in urban areas.

CHARACTERISTICS AND TYPES OF m-PFAL

An m-PFAL is a closed or semi-closed indoor plant growing system used for various purposes other than the commercial production and sale of plants. m-PFALs are characterized by (1) hydroponic or soilless culture, (2) plants are grown mostly under LED lamps, (3) well-designed as furniture or for greening of interiors (beautiful-looking, safe and robust), (4) easy to use and maintain, (5) no use of pesticides, and (6) lighting period, watering, temperature, and so forth can be controlled automatically or manually by the user. The size, i.e. the air volume of the plant-growing space, of most micro-PFALs ranges from 0.03 m^3 (e.g., $0.2 \times 0.5 \times 0.3$ m) to 1 m^3 (e.g., $1.0 \times 0.5 \times 2$ m) and that of most mini-PFALs ranges from 2 m^3 (e.g., $2 \times 0.5 \times 2$ m) to 30 m^3 (e.g., $2 \times 5 \times 3$ m).

Plant Factory. http://dx.doi.org/10.1016/B978-0-12-801775-3.00005-6

m-PFALs are classified into three types: Type (A) natural or forced ventilation through air gaps covered with fine mesh nets to prevent insects from entering, and without air conditioning; Type (B) air conditioned but some degree of natural ventilation and no CO_2 enrichment; and Type (C) an airtight structure equipped with an air conditioner and CO_2 enrichment unit. Types A and B produce vegetables that can be eaten fresh after washing with tap water. Type C produces vegetables that can be eaten fresh without washing because they are kept highly clean. Most m-PFALs belong to Type A or Type B, while most PFALs for commercial plant production belong to Type C.

In Types B and C, the air conditioner is used for cooling to remove the heat generated by lamps. Water vapor transpired from leaves condenses into liquid water at the cooling panel of the air conditioner, and is returned to the nutrient solution tank for recycling. Thus, less fresh water needs to be added by the user to the nutrient solution tank, and no drainage or drainpipe is required.

VARIOUS APPLICATIONS OF m-PFALs
HOMES

m-PFALs can be used in the home by individuals or the whole family (Figures 5.1A–5.1C). By using an m-PFAL instead of an outdoor plot, beginners can relatively easily grow a variety of leafy vegetables, herbs, small root vegetables, and edible flowers all year around. The user can choose the level of support

FIGURE 5.1A

Tabletop m-PFAL for personal use (The Plant Environment Designing Program by H. Hara).

FIGURE 5.1B

m-PFAL as green furniture available on the Internet. Estimated retail prices of units a, b, and c are about US$1500, 200, and 100. Products of U-ing Corporation (http://www.greenfarm.uing.u-tc.co.jp/tri-tower).

FIGURE 5.1C

Household m-PFAL connected with the Internet for SNS, designed by H. Hara and J. Hashimoto, Panasonic Corporation, Patent pending.

desired from m-PFAL software to grow the plants. These m-PFALs help create a green interior where people can gain pleasure from watching the plants grow and looking after them daily, and then enjoy eating the fresh produce.

Although people can enjoy growing and harvesting plants outdoors if the weather conditions are favorable, some people live in cold or hot regions, while others only have spare time at night. With an m-PFAL, anyone can enjoy growing plants indoors at any time, regardless of the weather.

The monthly electricity consumption of a micro-PFAL that produces sufficient leafy vegetables for fresh salads for two adults is about 50 kWh ($=0.1$ kW $\times$ 16 h/day $\times$ 30 days/month), costing about US$5 in the USA or about US$10 in Japan.

RESTAURANTS AND SHOPPING CENTERS

m-PFALs are sometimes placed at the entrance of a restaurant, café, or shopping center as an eye-catching feature (Figures 5.2A–5.2C and Figure 5.3). Vegetables, herbs, and medicinal plants grown at restaurants are served to customers. Typically, the chef picks the plants 5 min before serving them to the customer, who can see how the vegetables are grown in the m-PFAL before or after enjoying their meal.

FIGURE 5.2A

m-PFALs placed at the entrance of a restaurant in Kashiwa, Chiba, Japan. The vegetables are served to the customers. Café Restaurant 'Agora' designed by H. Hara and T. Hatta (left). Restaurant 'Come sta' at Mitsui Garden Hotel Kashiwa-no-ha (right).

FIGURE 5.2B

m-PFAL built adjacent to a French-style Chinese restaurant at Grand Front Osaka, Japan (left) and m-PFAL for growing medicinal plants (right).

FIGURE 5.2C

Eight m-PFALs under the service counter of Café 'Foodie Foodie' at Panasonic Center, Osaka.

SCHOOLS AND COMMUNITY CENTERS

m-PFALs are being used for education (Figure 5.4A), self-learning (Figure 5.4B), and communication (Figure 5.5). While experiencing the joy of raising plants empirically, school children gain an implicit and integrated understanding of the scientific principles at work in the environment, natural resources, food production, and ecosystems. m-PFALs could become practical tools for lifelong self-learning about diet, the environment, nature, energy, and natural resources. Using a m-PFAL reinforces awareness of the vital importance of foods and plants in our environment, of reducing our consumption of natural resources, and of preserving and protecting the environment.

FIGURE 5.3

m-PFAL at a shopping center as an eye-catching display (Lalaport Kashiwa-no-ha, Kashiwa City, Chiba).

Photos by Mitsui Fudosan Co., Ltd., Mirai Co., Ltd. and Sankyo Frontier Co., Ltd.

FIGURE 5.4A

m-PFAL used as an educational tool for extracurricular activities at Tomioka Elementary School in Fukushima after the earthquake and tsunami of March 11, 2011 (M. Takagaki).

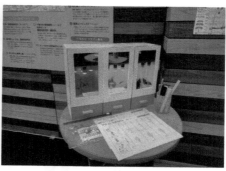

FIGURE 5.4B

m-PFAL at Kashiwa-no-ha Smart City Museum used as an educational tool for summer holiday activities by a volunteer group (H. Hara).

FIGURE 5.5

m-PFAL used as a communication tool in daily life at a meeting room for temporary housing residents in Miyagi after the earthquake and tsunami of March 11, 2011 (H. Hara).

HOSPITALS

m-PFALs can also be used in hospitals. The m-PFAL shown in Figure 5.6 is placed in the lobby of Sakakibara Memorial Hospital, Tokyo. The vegetables are grown by recuperating patients and served to other hospitalized patients; such growing of plants is a kind of horticultural therapy for recovering patients. The m-PFAL also facilitates communication among patients and hospital staff such as nurses, and there are plans to use it as a part of alternative integrated medicine at the hospital.

FIGURE 5.6

m-PFAL at the lobby of Sakakibara Memorial Hospital in Tokyo. The vegetables are grown by recuperating patients and served to other hospitalized patients. Designed by H. Hara and J. Hashimoto.

OFFICES

Watching a computer screen all day in an office with no green plants and little chance to chat with colleagues can be very stressful. The m-PFAL shown in Figure 5.7A serves as a partition between desks. The m-PFALs shown in Figure 5.7B are placed in meeting rooms and coffee-break areas, and can assist communication among colleagues and help staff to relax.

SMALL SHOPS AND RENTAL m-PFALs

A company in Kashiwa is displaying m-PFALs of size 3 or 6 m^2 for rent (the smaller one costs US$500 per month) at a nearby exhibition site (Figure 5.8). The m-PFAL can be lifted up by a crane and then transported by a small truck. The company is also producing and selling vegetables using a larger PFAL at the exhibition site.

DESIGN CONCEPT OF m-PFALs

The objective of the m-PFAL is to bring together the cultivation and use of plants for users, and to balance these two elements. Users' situations and needs vary depending on whether they are in the home, restaurant, school, community facility, or office. To create the ideal situation, it is necessary to integrate the design for users with the design for cultivation. By integrating not only the function but also the esthetics, the value of m-PFALs in society is increased. A good design emphasizes the beauty of the plants, encouraging users to use the m-PFAL and producers to grow better plants. This virtuous circle of users and plants is the basic design concept of the m-PFAL.

FIGURE 5.7A

m-PFAL as a partition between office desks (left), MTI Ltd., Tokyo.

FIGURE 5.7B

m-PFAL at a meeting room (left) and a café in the office (right), MTI Ltd., Tokyo (H. Hara).

m-PFALs CONNECTED TO THE INTERNET

A trial social experiment using m-PFALs was launched in 2012 in the Kashiwa-no-ha district of Kashiwa City in Chiba Prefecture, Japan (Figure 5.9; Kozai, 2013). This project was conducted by the Working Group of "In-town Plant Factory Consortium Kashiwa-no-ha," a consortium involving Chiba University, Mitsui Fudosan Co., Ltd., Panasonic Corporation, and Mirai Co., Ltd., with the collaboration of local residents acting as end-users in the project.

FIGURE 5.8

m-PFAL for rent (US$500/month), 2 m wide, 2 or 3 m long and 2.5 m high. A stall-type shop selling PFAL-grown vegetables at a production site (Sankyo Frontier Co., Ltd., Nagareyama City, Chiba) (right).

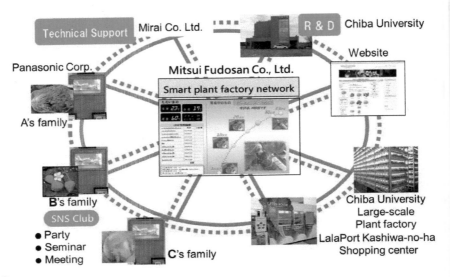

FIGURE 5.9

m-PFAL network as a social experiment.

The trial project examined (1) how people can grow vegetables in household spaces; (2) the effectiveness of providing cultivation advice; (3) the ease of using the m-PFAL; and (4) the value added from creating a network. It also assessed the usefulness and commercial feasibility of an exclusively web-based service where m-PFAL growers can ask experts questions on cultivation management, share information on their own circumstances and experiences, and arrange to exchange the vegetables they have grown.

The project showed that (1) linking m-PFALs by the Internet enabled users to gain the latest information on cultivation methods, (2) information on various plant species could be downloaded from cloud servers, (3) growers could swap tips about cultivation and food preparation using social network services such as Facebook and Twitter, and (4) users could check remotely how their plants are doing using built-in cameras. The users also enjoyed meeting each other at a party held every few months and bringing their own m-PFAL-grown vegetables (Figures 5.10A and 5.10B).

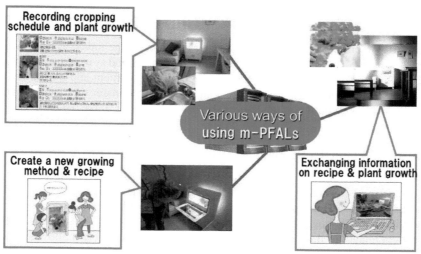

FIGURE 5.10A

Various ways of using m-PFAL networks.

FIGURE 5.10B

A gathering for m-PFAL users, May 11, 2013, Kashiwa, Chiba, Japan.

This project is an integral part of making Kashiwa-no-ha in Kashiwa City a "smart and streamlined" community, in line with its designation as an "Environmental-Symbiotic City" that strives to preserve green spaces and maintain harmony between cities and agriculture. In this district, there are also community gardens and rooftop gardens for growing organic vegetables. This project can be extended to networks linking schools and similar organizations.

ADVANCED USES OF m-PFALs CONNECTING WITH A VIRTUAL m-PFAL

A virtual m-PFAL system incorporating an e-learning function can be linked as a network with a real m-PFAL system involving the actual cultivation of plants. In the virtual system, users can operate an m-PFAL using a simulator that incorporates plant growth, an environment and a business model; after attaining a certain level of proficiency in the simulator, they can transfer their new-found knowledge to the real m-PFAL, or can do both simultaneously.

VISUALIZING THE EFFECTS OF ENERGY AND MATERIAL BALANCE ON PLANT GROWTH

Since m-PFALs are almost fully enclosed (airtight and thermally insulated) environments, it is possible to quantify the resources consumed and produced. The required inputs are light energy, water, CO_2, and fertilizer essential for the plants' photosynthesis. It is also possible to measure the exact amount of electricity required for lighting, air conditioning, and circulation of nutrient solution. Furthermore, analysis of camera images of the m-PFAL and measurements of plant weight allow the plants' growth to be estimated. In addition, by measuring the rates of consumption of CO_2, water, and electric energy and light energy, it is possible to estimate the plants' rate of photosynthesis, rate of transpiration, and increase of fresh weight (Li et al., 2012a,b,c), and hence to visualize the time course of plant growth as affected by the rates of inputs and outputs.

MAXIMIZING PRODUCTIVITY AND BENEFITS USING MINIMUM RESOURCES

As people become familiar with m-PFALs as a simplified model of an ecosystem, their ability to maximize food production while maximizing the benefits in terms of quality of life will become second nature. They will also minimize their environmental impact by reducing their electricity and water consumption, producing no waste water, absorbing carbon dioxide, and producing more oxygen for the atmosphere. Along with this increase in quality of life, they will enjoy helping living things to grow and protecting the ecological balance.

LEARNING THE BASICS OF AN ECOSYSTEM

The m-PFAL is a simple plant production system with a relatively well-controlled environment, so the effects of the environment on plant growth are straightforward. Accordingly, the ecological relationship between the environment and plants can be easily understood. The next step is to connect the m-PFAL with other biological systems to learn about and enjoy more complicated ecosystems.

CHALLENGES

Challenges for further applications of m-PFALs are (1) evidence-based experiments for verifying the benefits of m-PFAL, (2) collaboration among designers, citizens, and researchers, educators, community planners, and architects to design, construct and develop m-PFALs, and (3) development of "citizen-centered horticulture science"—a new science built from people's perspective.

m-PFAL CONNECTED WITH OTHER BIOSYSTEMS AS A MODEL ECOSYSTEM

Networks connecting mushroom culture, aquaculture (fish farming) and aquaponics to raise fish and aquatic plants together are also possible (Figures 5.11 and 5.12). m-PFAL is an inorganic plant production system, while m-aquaponics is an organic plant production system. By comparing the two, we can learn the principles of the two biological systems and their differences and similarities.

Small-scale experiments and results could also be applied to large-scale commercial and industrial PFALs. By connecting PFALs with local community energy management systems, these networks can form the infrastructure of a sustainable city. Thus, m-PFAL and PFAL will play an important role in the near future in improving the quality of our lives and communities while minimizing the consumption of nonrenewable natural resources and the emissions of environmental pollutants (Figure 5.13).

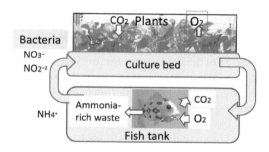

FIGURE 5.11

CO_2, O_2, and nitrogen circulation in aquaponics (=aquaculture+hydroponics). Both plants and fish can be harvested. Fish waste is decomposed by bacteria to nitrates and other inorganic nutrients for plant growth. Water from the fish tank is pumped into the culture bed and filtered by plants and substrate, and then returned to the fish tank.

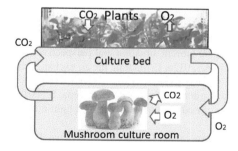

FIGURE 5.12

CO_2 and O_2 circulation in a plant-mushroom culture system. Both plants and mushrooms can be harvested. The plants and mushrooms grow better in their respective CO_2-enriched and CO_2-reduced air environments.

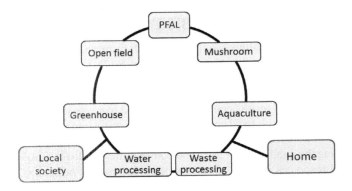

FIGURE 5.13

Combinations of PFAL with other eco-systems.

LIGHT SOURCE AND LIGHTING SYSTEM DESIGN

When choosing the type and design of the light source for m-PFALs, we need to consider the physiology and psychology of light perception by humans. This is because the m-PFAL is not only a plant production system but also part of a green interior. The color rendering index (CRI) of a light source is an important index used when designing lighting for humans (http://en.wikipedia.org/wiki/Color_rendering_index). CRI is a quantitative measure of the ability of a light source to reveal the colors of various objects faithfully in comparison with an ideal or natural (solar) light source (Wikipedia). The CRI is 100 when the light source functions exactly as natural light. The color of a plant perceived by the human eye is strongly affected by the CRI of the light source. For example, green lettuce plants under red-based LEDs tend to look dark red; the plants do not look fresh, tasty, or beautiful. The design of the m-PFAL lighting system is also important. Other design factors to be considered for the human eye include (1) the color temperature of the light source (cool white or warm white), (2) the spatial distribution of luminous intensity of the light source, and (3) the three-dimensional shape and optical characteristics of reflectors.

REFERENCES

Despommier, D., 2010. The Vertical Farm: Feeding the World in the 21st Century. St. Martin's Press, New York.

Kozai, T., 2013. Plant factory in Japan: current situation and perspectives. Chronica Hortic. 53 (2), 8–11.

Li, M., Kozai, T., Niu, G., Takagaki, M., 2012a. Estimating the air exchange rate using water vapour as a tracer gas in a semi-closed growth chamber. Biosyst. Eng. 113, 94–101.

Li, M., Kozai, T., Ohyama, K., Shimamura, S., Gonda, K., Sekiyama, S., 2012b. Estimation of hourly CO_2 assimilation rate of lettuce plants in a closed system with artificial lighting for commercial production. Eco-engineering 24 (3), 77–83.

Li, M., Kozai, T., Ohyama, K., Shimamura, D., Gonda, K., Sekiyama, T., 2012c. CO_2 balance of a commercial closed system with artificial lighting for producing lettuce plants. HortSci. 47 (9), 1257–1260.

Takagaki, M., Hara, H., Kozai, T., 2014. Indoor horticulture using micro-plant factory for improving quality of life in urban areas: design and a social experiment Approach, IHC 2014, Abstract book.

ROOFTOP PLANT PRODUCTION SYSTEMS IN URBAN AREAS

Nadia Sabeh

Building Performance Engineer, Guttmann & Blaevoet Consulting Engineers, Sacramento, California, USA

INTRODUCTION

Rooftop plant production (RPP) is the practice of growing plants on top of residential, commercial, and industrial buildings. Plants grown on roofs may be ornamental or edible, require regular pruning and harvesting, or be maintenance free. Many new buildings are being designed with "green roof" space to capitalize on the many proven and potential benefits (The City of London, 2008), including environmental—stormwater management, heat island reduction, carbon sequestration, and building insulation—and social—community connection, k-12 education, and plant therapy and rehabilitation (Mandel, 2013).

There are many added benefits for using the roof to grow food crops, including local food production and security, decreased fuel use for transporting food, agricultural jobs training, and improved local economy (Ganguly et al., 2011). Rooftop agriculture, as a subset of RPP, is gaining popularity within the urban landscape as city dwellers gain interest in local and sustainable food production. This interest is gaining traction as more and more cities lift historical bans on urban agriculture and pass ordinances that allow homeowners, communities, and businesses to grow food in open spaces for more than personal consumption (Engelhard, 2010).

The rooftop is an ideal landscape for growing plants in dense cities, as it typically has greater exposure to solar energy than the ground below. Further, rooftops are often vacant, except for the mechanical equipment used to condition the building. This combination of factors—sunlight and space—is also why rooftops are good locations for generating on-site solar electricity and solar hot water. The urban PFAL (Plant Factory with Artificial Light), like any city building, will have underutilized space that can be used to capture solar energy, either for food production or electricity generation.

ROOFTOP PLANT PRODUCTION

Rooftop production of plants for human consumption may be classified into three categories (Mandel, 2013): (1) Rooftop Gardening (small-scale, noncommercial); (2) Rooftop Farming (medium-scale, commercial); and (3) Rooftop Agriculture (large-scale, commercial). This chapter focuses on commercial plant production for food, and does not consider the larger context of Rooftop Agriculture, which may also include cultivation of animals, bees, fish, fungi, fiber, and feed. Rooftop Farming is often

Plant Factory. http://dx.doi.org/10.1016/B978-0-12-801775-3.00006-8

characterized by one of the following methods of cultivation: container and raised bed production, continuous row farming, and hydroponic greenhouse growing.

Depending on the method of RPP, there can be many proven benefits. These benefits include mitigation of storm water; reduction of the urban heat island effect; insulation of the roof; and increased biodiversity of pollinators, such as bees, birds, bats, and butterflies (The City of London, 2008). If carefully designed, RPP can also help to maximize resource use efficiency (RUE) by reusing waste water and waste heat from the building below. RPP has added benefits to PFAL by providing a space to grow crops that are less responsive to the low-light conditions and space constraints of vertical farming. Specifically, the rooftop can be used to cultivate crops that require high light intensities and greater vertical growth, as well as flowering plants that require pollination, root crops that require deeper soil beds, and plants that are more tolerant to heat and variations in temperature and humidity.

Some crop varieties suitable for RPP of PFAL include tomatoes, cucumbers, and peppers. Further, these plants are often cultivated as seedlings within the PFAL and could easily be moved to the roof for full-scale production, thereby reducing the energy and carbon footprint associated with transportation to remote facilities.

Although many cities are beginning to support urban agriculture and green roof construction, local policy and building codes can still be a barrier in many localities (Engelhard, 2010). Some of the other potential and perceived challenges to overcome when designing a RPP system include structural loads on the roof, accessibility to the roof, utility hookups, leakage and damage to waterproofing, and crop and building maintenance (The City of London, 2008). However, once implemented, the benefits of the green roof agriculture system will quickly outweigh the challenges that were overcome to build it.

RAISED-BED PRODUCTION

Raised-bed production results in moderate yields of food crops that would not be possible inside the PFAL. Raised beds are low-profile structures, typically built out of plastic, steel or wood, and filled with soil for growing plants (Mandel, 2013). The width and length of raised beds can vary and will depend on the roof dimensions and space availability. Raised beds can also be built to various heights and soil depths, influencing crop selection. Root vegetables and tomatoes require soil depths 12 in. (30.48 cm) or greater; while leafy greens can be grown in shallow depths of 6 in. (15.24 cm). The soil is often a combination of compost, peat moss, and vermiculite or perlite. The compost can be generated on site, from unused plant material (leaves, stems, etc.) at the PFAL, as well as any food waste. Plants may be irrigated by hand or using drip tubes. Raised beds are designed to allow excess water to drain through the soil and out of the bottom to the roof deck. Therefore, the roof should be protected with durable, water-proof material that is designed to channel excess water to a collection system or to the storm drain on the street below. The use of soil and heavy framing materials, such as steel or wood, results in relatively high structural loads, which should be carefully considered. Further, wood is susceptible to rot and should be treated prior to soil infill and planting.

CONTINUOUS ROW FARMING

Continuous row farming can produce relatively high yields of many crop varieties. Soil is placed directly on the roof structure with continuous row farming, which is most comparable to field farming. Because planting area is not limited by framing structures, the layout can be flexible and the soil depth varied to accommodate a diverse array of crops in a single growing cycle (Mandel, 2013). Like raised-bed production, soil may be comprised of compost, peat moss, and perlite or vermiculite. However, unlike raised beds, because continuous row farming does not containerize the soil, there may be challenges with soil loss and nutrient leaching during rain events. Providing "edging" around the roof and growing area can help prevent soil loss due to wind and rain. The roof must be designed similarly to a traditional "green roof," with greater structural capacity and a waterproof membrane that also acts as a root barrier.

HYDROPONIC GREENHOUSE GROWING

Of all the RPP systems, growing food in a hydroponic greenhouse will result in the overall highest yields per unit area for any given crop (Mandel, 2013). Hydroponic systems use soilless media to support the plants' root structure. This media is inert and can consist of rockwool, coconut coir, recycled glass, perlite, sand, and other alternatives. Drip tubes deliver water and nutrients in precise quantities, greatly increasing the water use efficiency (WUE) of the crop. The greenhouse structure itself may be made of glass, rigid plastic (polycarbonate), or plastic film (polyethylene, ETFE, etc.), and sometimes acrylic. The greenhouse is often provided with climate control systems, such as fans, heaters, evaporative cooling, and operable windows, to condition the indoor air and achieve the optimal air temperature, relative humidity, and carbon dioxide levels regardless of outside conditions. The greenhouse is a complex system with design criteria similar to PFAL. The greenhouse requires the largest initial investment and longest construction period of the three RPP growing methods. The structure can be heavy and susceptible to wind shear, but with close consultation by a structural engineer, these challenges can be managed. Because energy is required to operate the irrigation and climate control systems, the total electricity consumption of the PFAL will increase. However, depending on the location, the greenhouse may be constructed with solar glass, which collects specific wavelengths of sunlight for generating electricity, while transmitting and diffusing other wavelengths into the greenhouse.

BUILDING INTEGRATION

RPP presents several opportunities for increasing the sustainability and RUE of the PFAL, including storm water management, reduction of heating and cooling loads in the building below, reducing the building's heat island effect, and utilizing waste heat from the building and its equipment.

STORMWATER MANAGEMENT

Stormwater runoff poses many challenges to cities, including flooded streets, strain on sewage conveyance systems and waste water treatment plants, and groundwater pollution of nearby water bodies. Mitigating stormwater has become a prime directive of cities and states as they look to reduce the impact on aging and undersized sewage systems or simply want to drive sustainable city planning and the construction of green buildings. Several standards and building codes exist now that require or encourage stormwater management, including California's Green Building Code (CALGreen), the US Green Building Council (USGBC) green building certification program LEED (Leadership for Energy and Environmental Design), and the International Green Construction Code (IgCC), to name a few.

RPP systems, commonly known as "green roofs" in the building industry, have become an accepted strategy for mitigating stormwater in cities. Open planting structures, such as raised beds and row farming, help control the flow of rainwater from rooftop to street by absorbing rainwater (Figure 6.1). For example, an open green roof planting system that covers 80% of the roof can retain up to 40% of rainwater for any given location (Table 6.1). In large rain events, the water not absorbed by planting systems can be diverted to cisterns and tanks for storage. This water may be filtered for toilet flushing or evaporative cooling and humidification; or it may be stored for future irrigation of the green roof or landscaping. In PFAL with RPP, all stormwater will be used to support plant production, either for irrigation, evaporative cooling, or humidification as needed.

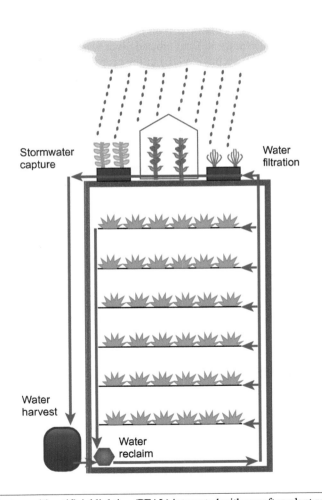

FIGURE 6.1

Water cycle of a plant factory with artificial lighting (PFAL) integrated with a rooftop plant production (RPP) system.

Table 6.1 Potential Stormwater Capture and Solar Electric Generation on a Standard 930 m^2 Roof in Selected International Cities. Stormwater Capture Assumes That Raised-Bed or Row Farming Occupies 80% of the Roof Surface. Solar Electric Generation Assumes 20% of the Roof is Available for a 28.1 kW DC-Rated Photovoltaic System with a Fixed Tilt of 35° and System Loss of 14%

Location	Maximum Daily Rainfall (kg m^{-2} h^{-1})	Maximum Stormwater Capture (kg m^{-2} h^{-1})	Daily Average Solar Radiation (kWh m^{-2} d^{-1})	Annual Solar Electric Generation (kWh)
Phoenix	1.06	0.42	6.56	48,600
London	3.29	1.32	3.17	26,000
Barcelona	3.90	1.56	5.17	40,800
San Francisco	4.63	1.85	5.45	44,100
Sydney	5.83	2.33	5.03	39,600
Mexico City	7.63	3.05	4.44	34,000
Tokyo	8.71	3.48	4.51	36,200
Seoul	14.50	5.80	4.01	32,100

ENERGY USE REDUCTIONS

Climate control accounts for approximately 15% of all energy consumption in PFAL (Kozai, 2013a). Although this energy consumption is much less than the typical commercial building (about 35%), strategies for reducing HVAC (heating, ventilation, and air conditioning) energy use can translate into significant operational cost savings (ASHRAE, 2013). In the U.S., the demand for electricity during peak periods of the day burdens many regional grids as they try to keep up with simultaneous uses of air conditioning, lights, and appliances. Utility companies and some building codes are responding by penalizing consumers for energy use during peak demand periods, typically between 14:00 and 20:00. Therefore, strategies for reducing energy use during these periods can greatly offset energy costs and reduce grid stress (ASHRAE, 2013). Additionally, on-site generation of electricity using rooftop solar panels can power lights, heating and cooling equipment, computers, and other equipments (Figure 6.2).

A well-designed PFAL will use high-efficiency equipment for operations (Kozai, 2013b), including heat pumps for cooling (and heating), variable speed motors on pumps and fans for moving water and air, and LEDs for illuminating the crop. The RPP can contribute to the RUE of the PFAL by insulating the roof and obstructing direct solar heat gain through the roof, two factors that will greatly reduce the heating and cooling loads of the grow space below. The plants themselves can mitigate the building's microclimate through evapotranspiration and by absorbing sunlight rather than reflecting it to the atmosphere above. This combination of factors mitigates the heat island effect caused by many urban buildings that reflect sunlight or emit heat from the building's exterior and dump heat from the building's interior during air conditioning.

Waste heat from inside the building can be further mitigated by recirculating it to the RPP (Figure 6.2). The waste heat from PFAL may be generated from primary airstreams (exhaust from

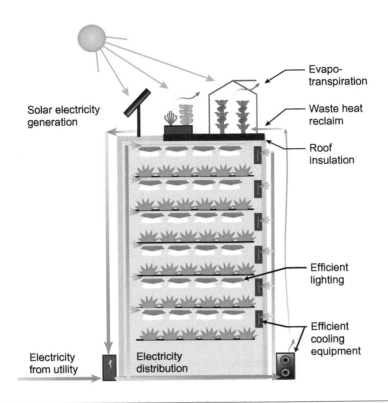

Solar electricity generation

Evapo-transpiration

Waste heat reclaim

Roof insulation

Efficient lighting

Efficient cooling equipment

Electricity from utility

Electricity distribution

FIGURE 6.2

Energy cycle of a plant factory with artificial lighting (PFAL) integrated with a rooftop plant production (RPP) system and rooftop solar electricity generation.

the building) or secondary airstreams (heat generated by motors and equipment). Primary airstreams may be ducted directly to a rooftop greenhouse for climate control. Secondary airstreams could be passed through a heat exchanger to produce hot water used for root zone heating of soil beds or hydroponic media, as well as perimeter heating in greenhouses. If a solar hot water system is designed into the building's program, the secondary airstream could also be used to preheat the water storage tank, reheat recirculating water, or provide the primary source of heating when ample sunlight is available.

REFERENCES

ASHRAE, 2013. ASHRAE Greenguide: Design, Construction and Operation of Sustainable Buildings, fourth ed. ASHRAE, Atlanta, GA.

Engelhard, B., 2010. Rooftop to Tabletop: Repurposing Urban Roofs for Food Production. University of Washington, Thesis.

Ganguly, S., Kujac, P., Leonard, M., Wagner, J., Worthington, Z., 2011. Lively'Hood Farm: Strategy Plan. Presidio Graduate School, In partnership with SF Environment.

Kozai, T., 2013a. Plant factory in Japan: current situation and perspectives. Chron. Horticult. 53 (2), 8–11.

Kozai, T., 2013b. Resource use efficiency of closed plant production system with artificial light: concept, estimation and application to plant factory. Proc. Jpn. Acad., Ser. B B89, 447–461.

Mandel, L., 2013. Eat Up: The Inside Scoop on Rooftop Agriculture. New Society Publishers, Canada.

The City of London, Greater London Authority, Design for London, London Climate Change Partnership, 2008. Living roofs and walls—technical report: supporting London plan policy. London

BASICS OF PHYSICS AND PHYSIOLOGY— ENVIRONMENTS AND THEIR EFFECTS

LIGHT

INTRODUCTION

Light is one of the most important environmental factors influencing plant growth and development. The selection of light sources can have a significant influence on the initial costs and production running costs of a PFAL, in addition to the effects on plant growth and development. This chapter describes the physical properties of light and its measurement. Then, light sources commonly used in PFALs are introduced with a simple explanation of the fundamentals necessary to understand the light sources. Particular emphasis is placed on light-emitting diodes (LEDs), which have received remarkable attention recently, and on fluorescent lamps, which are still widely used in PFALs. Lesser-known benefits of using LEDs and pulsed lighting effects with LEDs are described briefly along with relevant references.

PHYSICAL PROPERTIES OF LIGHT AND ITS MEASUREMENT

Genhua Niu

Texas AgriLife Research at El Paso, Texas A&M University, El Paso, Texas, USA

PHYSICAL PROPERTIES

Light is electromagnetic energy, which is also defined as electromagnetic radiation including both visible and invisible wavelengths (Figure 7.1). The shorter the wavelength, the greater the energy. The wavelength of visible light ranges from approximately 380 to 780 nm, which is what human eyes perceive. Visible light is very important to plants because it coincides roughly with photosynthetically active radiation (PAR, 400–700 nm). For solar radiation, 97% is within the 280–2800 nm range. Of this, 43% is visible light, which is useful for plant growth, 4% is ultraviolet, and 53% is infrared, which produces heat. However, only electric lights are used in PFALs. Wavelengths are classified as shown in Table 7.1; each waveband has its own importance or function.

Light possesses two contradictory properties: it can be observed as a wave phenomenon and yet it also acts as discrete particles called photons. A photon is the smallest particle of light, or a single quantum of light. Unlike other environmental factors such as temperature, humidity, and CO_2 concentration, light varies in at least three dimensions: quantity, quality, and duration. When electric lights are used in a PFAL, the lighting cycle, which affects plant growth and development, can be readily

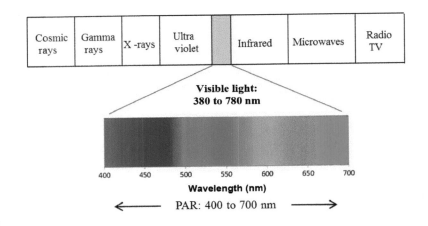

FIGURE 7.1

The electromagnetic spectrum.

Table 7.1 Wave Properties of Light and Their Importance		
Classification	**Wavelength (nm)**	**Importance**
Ultraviolet	100–380	
UV-C	100–280	Disinfecting
UV-B	280–320	Sun burn
UV-A	320–380	
Visible	380–780 (often 400–700)	Photosynthesis and morphology
Far red	700–800 (at the extreme red end of the visible spectrum, between red and infrared light)	Morphology
Near-infrared	780–2500	Heat
Infrared	2500+	Heat

changed. Basically, light influences plants in two ways: providing energy or a quantum source and acting as an information medium. As an energy source, the light energy, or photons of light, is captured by plants and a proportion (up to 10%) of photons captured by plants is converted to chemical energy (carbohydrates) through photosynthesis. Most of the light energy or photons captured by plants is converted into heat energy. As an information medium, light is involved in regulating various growth and development processes such as photomorphogenesis and photoperiodism. Photoreceptors in plants function as light sensors and provide plants with information on subtle changes in light composition in the growing environment and control physiological and morphological responses independent of photosynthesis.

LIGHT MEASUREMENT

There are three systems to measure light, as shown in Table 7.2. The preferred measurement system is quantum, since this system expresses a number of photons (or quanta) of light incident on a unit area (m^2) per unit time (second). A quantum sensor is designed to measure light only in the PAR waveband. The photometric system evaluates light according to its capacity to produce a visual sensation. A photometer measures the brightness of light to the human eye with its peak at 555 nm (Figure 7.2). The eye cannot see light in the infrared (>700 nm) or ultraviolet (<400 nm) range of the spectrum. Therefore, a photometer is not suitable for evaluating the light environment for plant production. The radiometric system evaluates light based on the absolute amount of energy per unit area per unit time ($W\,m^{-2}$ or $J\,m^{-2}\,s^{-1}$). Table 7.3 lists the approximate conversions for different lamp types, quoted from Thimijan and Heins (1983). These are approximate values because the spectral output of sources varies with individual luminaires, lamps, and ballasts, and their hours of use. Conversions for LEDs are not included in this table because the spectral output varies largely among different LEDs.

In addition to instantaneous light intensity, the total amount of light received by plants in a day or daily light integral (DLI) is closely related to plant growth, development, and quality. DLI is the

Table 7.2 Three Light Measurement Systems

Measurement System	Definition	Units
Quantum	Number of photons per unit area per unit time in a defined waveband, typically from 400 to 700 nm	$\mu mol\,m^{-2}\,s^{-1}$
Photometric	Capacity of producing a visual sensation	lx; footcandle 10.8 lx = 1 footcandle
Radiometric	Absolute amount of energy	$W\,m^{-2}$

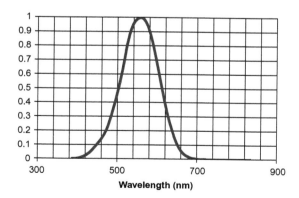

FIGURE 7.2

The human eye's response to different wavelengths of light. Green light generates a maximum response in the eye (peaks at 555 nm).

Table 7.3 Mixed Conversion Among Light Sources

Light Source	μmol m^{-2} s^{-1}	W m^{-2}	Footcandle	lx
Solar light	1	0.219[a]	5	54
High-pressure sodium	1	0.201	7.59	82
Metal halide	1	0.218	6.57	71
Cool-white fluorescent	1	0.218	6.85	74
Incandescent	1	0.200	4.63	50

[a]Values vary depending on light source, lamp, ballast and hours of use.
From Thimijan and Heins (1983).

amount of PAR received each day and is expressed as moles of light (mol) per square meter (m^{-2}) per day (d^{-1}), or: mol m^{-2} d^{-1}. In a PFAL, light intensity or PAR is usually kept relatively constant, and so DLI can be easily calculated from PAR and photoperiod.

In PFALs, various kinds of electric lights are used. The light spectral distribution or composition, also called light quality, affects plant growth and morphology. Light quality is often measured by a spectroradiometer, which is designed to measure the spectral power distribution of a light source. From the spectral power distribution, the characteristics of a light source can be determined in radiometric, photometric, and colorimetric quantities.

As light technology has advanced quickly in recently years, LEDs are increasingly being used in PFALs. When photosynthetic photon flux density (PPFD) under LEDs is measured using quantum sensors, especially for narrowband LEDs, spectral errors occur. This is because the quantum sensor is designed to measure the total number of photons between 400 nm and 700 nm. When measuring PPFD of a light source with a different spectrum from that used to calibrate the meter, spectral errors occur. Ideally, spectroradiometers should be used to measure PPFD for LEDs; however, quantum sensors are simple, inexpensive, and easy to use, provided measurement errors are considered.

LIGHT SOURCES

Kazuhiro Fujiwara

Graduate School of Agricultural and Life Sciences, The University of Tokyo, Tokyo, Japan

Until recently, most of the light sources used for PFALs were fluorescent lamps and high-intensity discharge (HID) lamps (especially high-pressure sodium lamps). According to the latest study by Shoji et al. (2013), 60% of PFALs in Japan use fluorescent lamps (including 6% using cold cathode fluorescent lamps), 27% use LEDs and 13% use HID lamps. Until just a decade ago, LEDs were used almost exclusively for research on plant cultivation, but are now being used as a light source for practical plant cultivation in PFALs because of their steady price decline and rapid improvement in luminous efficacy [lm W^{-1}], which is a measure of how efficiently an electrical lamp produces visible light.

This section examines the fundamentals necessary to understand light sources for PFALs, particularly LEDs, which are attracting increasing attention, and fluorescent lamps, which remain widely

Table 7.4 Luminosity-, Energy-, and Photon-Based Physical Quantities With SI Units Describing the Light Environment

Relations	Luminosity Basis	Energy Basis	Photon Basis
A	Luminous intensity [cd]	Radiant intensity $[W\ sr^{-1}]$	Photon intensity $[mol\ s^{-1}\ sr^{-1}]$
A sr =B	Luminous flux [cd sr] =[lm]	Radiant flux $[(W\ sr^{-1})\,sr]$ =[W]	Photon flux $[(mol\ s^{-1}\ sr^{-1})\,sr]$ =$[mol\ s^{-1}]$
A sr s =B s	Quantity of light [lm s]	Radiant energy [W s] =[J]	Photon number $[(mol\ s^{-1})\,s]$ =[mol]
A sr m^{-2} =B m^{-2}	Illuminance $[lm\ m^{-2}]$ =[lx]	Irradiance $[W\ m^{-2}]$	Photon flux density (Photon irradiance) $[(mol\ s^{-1})\,m^{-2}]$ =$[mol\ m^{-2}\ s^{-1}]$

used in PFALs. Pulsed lighting effects with LEDs are described briefly along with relevant references. The terminology used in this section is based on recently published books by Mottier (2009), Kitsinelis (2011), IEIJ (The Illuminating Engineering Institute of Japan) (2012), Khan (2013), and Japanese Industrial Standards (JIS) handbook No. 61 (2012).

Table 7.4 presents the relations among luminosity-based, energy-based, and photon-based physical quantities of light environments because some quantities are used in this section without explanation.

CLASSIFICATION OF LIGHT SOURCES

Electrical lamps used today for general illumination can be classified into three categories according to the principle of light emission: incandescence, discharge light emission, and electroluminescence (Table 7.5). The principle of incandescence is the same as that of the thermal radiation of a blackbody that radiates or emits light as it is heated. A traditional incandescent lamp is a typical example of one that emits light through incandescence. Discharge light emission refers to the emission of light by transitioning an electron of an atom which has been excited by an electric discharge, from a higher energy state to a lower energy state (mostly to ground state). Fluorescent lamps and high-pressure sodium lamps emit light through discharge light emission. The third mode, electroluminescence, refers to light emission caused by application of an electric field to a material. Of lamps in this category, LEDs have received the greatest attention for use in PFALs.

LIGHT-EMITTING DIODES

The following fundamentals must be understood when using LEDs for plant cultivation: (1) general benefits, (2) outline of the light-emitting mechanism, (3) configuration types, (4) basic terms expressing electrical and optical characteristics, (5) electrical and thermal characteristics in operation, (6) lighting and light intensity control methods, (7) lesser-known benefits and disadvantages related

Table 7.5 Light Emission Principles and Corresponding Major Electric Lamps and Devices

Light Emission Principle	Electrical Lamps/Devices
Incandescence	Incandescent lamps
	Halogen incandescent lamps
Discharge light emission	Low pressure discharge lamps
	Low pressure sodium lamps
	Fluorescent lamps
	Preheat fluorescent lamps
	Rapid-start fluorescent lamps
	High-frequency fluorescent lamps
	Cold cathode fluorescent lamps
	High pressure discharge lamps
	High pressure mercury lamps
	Metal halide lamps
	High pressure sodium lamps
	High pressure xenon lamps
Electroluminescence	Intrinsic electroluminescence devices
	Inorganic electroluminescence devices
	Injection electroluminescence devices
	Light-emit diodes
	Organic electroluminescence devices

to use, and (8) LED modules with different color LEDs for PFALs. These items are explained below, together with (9) pulsed light and its effects, because pulsed light is often included in discussions on light sources for PFALs.

General Benefits

Generally speaking, LEDs offer advantages over incandescent, fluorescent, and HID lamps: they are robust; produce a stable output (immediately after an electric current flows); are long-lived, compact, and lightweight; instantaneously turn on; and allow the light output to be easily controlled. Moreover, the spectral distribution of emitted light can be controlled with a light source comprising several color types of LEDs. The last point is described at greater length in section "Lesser-Known Benefits and Disadvantages Related to Use."

Outline of the Light-Emitting Mechanism

An LED is a semiconductor diode formed by contacting p-type and n-type materials. By applying an appropriate forward voltage to the diode to move holes (having a positive electrical charge) in the p-type material side and electrons (having a negative electrical charge) in the n-type material side towards the other side, the holes and electrons can recombine in the junction. The hole–electron recombination produces a photon, which has energy equivalent to the amount of energy that the electron released by transitioning when recombined. Light is emitted from LEDs in this manner (Figure 7.3).

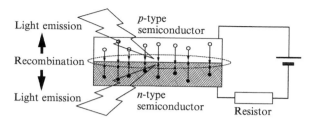

FIGURE 7.3

LED light emission from the junction of *p*-type and *n*-type semiconductors when electricity is supplied. Open and solid circles respectively denote holes (positively charged) and electrons (negatively charged).

Configuration Types

The lamp type (or round type, recently also called "through hole" or "thru-hole" type) and surface mount device (SMD) type (or surface mount type) are the major configurations for LEDs (Figure 7.4). The so-called "(4-pins) flux type LEDs" have two anodes and two cathodes, and are shaped like the four legs of a table. Figure 7.5 shows the arrangement of lamp type LED components.

Basic Terms Expressing Electrical and Optical Characteristics

Basic terms expressing electrical and optical characteristics of LEDs include the following:

Forward current [A]: The maximum forward current allowed for continuous operation. A similar term, "pulse forward current," refers to the maximum forward current allowed when producing pulsed light.

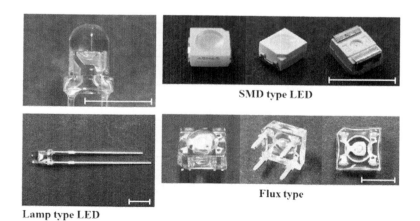

FIGURE 7.4

Lamp type (NSPB310B; Nichia Corp.), surface mount device (SMD) type (NESG064; Nichia Corp.), and flux type (NSPWR60CS-K1; Nichia Corp.) LEDs. White bars indicate a scale of approximately 5 mm.

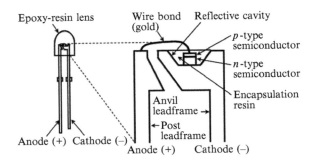

FIGURE 7.5

Schematic diagram of a lamp type LED.

Luminous intensity [cd]: This is defined as the luminous flux [lm] emitted into a solid angle [sr] of space in a specific direction (normally the optical axis direction as for LEDs, at which the value shows the maximum). Regarding LEDs with peak wavelength outside the visible wavelength range, radiant intensity [W sr^{-1}] is used.

Radiant flux/Total radiant power [W]: The amount of radiant energy [J] emitted per unit time [s].

Peak wavelength [nm]: The wavelength at which the spectral radiant flux [W nm^{-1}] in the spectral radiant flux distribution curve (spectrum) of an LED is the maximum (Figure 7.6).

Half width [nm]: The width of the spectral radiant flux distribution curve (spectrum) at 50% of the maximum spectral radiant flux. It is a measure of the light monochromaticity (Figure 7.6).

Viewing half angle [°]: The angle from the LED optical axis (at 0°), at which LED radiant intensity [W sr^{-1}] is the maximum, to which the radiant intensity is reduced to half the maximum value at 0°.

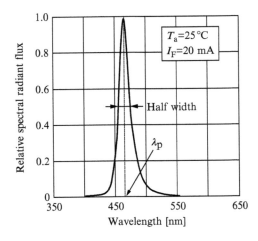

FIGURE 7.6

Peak wavelength (λ_p) and half width using a sample spectral distribution of a blue LED (NSPB310B; Nichia Corp).

The figure was redrawn from the specification sheet.

Electrical and Thermal Characteristics in Operation

LEDs have common operating characteristics that must be well understood when using LEDs for both plant production and research on the effects of the light environment. The major characteristics are as follows:

(a) The electric current flowing through an LED increases exponentially with increasing voltage applied to the LED (Figure 7.7a).

(b) The relative spectral radiant flux (and also relative luminous intensity) emitted from an LED is roughly proportional to the electric current flowing through the LED when the ambient temperature is constant (Figure 7.7b).

(c) Even if the electric current flowing through an LED is constant, the relative spectral radiant flux (and also relative luminous intensity) emitted from the LED decreases with increasing ambient temperature (Figure 7.7c).

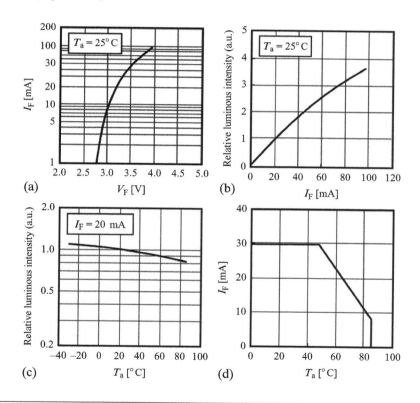

FIGURE 7.7

Common electrical and thermal characteristics of a white LED (NSPW310DS; Nichia Corp.) in operation: (a) relationship between forward voltage (V_F) and forward current (I_F), (b) effect of forward current on relative luminous intensity, (c) effect of ambient air temperature (T_a) on relative luminous intensity, and (d) allowable forward current (I_F) with ambient air temperature (T_a).

The figures were redrawn from the specification sheet.

(d) The maximum forward current allowed (allowable forward current) drops sharply when the ambient temperature increases beyond a certain temperature (typically around 40°C). A maximum ambient temperature exists for the use of LEDs (typically around 80°C) (Figure 7.7d).

Lighting and Light Intensity Control Methods

Two lighting methods are used for plant cultivation in respect of "truly" or "apparently" continuous light during the photoperiod. The most common method is "truly" continuous lighting, in which light is emitted continuously "in a strict sense" during the photoperiod. The other is intermittent lighting, in which light is emitted intermittently in a short time cycle. Although there is no clear definition, intermittent lighting is sometimes designated as pulsed lighting, especially when the one-cycle time (light-on time period plus light-off time period) is less than or equal to 1 s. An LED can produce pulsed light with a light-on and light-off cycle of less than 10 µs.

Simple and general methods for controlling light intensity (radiant flux in the strict terminology) or dimming an LED include controlling the electric current flowing through the LED. This is called constant-current operation. Constant-voltage operation is a similar method, but it can be done only under the condition of constant temperature of both the LED chip and ambient air. Another method of controlling light intensity is so-called "pulse forward current". In this method, an LED is repeatedly turned on and off at extremely short intervals by applying and breaking a constant electric current flowing through the LED, causing the LED to produce light intermittently. The pulse width is the time period during which current is applied and the light is on. The percentage of the light-on time period to the one-cycle time period is called the duty ratio [%]. A change in the duty ratio results in an increase or decrease in radiant flux from the LED. At 50% duty ratio, the LED will emit half the radiant flux that it does at 100% duty ratio. Intermittent light with a light-off time period shorter than a certain level is perceived by the human eye as continuous light.

Lesser-Known Benefits and Disadvantages Related to Use

An important but less widely recognized benefit of using LEDs as a light source for PFALs is that LEDs offer great flexibility for producing various light environments compared to conventional lamps. For instance, a light source having several types of LEDs with different peak wavelengths can produce light of which the spectral radiant flux can be varied with time. Such light source systems have been developed for research purposes with 5 (Fujiwara et al., 2011; Yano and Fujiwara, 2012), 6 (Fujiwara and Yano, 2013), and 32 (Fujiwara and Sawada, 2006; Fujiwara and Yano, 2011; Fujiwara et al., 2013) types of LEDs for advanced control of the light environment. Such LED light sources are ideal for conducting research on controlling the light environment to improve plant production in PFALs.

The greatest disadvantage of using LEDs for PFALs is the high initial cost for a set of LED light sources and the luminaire ("luminaire" refers to the entire electric light fitting, including all components needed to mount, operate and protect the lamps) compared to conventional lamps. This disadvantage will take time to overcome.

LED Modules With Different Color LEDs for PFALs

LED modules with different color LEDs (strictly speaking, LEDs that emit light with different relative spectral power distributions) are used in some PFALs. The most common combination is a blue LED with a peak wavelength (λ_p) of around 460 nm and a red LED with λ_p of around 660 nm. A combination of a red LED and white LED (made of a blue LED chip coated with a yellow phosphor) is also used for

LED modules in a PFAL. An LED module incorporating blue, red, and far-red LEDs is already being used for commercial production of lettuce seedlings in the Netherlands. A multicolor LED module that has green, violet, or ultraviolet LEDs in addition to blue and red LEDs could be used in a PFAL if such a module is shown to improve plant growth and/or increase nutritional and functional ingredients that are beneficial to human health. If asked to choose a single color LED for plant production, the author would select an LED with a spectral power distribution similar to that of sunlight at ground level.

Pulsed Light and Its Effects

An LED can be turned fully on and fully off extremely rapidly; the response time is less than 100 ns. For that reason, LEDs are widely used for investigating the effects of pulsed light on plant photosynthesis and growth. Tennessen et al. (1995) measured the net photosynthetic rate (NPR) of tomato leaves in continuous light and compared the results with those for pulsed light with the same average PPFD (50 μmol m^{-2} s^{-1}) at a duty ratio of 1% from red LEDs. The results showed that NPRs were similar during continuous light and pulsed light at frequencies (reciprocal of the cycle time period) greater than 0.1 kHz. However, at 5 Hz, NPR was half that measured in continuous light. Jishi et al. (2012) reported a similar result: no combination of frequency (0.1–12.8 kHz) and duty ratio (25–75%) showed significantly greater NPR in cos lettuce plants than in continuous light when irradiated from white LEDs at an average PPFD of 100 μmol m^{-2} s^{-1} (Figure 7.8). Sims and Pearcy (1993) and Pinker and Oellerich (2007) reported no positive effect of pulsed light on growth rate in *Alocasia* plants and *Amelanchier* and *Tilia* shoots *in vitro*, respectively. In contrast to those studies, Mori et al. (2002) observed 1.23 times greater NPR and relative growth rate (weight increase rate divided by weight) in lettuce plants under a combination of 2.5 kHz frequency and 50% duty ratio than in continuous light when irradiated from white LEDs at an average PPFD of 50 μmol m^{-2} s^{-1}. It seems that pulsed light generally does not

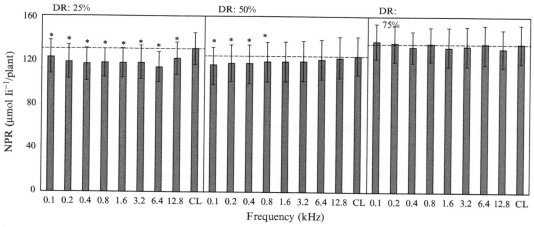

FIGURE 7.8

Net photosynthetic rate (NPR) of the aerial part of cos lettuce with different combinations of duty ratio (DR) and frequency of pulsed light from white LEDs. Dashed lines represent NPRs in continuous light (CL) from white LEDs. Bars represent S.E. ($n = 6$). Means with an asterisk are significantly different from that obtained on the same day in continuous light (paired t test, $P < 0.05$) (Jishi et al., 2012).

promote photosynthesis or plant growth as long as the spectral photon flux density (or spectral irradiance) is constant with time when irradiated, and the average PPFD is the same for pulsed light and continuous light.

FLUORESCENT LAMPS

Shoji et al. (2013) reported that 60% of PFALs in Japan use fluorescent lamps (including cool cathode fluorescent lamps) as a light source. However, newly constructed PFALs in Japan tend to use LEDs rather than fluorescent lamps. Regarding shape, tubular fluorescent lamps are exclusively used in PFALs. There is no report on the starting types (preheat, rapid-start, high-frequency, or cold cathode) of fluorescent lamps used, or their relative percentages. Presumably rapid-start and high-frequency types account for the greatest percentage of fluorescent lamps used in PFALs.

The (1) general benefits, (2) tubular fluorescent lamp configurations, (3) outline of the light emission mechanism and process, and (4) relative spectral radiant flux of light emitted from a fluorescent lamp are outlined below.

General Benefits

Fluorescent lamps offer no clear single advantage over other lamps including LEDs. Nevertheless, tubular fluorescent lamps are the most appropriate light source at present for PFALs when considering all the factors of lamp and luminaire prices, rated life, luminous efficacy, ready availability, and lighted lamp surface temperature.

Configuration of Tubular Fluorescent Lamps

A tubular fluorescent lamp generally comprises a glass tube (mostly 25 mm diameter, several lengths available according to the required luminous flux) coated inside with a fluorescent material (phosphor), two tungsten electrodes at the two inside ends that are coated with an electron-emissive material, a small amount of mercury, and low vapor-pressure inert gas (mostly argon) enclosed in the glass tube.

Outline of the Light Emission Mechanism and Process

The light emission from a fluorescent lamp is the result of electron-transitioning of mercury atoms that have been excited by an electric discharge generated by the application of an alternating voltage to the electrodes. Therefore, the principle of light emission of a fluorescent lamp is classified as discharge light emission. Incidentally, based on the fluorescence process taking place on the fluorescent material, the light emission of a fluorescent lamp is classified as photoluminescence.

Light is produced by a fluorescent lamp through the following processes: (1) electricity is supplied to the electrodes through ferrule pins (metal fittings), (2) thermal electrons are emitted from the electrodes, (3) the thermal electrons collide with mercury atoms, (4) the mercury atoms radiate ultraviolet light (emission line: 254 nm), (5) the ultraviolet light reaches the fluorescent material coated on the inside of the glass tube, (6) the fluorescent material converts some ultraviolet light into visible light, and (7) the visible light is transmitted through the lamp glass to the outside of the glass tube.

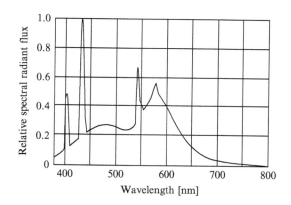

FIGURE 7.9

Spectral distribution of a tubular daylight white fluorescent lamp (FLR110HN/A/100; Toshiba Lighting & Technology Corp.).

The figure was redrawn from the specification sheet.

Relative Spectral Radiant Flux of Light Emitted from a Fluorescent Lamp

A fluorescent lamp emits broad-wavelength light ranging from ultraviolet to visible light with the emission lines of mercury (365, 405, 436, 546, and 577 nm within visible light) in its spectrum (Figure 7.9). The continuum spectrum shows light emission from the fluorescent material. Fluorescent lamps with various spectra or relative spectral radiant flux, in which different types of fluorescent materials are used to coat the inside of the glass tube, are commercially available. Several types of fluorescent lamps developed for plant production are also available, although they cost more than general white fluorescent lamps.

REFERENCES

Fujiwara, K., Sawada, T., 2006. Design and development of an LED-artificial sunlight source system prototype capable of controlling relative spectral power distribution. J. Light. Vis. Environ. 30 (3), 170–176.

Fujiwara, K., Yano, A., 2011. Controllable spectrum artificial sunlight source system using LEDs with 32 different peak wavelengths of 385–910 nm. Bioelectromagnetics 32 (3), 243–252.

Fujiwara, K., Yano, A., 2013. Prototype development of a plant-response experimental light-source system with LEDs of six peak wavelengths. Acta Horticult. 970, 341–346.

Fujiwara, K., Yano, A., Eijima, K., 2011. Design and development of a plant-response experimental light-source system with LEDs of five peak wavelengths. J. Light. Vis. Environ. 35 (2), 117–122.

Fujiwara, K., Eijima, K., Yano, A., 2013. Second-generation LED-artificial sunlight source system available for light effects research in biological and agricultural sciences. In: Proceedings of 7th LuxPacifica, Bangkok, Thailand, pp. 140–145.

IEIJ (The Illuminating Engineering Institute of Japan) (Ed.), 2012. Illuminating Engineering, Ohmsha, Ltd, Tokyo, Japan (in Japanese).

Japanese Standard Association (Ed.), 2012. JIS (Japanese Industrial Standards) Handbook No. 61 Color. Japanese Standard Association, Tokyo, Japan (in Japanese).

Jishi, T., Fujiwara, K., Nishino, K., Yano, A., 2012. Pulsed light at lower duty ratios with lower frequencies is less advantageous for CO_2 uptake in cos lettuce compared to continuous light. J. Light. Vis. Environ. 36 (3), 88–93.

Khan, M.N., 2013. Understanding LED Illumination. CRC Press, Boca Raton, FL, USA.

Kitsinelis, S., 2011. Light Sources: Technologies and Applications. CRC Press, Boca Raton, FL, USA.

Mori, Y., Takatsuji, M., Yasuoka, T., 2002. Effects of pulsed white LED light on the growth of lettuce. J. Soc. High Technol. Agric. 14 (3), 136–140 (in Japanese with English abstract and captions).

Mottier, P. (Ed.), 2009. LEDs for Lighting Applications. ISTE Ltd., John Wiley & Sons, Inc, Great Britain, UK.

Pinker, I., Oellerich, D., 2007. Effects of chopper-light on *in vitro* shoot cultures of *Amelanchier* and *Tilia*. Propag. Ornam. Plants 7 (2), 75–81.

Shoji, K., Moriya, H., Goto, F., 2013. Surveillance study of the support method to the plant factory by electric power industry: development trend of plant factory technology in Japan. Environment Science Research Laboratory Report No. 13002, Central Research Institute of Electric Power Industry, Tokyo, pp. 1–16.

Sims, D.A., Pearcy, R.W., 1993. Sunfleck frequency and duration affects growth rate of the understorey plant *Alocasia macrorrhiza* (L.) G. Don. Funct. Ecol. 7, 683–689.

Tennessen, D.J.R., Bula, J., Sharkey, T.D., 1995. Efficiency of photosynthesis in continuous and pulsed light emitting diode irradiation. Photosynth. Res. 44, 261–269.

Thimijan, R.W., Heins, R.D., 1983. Photometric, radiometric, and quantum light units of measure: a review of procedures for interconversion. HortSci. 18, 818–822.

Yano, A., Fujiwara, K., 2012. Plant lighting system with five wavelength-band light-emitting diodes providing photon flux density and mixing ratio control. Plant Methods 8, 46.

PHYSICAL ENVIRONMENTAL FACTORS AND THEIR PROPERTIES

Genhua Niu[1], Toyoki Kozai[2], Nadia Sabeh[3]

Texas AgriLife Research at El Paso, Texas A&M University, El Paso, Texas, USA[1] Japan Plant Factory Association, c/o Center for Environment, Health and Field Sciences, Chiba University, Kashiwa, Chiba, Japan[2] Building Performance Engineer, Guttmann & Blaevoet Consulting Engineers, Sacramento, California, USA[3]

INTRODUCTION

In order to provide the optimum environment for plants to grow in a plant factory with artificial lighting (PFAL), it is essential to understand the nature of each of the environmental factors and how to measure and quantify them. This chapter describes the physical and chemical properties of the following environmental factors and their measurement: temperature, humidity, CO_2 concentration, air current speed, and number of air exchanges per hour. In addition, the basic concepts of energy balance, radiation, and heat conduction and convection are described in detail. Furthermore, the concept and use of the psychrometric chart are described to introduce its importance in understanding environmental control of PFALs. Light, one of the most important environmental factors in the PFAL, is described in Chapter 7. The response of plant growth and development to these environmental factors is described in Chapter 10.

TEMPERATURE, ENERGY, AND HEAT

Temperature is an indication of the sensible heat energy content of an object or a substance. Many plant physiological processes are affected by plant temperature, which is determined by the transfer of heat between plant tissues and the surrounding environment. Therefore, monitoring and controlling the air temperature is critical for managing plant physiological activity and response. In a PFAL, air temperature is often controlled at a relatively constant level, resulting in constant plant temperature and, therefore, consistent physiological activity.

ENERGY BALANCE

Any object with a temperature above 0 K (absolute zero) emits thermal radiation, including the plants themselves and their surroundings. Energy received by plants includes absorbed radiant energy from lamps and the absorbed infrared irradiation from surroundings. Energy leaving plants includes energy lost through emitting infrared radiation, heat convection, heat conduction and heat loss through evaporation. The heat by conduction and convection from leaves is referred to as sensible heat, and that associated with the evaporation or condensation of water is known as latent heat. Plant leaves have high absorptivity in the photosynthetically active radiation (400 to 700 nm), but the chemical energy fixed by photosynthesis is negligibly small compared to the total energy budget of the plant. Leaves of most species have low absorption in the near infrared range (700 to 1500 nm) because those wavelengths are transmitted through or reflected from the leaf. In contrast, absorption is high (approximately 95%) in the far infrared waveband (1500 to 30,000 nm), which can contribute significantly to the thermal energy load on the plant.

RADIATION

Radiation in the far infrared waveband is essentially blackbody radiation emitted by surrounding objects. Objects of higher temperature emit larger quantities of far infrared radiation than objects at a lower temperature.

The primary sources of radiant energy in PFALs are lamps and reflectors. Traditional lamps for growth chambers and greenhouses, such as high pressure sodium lamps and metal halide lamps, have surface temperatures of over 100°C and emit large quantities of far infrared radiation. This radiation is absorbed by plants, causing increased plant temperature regardless of the surrounding air temperature, thereby impeding control over plant physiological activity. In PFALs, this challenge is compounded by the short distance between lamps and plants that is desirable for maximizing space use efficiency and plant productivity. Therefore, it is preferable to use light sources that emit much less far infrared radiation, such as LED (surface temperature: about 30°C) and fluorescent lamps (surface temperature: about 40°C).

HEAT CONDUCTION AND CONVECTION

Energy is conducted between a plant and its environment at the molecular level. Energy is transferred by conduction from the leaf cells to the air molecules in contact with the leaf. Conductive heat transfer at the interface between leaf and air is limited without convective movement due to the low thermal conductivity of air. Conductive heat exchange can also occur between plant parts and other solid or liquid media. However, the impact of this conductive heat exchange on the plant's energy budget is small, because plants do not have much physical contact with solid objects or liquid media.

Convective heat transfer occurs when air moves across the plant. Temperature gradients between the leaves and air result in corresponding gradients in air density and pressure, which cause turbulence and the initiation of convection. There are two types of convection: free and forced. Free (natural) convection occurs when the heat transferred from leaves causes the air outside the unstirred layer to warm, expand, and thus to decrease in density. This more buoyant warmer air then moves upward and thereby transfers heat away from the leaf.

Forced convection, caused by wind or fans, is more effective in heat exchange. As air speed increases, more and more heat is dissipated by forced air movement. An air current speed of more than 0.5 m s^{-1} at the leaf canopy is required to promote gas exchange; however, air current speeds greater than 1.0 m s^{-1} do not significantly increase gas exchange (Kitaya et al., 2000). Also, higher air current speeds are not recommended due to possible mechanical stress on the plant. Therefore, in PFALs, the air current speed is usually controlled between 0.5 to 1.0 m s^{-1} using air circulation fans to promote gas exchange. Further, when lamps are placed close to plants, leaves absorb significant amounts of thermal energy from lamps, which increases leaf temperatures. Air circulation can help mitigate leaf temperatures by facilitating convective heat transfer away from the plants.

Controlling leaf and air temperatures uniformly at every growing shelf is important in a PFAL. If air circulation in a PFAL is insufficient, air temperatures at the upper growing shelves will be warmer than lower shelves, causing the leaves in the upper canopy to also be warmer. By providing air movement in the whole PFAL system, the vertical air and leaf temperature gradients can be minimized, as well as differences within each horizontal canopy.

LATENT HEAT—TRANSPIRATION

Evaporation of water is a cooling process, caused by water absorbing sensible heat from its surroundings when it changes from liquid to vapor. During plant transpiration, water evaporates at the air–liquid interfaces along the pores in the cell walls of mesophyll, epidermal, and guard cells and then diffuses out of a leaf. Thus, transpiration represents a means of latent heat loss by a leaf. A leaf can also gain latent heat if dew or frost condenses onto it.

When 1 g of free water evaporates at 25°C, 2436 J of energy is transferred between the leaf and the air. The rate of transpiration, and thus the amount of latent heat transferred from a leaf via evaporation, depends on the difference in vapor pressure between the leaf's surface and the air immediately surrounding it. Therefore, latent heat transfer increases as the leaf/air temperature difference increases. During unstressed conditions (adequate water supply and fully open stomata), transpiration increases with increasing leaf temperature. As water evaporates from the leaves, leaf temperature decreases, often below the surrounding air temperature.

MEASUREMENT OF TEMPERATURE

Temperature is one of the most critical variables to measure in PFALs. Air temperature, leaf temperature, and even media and nutrient solution temperature can all impact plant physiological activity. Additionally, measuring plant temperature and comparing it to the air temperature can indicate if a plant is stressed. In a PFAL, multiple growing shelves are arranged in close proximity to each other both vertically and horizontally. To achieve the greatest control over air and leaf temperatures, it is ideal to monitor air temperature at several locations, both vertically and horizontally along the growing shelves.

Although measuring temperature is relatively easy and inexpensive, selecting the correct temperature-sensing device is important, regardless of the capability and sophistication of the environmental control system. Selection of temperature sensors should be based on operating temperature range, reliability, sensitivity to temperature change, and long-term stability. Thermistors often are chosen because they fulfill these criteria, are relatively inexpensive, and respond quickly to temperature changes. Thermistors are essentially ceramic resistors that change resistance with changing temperature. They are regularly used for climate control applications and are commonly found in digital thermometers.

The thermocouple is another commonly used device for air temperature measurement. A thermocouple consists of two different metals that are combined to produce a voltage related to their temperature difference. There are several types of thermocouples that contain different metals, wire thickness and thus different costs and accuracy. The most common type for greenhouses, growth chambers and PFALs is type T (copper-constantan) welded thermocouple. The primary challenge of using thermocouples to measure air temperature is that they are sensitive to sources of heat other than the air.

Specifically, thermocouples can produce erroneously high temperature readings if exposed to direct radiation from the sun or nearby objects. They can also produce erroneously low temperature readings if condensation forms on the metal wires. Therefore, to increase the accuracy of thermocouple measurements, they should be mounted in an enclosed box with a small fan that blows air from the surroundings over the coupled wires.

WATER VAPOR
HUMIDITY

Water vapor is the gaseous state of water and humidity is a measure of its content in the air. The amount of atmospheric water vapor can range from nearly zero up to 4% of the total mass of air. Absolute humidity (AH), or humidity ratio, is a measure of the actual water vapor content in the air and is expressed as the ratio of mass of water vapor to the mass of dry air for a specified volume of air. The air can hold more water vapor at higher temperatures than at lower temperatures. Relative humidity (RH) is temperature-dependent and used to express the water vapor content of air based on the maximum amount of water the air can hold for a given temperature and pressure. It is most often expressed as a percentage or ratio of the given water vapor content to the maximum at a given temperature. As an example, if the air temperature decreases with no change in water vapor content (or AH), the maximum water holding capacity of the air drops, resulting in a higher RH.

Water vapor is produced by evaporation from open water surfaces (lakes, puddles, rivers, and oceans) and evaporation from wet surfaces such as soil and plants. In PFALs, plants are constantly adding water vapor to the air through transpiration, which is the evaporation of water from plant surfaces to the atmosphere. Healthy, actively growing plants can transpire a lot of water, resulting in a rapid increase in the water vapor content and humidity in a semi-closed PFAL. When the air-conditioning system is operating, humidity is kept under control because water vapor condenses on the cooling coils, dropping the moisture content, and thus humidity, of the air. Therefore, one approach to controlling humidity in a PFAL is to alternate the operation of lights to generate heat and cause the air conditioner to run, resulting in simultaneous cooling and dehumidification of the space. Alternatively, stand-alone dehumidifiers can be installed in the PFAL that do not rely on the operation of air conditioners. These units may be used in PFAL applications that require distinct day/night cycles, when turning on the lights for dehumidification would be undesirable. They can also be used to avoid operating lights and air conditioners during peak energy use periods, lowering energy costs.

In a PFAL, the amount of condensation can be significant and recycling the condensation is desirable for conserving water. The safety of this condensed water for irrigation should be confirmed prior to use due to the possibility of contaminants such as heavy metal. If used for irrigation, the condensate should be filtered and treated prior to being delivered to plants.

VAPOR PRESSURE DEFICIT

Although RH is commonly used as a measure of air humidity, it provides no direct information about the driving force of transpiration and evaporation. Instead, the vapor pressure deficit (VPD) is a measure of the driving force, meaning that transpiration and evaporation rates are proportional to VPD.

VPD is the difference (deficit) between the amount of moisture in the air and how much moisture the air can hold when it is saturated at the same air temperature and is expressed in units of pressure. The SI unit of VPD is kPa (kilopascal), but $kg\ m^{-3}$ is still used in some literature. As water vapor content increases, water molecules exert more force on each other, resulting in a higher vapor pressure. Therefore, because air can hold more water vapor at higher temperatures, the maximum water vapor pressure is higher at higher temperatures.

When the VDP is too low, transpiration will be inhibited and can lead to condensation on leaves and surfaces inside the PFAL. On the other hand, when the VPD is high, the plant will draw more water from its roots in an effort to avoid wilting. If the VPD gets too high, plants close stomata and shut down the transpiration altogether in an effort to prevent excessive water loss. In a PFAL, the ideal range for VPD is from 0.8 kPa to 0.95 kPa, with an optimal setting of around 0.85 kPa.

MEASUREMENT OF HUMIDITY

As previously discussed, humidity is the amount of water vapor in the air. RH is the most common measure of water vapor content because it relates water vapor content to the maximum water vapor content the air could hold at saturation. For example, 50% RH indicates that the air has half the water vapor that it could hold if it were completely saturated. As the air temperature rises, more water vapor must be added to the air to maintain the same RH.

The traditional method of measuring humidity is to measure both wet bulb and dry bulb temperatures, and then convert to RH using a psychrometric chart. The dry bulb temperature is commonly measured with a standard thermometer. The wet bulb temperature is determined from a standard thermometer modified with a wetted fabric wick covering the sensor bulb. Sufficient air flow is provided over the wick material so that as water evaporates from the wet wick, the temperature falls and the thermometer reading reflects the wet bulb temperature. Since wet bulb needs to be wet all the time, this method is not suitable for continuous measurement of humidity.

RH can also be measured directly using a hygrometer. There are many types of hygrometers for direct RH measurement with different degrees of accuracy and at various prices. Modern humidity sensors use electronic devices, which are based on changes in electrical capacitance or resistance to measure humidity differences. Most humidity sensors need calibration every year or two to maintain accuracy. Even with frequent calibration, most humidity sensors have low accuracy under very humid (above 95%) and dry environment (below 10%).

MOIST AIR PROPERTIES
COMPOSITION OF AIR

Atmospheric air comprises approximately 78% nitrogen, 21% oxygen, 0.93% argon, 0.04% carbon dioxide (CO_2), a variable amount of water vapor, and other gases. The density of dry air (assuming zero water vapor) under standard pressure (101 kPa) is $1.293\ kg\ m^{-3}$ at 0°C, $1.205\ kg\ m^{-3}$ at 20°C, and $1.185\ kg\ m^{-3}$ at 25°C. Therefore, the higher the temperature, the lighter is the air. Compared with the density of pure water ($1000\ kg\ m^{-3}$), the density of air is very small.

FIGURE 8.1

Schematic diagram of dry and moist air.

In reality, air is never completely dry, even in a desert, which means the air is always a mixture of dry air and water vapor (moist air, Figure 8.1). Although the water vapor content is often less than 1% of the total, it has a large impact on the condition and overall behavior of air. Water vapor plays a major role in earth's water cycle, influences local climate conditions, and both stores and transfers energy in the form of sensible and latent heat. Sensible heat is energy that a person can "feel" and can be measured directly with a thermometer. Latent heat is energy that cannot be felt, and is contained within water until it is transferred to a substance that has less energy.

The process of evaporative cooling demonstrates latent heat transfer. When liquid water evaporates, it absorbs heat from the air around it, causing the sensible air temperature to decrease and the water content (humidity) of air to increase. However, this energy does not disappear. It is "captured" by water vapor in the form of latent heat. Therefore, the energy content of the mixed air does not change, even though the dry bulb temperature has decreased.

PSYCHROMETRIC CHART

Temperature and humidity are often used to characterize the properties of moist air for climate control applications. The psychrometric chart provides a graphical representation of the relationships between water vapor, temperature, and energy (Figures 8.2 and 8.3). This chart allows all the parameters of moist air to be determined when only two of the following independent properties are known: dry bulb temperature (T_{db}), wet bulb temperature (T_{wb}), dewpoint temperature (T_d), AH (also called humidity ratio), RH, and water vapor pressure (P_v). When any two of these six independent variables are known, saturated vapor pressure ($P_{v,sat}$), enthalpy (h), and specific volume (v) can be determined. By knowing the relationships of these moist air properties, one can predict the effectiveness of various environmental control strategies to influence plant behavior and productivity. Table 8.1 lists the definition and units of psychrometric properties.

Dry bulb temperature, T_{db}, which is commonly referred to simply as "air temperature," is the air property that is most familiar to people. Dry bulb temperature can be measured using a standard thermometer and indicates the quantity of "sensible heat." Dry bulb temperature is shown along the horizontal axis of the psychrometric chart. The vertical lines extending upward from this axis represent constant T_{db} (Figure 8.2).

Wet bulb temperature, T_{wb}, is the temperature of air if it were saturated with water vapor. It is measured with a thermometer that has a bulb covered with a water-moistened wick. The T_{wb} is measured

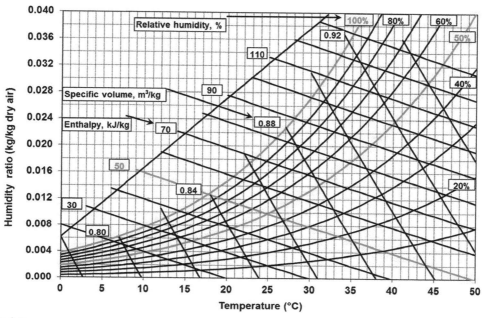

FIGURE 8.2

Psychrometric chart.

Figure courtesy of Dr. A.J. Both, Rutgers University.

when a constant, known velocity of air is passed over the wet wick. On the chart, the constant wet-bulb lines slope slightly up and to the left. The T_{wb} is read at the saturation line (100% RH). The wet bulb temperature will always be lower than the dry bulb temperature, unless the air is completely saturated with water vapor (100% RH).

Dewpoint temperature, T_d, is the temperature at which the water vapor in air at constant barometric pressure condenses into liquid water at the same rate at which it evaporates. If water vapor comes into contact with solid surfaces that are lower than T_d, liquid water will form into condensation, or dew. Likewise, if the dry bulb air temperature drops below T_d, water vapor will condense within the air itself and produce fog. The dewpoint is associated with RH. A high RH indicates that the dewpoint temperature is close to the current dry bulb air temperature. At RH of 100%, the dry bulb, wet bulb, and dewpoint temperatures are all the same, indicating that the air is maximally saturated with water vapor. When the absolute moisture content remains constant and temperature increases, RH decreases.

Understanding the psychrometric chart can help visualize the properties and behavior of moist air for environmental control of a PFAL or greenhouse. For example, looking at the psychrometric chart, one can see that warm air can hold more water vapor than cold air. Yet, as that air is cooled, the RH increases; and if it is cooled enough, water vapor will condense into liquid when the RH is 100%.

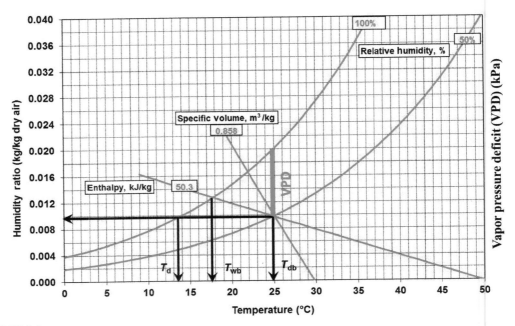

FIGURE 8.3

Simplified psychrometric chart showing dry bulb temperature (T_{db}), wet bulb temperature (T_{wb}), and dewpoint (T_d) of moist air on a psychrometric chart.

Figure courtesy of Dr. A.J. Both, Rutgers University.

Table 8.1 Psychrometric Properties Related to Psychrometric Chart and Their Definition

Psychrometric Property	Unit	Definition
Dry bulb temperature, T_{db}	°C	Air temperature measured with a regular thermometer
Wet bulb temperature, T_{wb}	°C	The temperature reading from a wetted bulb
Relative humidity, RH	%	The amount of water vapor in the air divided by the amount of water vapor the air can hold at the same temperature and pressure
Dew point temperature, T_d	°C	The saturation temperature of the moisture present in the sample of air. It can also be defined as the temperature at which the vapor changes to liquid (condensation)
Absolute humidity, AH	kg kg^{-1} (dry air)	The mass of water vapor per unit mass of dry air (kg kg^{-1}). This quantity is also known as the water vapor density
Specific volume, ν	m^3 kg^{-1}	The volume of moist air per kg of dry air. Warm air is less dense than cool air which causes warmed air to rise
Enthalpy, h	kJ kg^{-1}	The total heat (sensible and latent heat) energy per kg of dry air, which increases with increasing temperature and absolute humidity

CO$_2$ CONCENTRATION

NATURE

CO$_2$ is a naturally occurring chemical compound. It is a linear covalent molecule and is an acidic oxide, and it reacts with water to give carbonic acid. CO$_2$ is a nonflammable, colorless, odorless gas at standard temperature and pressure and exists in Earth's atmosphere at this state as a trace gas. As of September 2014, the average atmospheric CO$_2$ concentration was 395.28 ppm (parts per million) (http://www.esrl.noaa.gov/). Atmospheric CO$_2$ concentration varies with time of day and location depending on absorption and respiration of plants and animals, and human activity. With a molecular weight of 44.01 g mol^{-1}, CO$_2$ is heavier than air. CO$_2$ has a melting point of $-55.6°C$, boiling point (sublimation) of $-78.5°C$, and gas density (at 1.013 bar and 15°C) of 1.8714 kg/m^3. It freezes at $-78.5°C$ to form CO$_2$ snow. CO$_2$ is produced from the combustion of coal or hydrocarbons, the fermentation of liquids, and the respiration of humans, animals, and fungi.

DYNAMIC CHANGES OF CO$_2$ CONCENTRATION IN PFAL

Plants assimilate CO$_2$ during photosynthesis and release CO$_2$ during respiration. Small changes in CO$_2$ concentration can have significant impact on the rate of photosynthesis of plants. Although essential for plant photosynthesis and growth, CO$_2$ concentration is probably the least controlled factor in the traditional controlled environment of greenhouses and growth chambers. In a semi-closed PFAL, CO$_2$ concentration varies not only with "plant activity" (photosynthesis and respiration) but also with human activities. The air exhaled by the typical human is composed of 78.04% nitrogen, 13.6 to 16% oxygen, 4 to 5.3% CO$_2$, and 1% argon and other gases (en.wikipedia.org/wiki/breathing). With several people inside a PFAL, the CO$_2$ concentration can increase to several times higher than atmospheric levels without sufficient ventilation. On the other hand, during the photoperiod, when there is no CO$_2$ enrichment and no human activity, CO$_2$ concentration can decrease to a very low level that is close to the CO$_2$ compensation point, if plants are actively photosynthesizing and the culture room is not actively ventilated. The degree of fluctuation in CO$_2$ concentration inside the culture room depends on the volume of the culture room (buffering capacity), human occupancy, and plant activity. It is often difficult to detect "CO$_2$ starvation" because plants grow slowly without any symptoms under low CO$_2$ concentration. Similarly, a too high CO$_2$ concentration is equally hard to detect from plant symptoms in most cases. Therefore, to maintain optimal plant growth, it is important to continuously monitor, and preferably control, the CO$_2$ concentration in PFAL. At minimum, the CO$_2$ concentration in PFAL should be maintained at or around atmospheric levels.

MEASUREMENT OF CO$_2$ CONCENTRATION

Traditionally, CO$_2$ concentration is reported in the United States as parts per million (ppm) and in the United Kingdom and Europe as volume per million (vpm or ppmv). The International System of Units (SI) unit for CO$_2$ concentration is µmol mol^{-1}. Because the absolute values in µmol mol^{-1} are the same as those for µl L^{-1} or ppm on a volume basis, there is no need for interconversions. However, it should be noted that CO$_2$ is often purchased by weight and supplied in pressurized cylinders. The conversion factor from volume to mass (weight) is 1.80 kg m^{-3} at 25°C (Note: 1000 L $=$ 1 m^3).

Continuous measurement of CO_2 concentrations in PFAL is very important. The most commonly used CO_2 sensor is the nondispersive infrared (NDIR) CO_2 sensor. NDIR sensors are spectroscopic sensors that detect CO_2 in a gaseous environment. The key components are an infrared source, a light tube, an interface (wavelength) filter, and an infrared detector. The gas is pumped or diffuses into the light tube and the electronics measures how much infrared light is absorbed by the gas pumped into the light tube. Cost for a typical NDIR sensor ranges from US $100 to $1000, depending on accuracy. New developments include using microelectromechanical systems to bring down the costs of this sensor and to create smaller devices.

Avoid placing CO_2 sensors in locations where people may breathe directly onto the sensor. Also avoid places close to intake or exhaust ducts, or near windows and doorways, as they may give readings that do not accurately reflect the plant canopy conditions.

AIR CURRENT SPEED

NATURE AND DEFINITION

Air current speed is defined as the distance air travels over a given period of time, such as one meter per second ($m\ s^{-1}$). Air velocity is the term used when the direction of air current speed is specified. In this book, air current speed will be used to describe the air movement and air flow in the PFAL growing space.

Insufficient air current speed around plants suppresses gas diffusion in the leaf boundary layer, which subsequently reduces rates of photosynthesis and transpiration and hence plant growth (Kitaya et al., 2000). Maintaining appropriate air current speeds in PFAL creates small turbulent eddies around the leaf surface that facilitates gas exchange between the plants and the surrounding environment, thereby promoting plant growth. Conversely, low air current speeds can cause spatial variations in air temperature, CO_2 concentration, and humidity inside the plant canopy (Kitaya et al., 1998), resulting in inconsistent plant growth. Further, air movement helps to prevent condensation on leaves and other surfaces in the PFAL, helping to prevent unwanted growth of bacteria and molds.

Mechanical fans can be used to generate air movement and control air current speed within the plant canopy in PFAL. They may be strategically located in the PFAL and around shelves to promote consistent gas exchange and uniform plant growth. Without mechanical fans, the PFAL will rely on the natural convection and buoyancy of hotter air to produce air movement around plants. However, these convective air current speeds are extremely low and cannot be controlled. To achieve precise air speed control, special calculations and design strategies regarding the location, number, and capacity of fans are necessary when a PFAL is built.

MEASUREMENT

Air current speed is typically measured using an anemometer. When choosing anemometers for use in PFALs, high sensitivity at low air current speeds should be prioritized. Because of the limited space between growing shelves, smaller sizes are preferred.

NUMBER OF AIR EXCHANGES PER HOUR

NATURE AND DEFINITION

Number of air exchanges per hour (N, h^{-1}) is a measure of how many times the air within a defined space is replaced by new air, which is defined as the ratio of hourly ventilation rate divided by the volume of room air. For PFAL, ideally N should be small for the purpose of controlling the environment and preventing entry of pathogens and pests. However, a minimum air exchange rate should be maintained to prevent the accumulation of ethylene in PFAL, which can cause damage to the plants. Among the commercial PFALs, N varies largely. For relatively airtight, well-insulated PFALs, the N is approximately 0.01 to 0.02 h^{-1} (Kozai, 2013).

MEASUREMENT OF AIR EXCHANGE

Accurate estimation of the number of air exchanges (N) for a PFAL is important because the photosynthetic rate of plants can be predicted continuously based on N. For greenhouses, various techniques have been used to measure and predict ventilation and leakage rates such as tracer gas techniques, energy balance equations, and measurements of pressure differences between inside and outside (Baptista et al., 1999).

The energy balance method has been used to predict ventilation rates by Kozai et al. (1980) and Fernandez and Bailey (1993). The energy balance method is based on the fact that the ventilation removes energy from the greenhouse as a way of preventing excessively high temperatures. This method requires a large number of variables to be measured; and a single inaccuracy can have a large effect on the final result (Baptista et al., 1999).

Tracer gas technique has a greater accuracy than the energy balance method and uses CO_2, N_2O, or SF_6 as tracer gases. CO_2 is an inexpensive gas and is commonly used for estimating the air exchange rate of greenhouses and growth chambers. However, when plants are inside the room, CO_2 cannot be used reliably because it will be assimilated by plants. Unfortunately, N_2O and SF_6 are too expensive to be used for large PFALs and for continuous measurement. Water (H_2O) has also been studied as a potential tracer gas. Li et al. (2012) used H_2O as a tracer gas to continuously measure the N of a semiclosed growth chamber based on CO_2 and H_2O balance. When H_2O is used as a tracer gas, the key is to accurately measure the H_2O consumption of the plants and condensation amount, if any. For relatively airtight PFALs with uniform air circulation, condensation occurs at the cooling unit, which can be easily collected and quantified. Gas and liquid flow rates and their balances in the PFAL are described in detail in Chapter 4.

REFERENCES

Baptista, F.J., Bailey, B.J., Randall, J.M., Meneses, J.F., 1999. Greenhouse ventilation rate: theory and measurement with tracer gas techniques. J. Agric. Eng. Res. 72, 363–374.

Fernandez, J.E., Bailey, B.J., 1993. Predicting greenhouse ventilation rates. Acta Horticult. 328, 107–111.

Kitaya, Y., Shibuya, T., Kozai, T., Kubota, C., 1998. Effects of light intensity and air velocity on air temperature, water vapor pressure and CO_2 concentration inside a plant stand under artificial lighting conditions. Life Support Biosphere Sci. 5, 199–203.

Kitaya, Y., Tsuruyama, J., Kawai, M., Shibuya, T., Kiyota, M., 2000. Effects of air current on transpiration and net photosynthetic rates of plants in a closed plant production system. In: Kubota, C., Chun, C. (Eds.), Transplant Production in the 21st Century. Kluwer Academic Publishers, Dordrecht, The Netherlands.

Kozai, T., 2013. Resources use efficiency of closed plant production system with artificial light: concept, estimation and application to plant factory. Proc. Jpn. Acad., Ser. B 89, 447–461.

Kozai, T., Sase, S., Nara, M.A., 1980. A modeling approach to greenhouse ventilation control. Acta Horticult. 106, 125–136.

Li, M., Kozai, T., Niu, G., Takagaki, M., 2012. Estimating the air exchange rate using water vapor as a tracer gas in a semi-closed growth chamber. Biosyst. Eng. 113, 94–101.

PHOTOSYNTHESIS AND RESPIRATION

Wataru Yamori[1,2]

Center for Environment, Health and Field Sciences, Chiba University, Kashiwa, Japan[1] PRESTO, Japan Science and Technology Agency (JST), Kawaguchi, Japan[2]

INTRODUCTION

Crop yield is determined by plant growth and more than 90% of crop biomass is derived from the products of photosynthesis. Therefore, photosynthesis is the basic process underlying plant growth and food production. Although respiratory reaction is essentially the opposite of photosynthetic reaction, respiration as well as photosynthesis is also a metabolic pathway that produces chemical energy (i.e., adenosine triphosphate (ATP) and redox equivalents) to meet cell energy demands for growth and maintenance. Thus, plant growth is closely related to both photosynthesis and respiration. There is no growth without photosynthesis and respiration. Therefore, an understanding of the physiological processes of photosynthesis and respiration is necessary for a basic understanding of maximizing crop yield. In addition, plant growth rates are not simply determined by rates of photosynthesis and respiration at the single-leaf level, but are determined by rates of photosynthesis and respiration at the canopy level. In this chapter, the basic reactions of photosynthesis and respiration at the single-leaf level as well as the canopy level are summarized.

PHOTOSYNTHESIS

Green plants have leaves that contain chloroplasts. The primary processes of photosynthesis occur in the chloroplast in the mesophyll cells of the leaves. Three main processes for photosynthesis can be distinguished: (a) light absorption by photosynthetic pigments, (b) electron transport and bioenergetics, and (c) carbon fixation and metabolism (Figure 9.1).

LIGHT ABSORPTION BY PHOTOSYNTHETIC PIGMENTS

To convert the transient energy of light into stable chemical energy, the photosynthetic apparatus performs a series of energy-transforming reactions in the thylakoid membranes of the chloroplasts (Figure 9.1). In higher plants, the process is initiated by light capture by an antennae array containing two classes of pigments, chlorophylls and carotenoids, which are responsible for the absorption of light that drives photosynthesis. Chlorophylls are the dominant pigment and strongly absorb red and blue light (Figure 9.2a). Carotenoids are accessory pigments and absorb blue light strongly, enabling the

Plant Factory. http://dx.doi.org/10.1016/B978-0-12-801775-3.00009-3

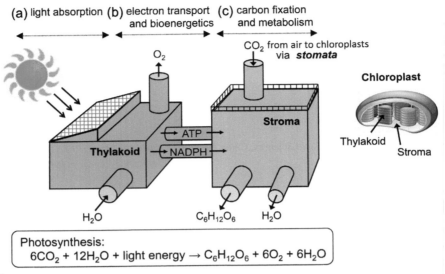

(a) light absorption **(b)** electron transport **(c)** carbon fixation
and bioenergetics and metabolism

FIGURE 9.1

Simplified view of photosynthetic reactions in the chloroplasts. Chloroplasts are the site of photosynthesis. Chloroplasts contain membranes arranged in disks referred to as thylakoids, and the region of the chloroplast outside of the thylakoid membranes is referred to as the stroma. The photosynthesis reactions consist of three processes: (a) light absorption by photosynthetic pigments in the thylakoid membranes, (b) electron transport and bioenergetics in the thylakoid membranes, and (c) carbon fixation and metabolism in the stroma.

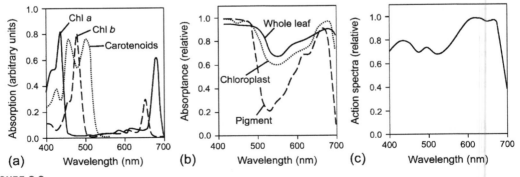

FIGURE 9.2

Absorption spectra of photosynthetic pigments from higher plants (a), absorptance spectra of isolated photosynthetic pigments, intact chloroplasts and whole leaf (b) and action spectra for photosynthesis (c).

The data are modified from Moss and Loomis (1952), McCree (1971/1972) and Govindjee (2004).

chloroplast to trap a larger fraction of the light energy (Figure 9.2a). Isolated chlorophylls and carotenoids transmit green light (Figure 9.2b). However, it is obvious that intact chloroplasts and whole leaves absorb most of the visible light spectrum, including green light (Figure 9.2b). Light scattering is an optical phenomenon that influences the global absorption properties of a leaf. Light scattering in leaves is caused primarily by reflection between the intercellular air spaces and cells (Evans et al., 2004). Internal reflection of light increases path-length, which increases the probability of light capture and absorption. Leaves absorb more than 90% of red and blue light, and also about 70% of green light (Figure 9.2b). Red and blue light are mostly absorbed in the illuminated side of a leaf due to their high absorption, whereas green light penetrates deeper into the leaf, increasing its chance to be absorbed as it travels a longer path-length with multiple reflections at cell wall/air interfaces (Terashima et al., 2009). An action spectrum, which shows what wavelengths of light are most effective for photosynthesis, clearly indicates that red light is more effective for photosynthesis, but most light within the photosynthetically active region of the spectrum (400 to 700 nm) is utilized in leaves (Figure 9.2c).

ELECTRON TRANSPORT AND BIOENERGETICS

The energy captured by the photosynthetic pigments is first transferred to the reaction centers of Photosystem I and Photosystem II in the thylakoid membranes of chloroplasts (Nelson and Yocum, 2006). The electron moves along the electron-transport chain from Photosystem II to Photosystem I and is finally used for the reduction of $NADP^+$ to NADPH (nicotinamide adenine dinucleotide phosphate reduced form). The primary electron donor of Photosystem II obtains electrons from the hydrolysis of H_2O (water), leading to the production of O_2 (oxygen). During electron transport, a H^+ gradient is formed across the thylakoid membrane, which is used to drive ATP production by ATP synthase. The overall process, which involves both Photosystem II and Photosystem I, produces ATP and NADPH, which are mainly used for carbon fixation and subsequent carbon metabolism.

CARBON FIXATION AND METABOLISM

ATP and NADPH are used in the photosynthetic carbon-reduction cycle (i.e., Calvin–Benson cycle, or reductive pentose phosphate pathway) in the chloroplast stroma (Figure 9.1). In this cycle, CO_2 (carbon dioxide) is assimilated leading to the synthesis of carbohydrates and finally sucrose and/or starch. To fix CO_2 into carbohydrate form, CO_2 needs to be transported from the air to chloroplasts in a leaf. CO_2 in the air diffuses through the stomatal pores in the leaf and then through intercellular airspaces into the cells and ultimately chloroplast as a result of a concentration gradient. Each stomatal pore is surrounded by a pair of guard cells, and these guard cells can change their turgor and hence aperture in response to various environmental stimuli. The most important variables are light, water status, and CO_2 concentration (i.e., stomata close under low light, water stress, and high CO_2 concentration). Stomatal pores cover only 1% of the total leaf surface, but mainly determine the rate of CO_2 diffusion from the leaf (Evans et al., 2004). Recently, it has been shown that the resistance to CO_2 diffusion from the intercellular airspaces within the leaf through the mesophyll to the chloroplast during photosynthesis is large, and its limitation to photosynthesis is of a similar magnitude as stomatal limitations (Evans et al., 2009). The magnitude of CO_2 diffusion through the mesophyll is considered to be determined by leaf anatomical features such as cell wall thickness and the surface areas of chloroplasts exposed to intercellular air spaces, and by protein-facilitated processes (Evans et al., 2009).

The fixation of CO_2 into carbohydrates is catalyzed by RuBisCO (Ribulose 1,5-bisphosphate carboxylase/oxygenase) in the chloroplast stroma. This enzyme makes up almost 50% of the chloroplast protein and is considered to be the most abundant protein on earth. Through a series of Calvin–Benson cycle reactions, which consume the ATP and NADPH produced in the thylakoid membranes, triose phosphate is produced. Some triose phosphate is transported out of the chloroplast and used in sucrose synthesis, while some remains in the chloroplast and is used in starch synthesis, which is stored in the chloroplast.

Sucrose synthesized within the cytosol of photosynthesizing cells is then available for general distribution and is commonly translocated to other carbon-demanding organs via the phloem (see Section 10.5 of Chapter 10). Starch is the principal storage carbohydrate in both photosynthetic and nonphotosynthetic tissues of higher plants (Beck and Ziegler, 1989; Smith et al., 2006). In the light, up to 30% of the CO_2 fixed by leaves is incorporated into starch. During the night, starch in chloroplasts is actively degraded and the product of this degradation, sucrose, is exported to the other parts (sinks) of the plant via the phloem (Patrick, 1997). There is substantial variation among plants in the ratio of starch to sucrose during primary partitioning. For example, spinach leaves store significant quantities of sucrose in their vacuoles before starch synthesis begins (Gerhardt et al., 1987) while bean makes starch as soon as photosynthesis exceeds a certain threshold rate (Sharkey et al., 1985). In addition, it should be noted that some plants store other compounds (e.g., sorbitol in the Rosaceae, fructan in asparagus, inulin in artichoke, and so on) instead of starch.

C$_3$, C$_4$ AND CAM PHOTOSYNTHESIS

In C_3 plants (most vegetables and trees, and several major cereals), CO_2 diffuses through the stomata and the intercellular air spaces, and eventually arrives in the chloroplast. In the chloroplast, CO_2 is fixed by RuBisCO and the first product is a compound with 3-carbon acid, hence, the name C_3 photosynthesis. Triose phosphates are produced through a series of reactions known as the Calvin–Benson cycle, leading to the production of sugars and starch using ATP and NADPH produced by photosynthetic electron transport in the thylakoid membranes.

In contrast, C_4 plants (e.g., some cereals and crops such as maize and sugarcane) have a CO_2 concentrating mechanism that increases CO_2 concentration by 10–100 fold at the catalytic sites of RuBisCO in specialized leaf cells compared to ambient air (Furbank and Hatch, 1987; Jenkins et al., 1989). As a result, the oxygenation reaction of RuBisCO in C_4 plants is greatly suppressed (see Section "Photorespiration"). These species are known as C_4 plants because the initial carboxylation reaction produces a 4-carbon acid. A third type of photosynthesis is represented by the so-called CAM (Crassulacean acid metabolism) plants (e.g., pineapple, agave and ice plant). These plants also possess a CO_2 concentrating mechanism, but it is characterized by a temporal separation of the C_3 and C_4 components, compartmentalized within a common cellular environment (Yamori et al., 2014). The common ice plant (*Mesembryanthemum crystallinum* L.) is known as an inducible (facultative) CAM species, because CAM can be induced in this species by water deficit or salinity stress.

RESPIRATION

Sucrose and starch are prime sources of respiratory substrates in plants, although other carbohydrates such as fructans and sugar alcohols can also be utilized. Carbon is exported from the chloroplasts to the cytosol and mitochondria, and used to generate redox equivalents (in particular, NADH), ATP and

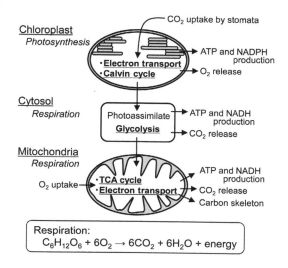

FIGURE 9.3

Simplified scheme involved in photosynthesis in chloroplast, glycolysis in cytosol and respiration in mitochondria. CO_2 assimilated by the Calvin–Benson cycle and electron transport in the chloroplasts is used to produce photoassimilates (e.g., sugar and starch). The carbon is subsequently exported from chloroplasts to the cytosol and mitochondria, and used to generate ATP, redox equivalents (in particular, NADH) and carbon skeletons (which are necessary for protein synthesis) via glycolysis, TCA (mitochondrial tricarboxylic acid) cycle and mitochondrial electron transport.

carbon skeletons via glycolysis, mitochondrial tricarboxylic acid cycle (TCA cycle) and mitochondrial electron transport (Figure 9.3). Generation of these respiratory products leads to CO_2 loss during glycolysis and during reactions in the TCA cycle.

Respiration can be separated into three major components: (1) growth respiration, (2) maintenance respiration and (3) active uptake of ions by roots. Growth respiration includes the synthesis of new structures in a growing plant and is regarded as proportional to the rate at which new plant material is being formed. Maintenance respiration covers the production of necessary energy for repair and maintenance of the plant. Maintenance respiration includes the energy needed for the turnover of metabolites in a nongrowing plant. Energy needed for maintenance respiration is generally considered to be proportional to tissue mass. It is no longer believed that minimization of respiration would be a feasible way to maximize production of plant organic matter. However, it is important to regulate air temperature in the plant factory with artificial lighting (PFAL), since the maintenance respiration for repair and maintenance of the plant is more stimulated than the growth respiration at high temperature (Rachmilevitch et al., 2006). Energy costs associated with nutrient acquisition by plant roots are often very high because ions have to be transported across root cell membranes using active transport systems that require substantial amounts of ATP. It has been reported that, in young growing plants, respiration for growth and ion uptake together account for about 60% of root respiration, while the maintenance respiration is relatively low (Figure 9.4). With increasing plant age, the maintenance respiration accounts for an increasing proportion of total respiration (over 85%).

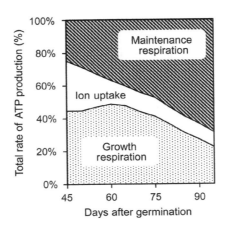

FIGURE 9.4

Time course of the relative contribution of the three major components for root respiration in *Carex* species: (1) growth respiration, (2) maintenance respiration, and (3) active uptake of ions by roots.

(modified from Van der Werf et al., 1988)

It should be noted that respiration does not directly depend on light, since cells continuously need energy to live. However, in general, the rate of mitochondrial respiration in daytime is 50% or less of that at night, since the light inhibition of respiration seems to be due to the rapid light inactivation of mitochondrial enzymes (Atkin et al., 1998, 2000). Photosynthesis and respiration are metabolic pathways that produce ATP and redox equivalents (e.g., NADPH and NADH) to meet cell energy demands for growth and maintenance. In this way, plant growth is closely related to respiration as well as photosynthesis (Noguchi and Yoshida, 2008).

PHOTORESPIRATION

RuBisCO facilitates CO_2 fixation (carboxylation) in the Calvin–Benson cycle, but at the same time, it also fixes O_2 (oxygenation). The latter reaction is the first step of photorespiration. Photorespiration involves the movement of metabolites among chloroplasts, peroxisomes and mitochondria in a metabolic pathway that fixes O_2, releases CO_2, and consumes ATP. Therefore, it reduces photosynthetic efficiency (Sharkey, 1988). Respiration and photorespiration both release CO_2 in the mitochondria. However, respiratory and photorespiratory processes are well separated with respect to both carbon intermediates and compartmentation.

Similar to photosynthesis, photorespiration is a dynamic process, the rate of which can be modulated by light intensity, internal CO_2/O_2 concentrations or leaf temperature (Jordan and Ogren, 1984). Generally, photorespiratory rate increases under strong light, high temperature and low humidity when stomata close and the CO_2 concentration in the leaf decreases. It has been proposed that photorespiration could prevent over-reduction of the chloroplast stroma under conditions of high light and low CO_2 when stomata are closed at midday to prevent water loss (Takahashi and Badger, 2011).

Photorespiration is inherently difficult to measure, although its rate is substantially higher than that of mitochondrial respiration. Net photosynthesis would be estimated by the difference between gross photosynthesis and the sum of photorespiration and dark respiration in the mitochondria: net photosynthesis = gross photosynthesis − (photorespiration + dark respiration). The rate of photorespiration is about 25% of the photosynthetic rates in moderate nonstressed conditions, but under stress conditions, it could be 50% or more of the photosynthetic rate. Due to the substantial reductions in photosynthetic efficiency by photorespiration and the fact that increases in CO_2 concentration reduce photorespiration, increasing CO_2 concentration can enhance plant growth both by its effect as a substrate for photosynthesis and by its effect on suppressing photorespiration (Ainsworth and Long, 2005), if other environmental factors are not limiting.

LAI AND LIGHT PENETRATION

Biomass is ultimately derived from photosynthesis under light. Crop yield commonly depends on the total amount of light intercepted, and the interception of light by a crop canopy is closely related to the total leaf area. The leaf area index (LAI) is the ratio of total projected leaf area per unit ground area, and is widely used to characterize canopy light conditions. A canopy with an LAI of 1.0 has a leaf area equal to the soil surface area, but this does not mean that all light is intercepted, since some leaves overlap and there are gaps between leaves in the canopy. Generally, light interception increases sharply with increase in LAI to about 90% once LAI exceeds 4.0, and approaches an asymptote at a higher LAI (Figure 9.5a).

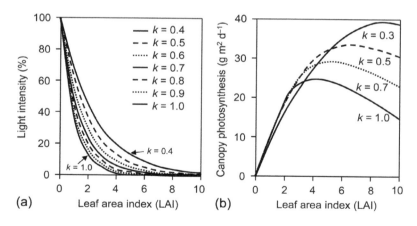

FIGURE 9.5

Relationship between light intensity within the canopy at a height and LAI from the canopy top (a) and between LAI and canopy photosynthesis (b). The data are modified from Nátr and Lawlor (2005) and Saeki (1960). k is an extinction coefficient. A canopy with a large k value implies that light intensity decreases rapidly with depth of canopy, whereas a canopy with a small k value would allow light irradiation to penetrate deeply. Canopies with more horizontal leaves have large k values (e.g., tomato, sunflower: 0.7–1.0), whereas canopies with more erect leaves have small k values (e.g., rice, sorghum: 0.3–0.6).

Young plants in a crop rarely compete with each other for light irradiation. At this stage, LAI is low (<1.5) and increases in canopy photosynthesis rate are almost proportional to increases in LAI. Thus, new leaf area expansion contributes to the growth rate to the same extent as the existing leaf area, because there are few leaves shading each other. However, in a closed canopy with high LAI (>3.0), leaf area expansion is of minor importance compared with a young crop, because most of the incident radiation is absorbed by the upper leaves of the canopy and thus further increase of LAI has only a marginal effect on canopy photosynthesis. LAI cannot be too large, because the lower (shaded) leaves would not be sufficiently irradiated and loss of assimilates due to respiration would be higher than gains as a result of photosynthesis (Figure 9.5b). Values for LAI between 3.0 and 4.0 are typical for horizontal-leafed species; values between 5 and 10 can occur in vertical-leafed species such as cereals and grasses.

SINGLE LEAF AND CANOPY

Photosynthesis is the basic process underlying plant growth and food production. It has been clarified that leaf photosynthesis is affected by various environmental conditions such as light (Yamori et al., 2010a), temperature (Yamori et al., 2005, 2006, 2010b), CO_2 concentration (Yamori et al., 2005, Yamori and von Caemmerer, 2009) and so on. However, growth rates are not simply determined by rate of photosynthesis per unit leaf area of a single leaf, nor by the total leaf area per plant. As light penetrates a crop canopy, it is intercepted by leaves with light intensity declining exponentially with cumulative leaf area (Figure 9.5a). Canopy photosynthesis can be described as the product of canopy light interception and light use efficiency (Loomis and Connor, 1992). Canopy structure (i.e., LAI and spatial distribution of leaf angles) is important for the canopy light climate and photosynthetic energy conversion. Large leaf angles with leaves close to the vertical ensure good light penetration, leading to a more uniform light environment throughout the canopy. There are positive linear relationships between crop growth rates and canopy light interception for most field crops (e.g., Wells, 1991; Board and Harville, 1993; Heitholt, 1994). Thus, the management of LAI, spatial distribution of leaf angles and environmental conditions are important for improving photosynthetic efficiency. The balance of daytime carbon gain and nighttime carbon loss from the whole plant determines the rate of daily carbon accumulation, which subsequently controls plant growth and development.

In a PFAL, appropriate plant densities are also important for improving the photosynthetic efficiency of crop plants, and consequently yield and quality. Improper plant spacing may cause either too dense or too sparse populations, resulting in a reduction of yield. It is obvious that increasing planting densities would reduce individual weights, leading to reductions in yields. However, the optimum plant density ensures that plants grow uniformly and properly through efficient utilization of moisture, nutrients, and light and thus results in a maximum, high-quality yield.

REFERENCES

Ainsworth, E.A., Long, S.P., 2005. What have we learned from 15 years of free-air CO_2 enrichment (FACE)? A meta-analytic review of the responses of photosynthesis, canopy properties and plant production to rising CO_2. New Phytol. 165, 351–372.

Atkin, O.K., Evans, J.R., Siebke, K., 1998. Relationship between the inhibition of leaf respiration by light and enhancement of leaf dark respiration following light treatment. Aust. J. Plant Physiol. 25, 437–443.

Atkin, O.K., Evans, J.R., Ball, M.C., Lambers, H., Pons, T.L., 2000. Leaf respiration of snow gum in the light and dark: interactions between temperature and irradiance. Plant Physiol. 122, 915–923.

Beck, E., Ziegler, P., 1989. Biosynthesis and degradation of starch in higher plants. Annu. Rev. Plant Physiol. Plant Mol. Biol. 40, 95–117.

Board, J.E., Harville, B.G., 1993. Soybean yield component responses to a light interception gradient during the reproductive period. Crop Sci. 33, 772–777.

Evans, J.R., Vogelmann, T.C., Williams, W.E., Gorton, H.L., 2004. Chloroplast to leaf. In: Smith, W., Vogelmann, T.C., Critchley, C. (Eds.), Photosynthetic Adaptation: Chloroplast to Landscape, vol. 178. Springer, Berlin Heidelberg, New York, pp. 15–41.

Evans, J.R., Kaldenhoff, R., Genty, B., Terashima, I., 2009. Resistances along the CO_2 diffusion pathway inside leaves. J. Exp. Bot. 60, 2235–2248.

Furbank, R.T., Hatch, M.D., 1987. Mechanism of C_4 photosynthesis. The size and composition of the inorganic carbon pool in the bundle sheath cells. Plant Physiol. 85, 958–964.

Gerhardt, R., Stitt, M., Heldt, H.W., 1987. Subcellular metabolite levels in spinach leaves. Plant Physiol. 83, 399–407.

Govindjee, 2004. Chlorophyll *a* fluorescence: a bit of basics and history. In: Papageorgiou, G.C., Govindjee, (Eds.), Chlorophyll *a* Fluorescence: A Signature of Photosynthesis. Kluwer Academic Publishers, Dordrecht, The Netherlands, pp. 1–42.

Heitholt, J.J., 1994. Canopy characteristics associated with deficient and excessive cotton plant population densities. Crop Sci. 34, 1291–1297.

Jenkins, C.L.D., Furbank, R.T., Hatch, M.D., 1989. Inorganic carbon diffusion between C_4 mesophyll and bundle sheath cells. Plant Physiol. 91, 1356–1363.

Jordan, D.B., Ogren, W.L., 1984. The CO_2/O_2 specificity of ribulose 1,5-bisphosphate carboxylase/oxygenase. Dependence on ribulose bisphosphate concentration, pH and temperature. Planta 161, 308–313.

Loomis, R.S., Connor, D.J., 1992. Crop Ecology: Productivity and Management in Agricultural Systems. Cambridge University Press, Cambridge.

McCree, K.J., 1971/1972. The action spectrum, absorptance, and quantum yield of photosynthesis in crop plants. Agric. Meteorol. 9, 191–216.

Moss, R.A., Loomis, W.E., 1952. Absorption spectra of leaves. 1. The visible spectrum. Plant Physiol. 27, 370–391.

Nátr, L., Lawlor, D.W., 2005. Photosynthetic plant productivity. In: Pessarakli, M. (Ed.), Hand Book of Photosynthesis. second ed. CRC Press, New York, pp. 501–524.

Nelson, N., Yocum, C.F., 2006. Structure and function of photosystems I and II. Annu. Rev. Plant Biol. 57, 521–565.

Noguchi, K., Yoshida, K., 2008. Interaction between photosynthesis and respiration in illuminated leaves. Mitochondrion 8, 87–99.

Patrick, J.W., 1997. Phloem unloading: sieve element unloading and post-sieve element transport. Annu. Rev. Plant Physiol. Plant Mol. Biol. 48, 191–222.

Rachmilevitch, S., Lambers, H., Huang, B., 2006. Root respiratory characteristics associated with plant adaptation to high soil temperature for geothermal and turf-type Agrostis species. J. Exp. Bot. 57, 623–631.

Saeki, T., 1960. Interrelationships between leaf amount, light distribution and total photosynthesis in a plant community. Bot. Mag. 73, 55–63.

Sharkey, T.D., 1988. Estimating the rate of photorespiration in leaves. Physiol. Plant. 73, 147–152.

Sharkey, T.D., Berry, J.A., Raschke, K., 1985. Starch and sucrose synthesis in *Phaseolus vulgaris* as affected by light, CO_2, and abscisic acid. Plant Physiol. 77, 617–620.

Smith, A.M., Zeeman, S.C., Smith, S.M., 2006. Starch degradation. Annu. Rev. Plant Biol. 56, 73–97.

Takahashi, S., Badger, M., 2011. Photoprotection in plants: a new light on photosystem II damage. Trends Plant Sci. 16, 53–60.

Terashima, I., Fujita, T., Inoue, T., Chow, W.S., Oguchi, R., 2009. Green light drives leaf photosynthesis more efficiently than red light in strong white light: revisiting the enigmatic question of why leaves are green. Plant Cell Physiol. 50, 684–697.

van der Werf, A., Kooijman, A., Welschen, R., Lambers, H., 1988. Respiratory energy costs for the maintenance of biomass, for growth and for ion uptake in roots of *Carex diandra* and *Carex acutiformis*. Physiol. Plant. 72, 483–491.

Wells, R., 1991. Response of soybean growth to plant density: relationships among canopy photosynthesis, leaf area and light interception. Crop Sci. 31, 755–761.

Yamori, W., von Caemmerer, S., 2009. Effect of Rubisco activase deficiency on the temperature response of CO_2 assimilation rate and Rubisco activation state: insights from transgenic tobacco with reduced amounts of Rubisco activase. Plant Physiol. 151, 2073–2082.

Yamori, W., Noguchi, K., Terashima, I., 2005. Temperature acclimation of photosynthesis in spinach leaves: analyses of photosynthetic components and temperature dependencies of photosynthetic partial reactions. Plant Cell Environ. 28, 536–547.

Yamori, W., Noguchi, K., Hanba, Y.T., Terashima, I., 2006. Effects of internal conductance on the temperature dependence of the photosynthetic rate in spinach leaves from contrasting growth temperatures. Plant Cell Physiol. 47, 1069–1080.

Yamori, W., Evans, J.R., von Caemmerer, S., 2010a. Effects of growth and measurement light intensities on temperature dependence of CO_2 assimilation rate in tobacco leaves. Plant Cell Environ. 33, 332–343.

Yamori, W., Noguchi, K., Hikosaka, K., Terashima, I., 2010b. Phenotypic plasticity in photosynthetic temperature acclimation among crop species with different cold tolerances. Plant Physiol. 152, 388–399.

Yamori, W., Hikosaka, K., Way, D.A., 2014. Temperature response of photosynthesis in C_3, C_4 and CAM plants: temperature acclimation and temperature adaptation. Photosynth. Res. 119, 101–117.

GROWTH, DEVELOPMENT, TRANSPIRATION AND TRANSLOCATION AS AFFECTED BY ABIOTIC ENVIRONMENTAL FACTORS

Chieri Kubota

School of Plant Sciences, The University of Arizona, Tucson, Arizona, USA

INTRODUCTION

Understanding of environmental factors affecting various aspects of plant growth and development is crucial in plant factory design and operation. Use of artificial lighting would allow growers to modify one or more selected abiotic environmental factors almost independently, giving opportunities for inducing desired responses of plants (growth, development, flowering, etc.). This chapter first defines vegetative growth (shoots and root growth) and then discusses the typical abiotic environmental factors (temperature, light intensity, light quality, photoperiod, humidity, carbon dioxide concentration, air current speed, and nutrient and root-zone environment) affecting plant growth and development. Then transpiration and translocation of sugar are briefly described using the general physiological understanding of water potential and the sink–source relationship.

SHOOT AND ROOT GROWTH
GROWTH: DEFINITION

Plant growth is generally expressed by biomass (fresh or dry mass), which is essentially determined by the cumulative amount of gas exchange (photosynthesis and respiration) over time. Environmental factors affecting plant photosynthesis and respiration are summarized elsewhere in this book. While studies on photosynthesis as affected by environmental factors are often conducted at a single leaf level, quantitatively understanding photosynthesis and growth of the whole plant and canopy is critical in crop production. Leaf expansion and light interception are major determinants of whole plant biomass production, in addition to leaf gas exchange. The relationship between plant growth rate, photosynthesis (assimilation) and leaf area (leafiness) is expressed in the following formula well adopted in plant growth analysis (Hunt, 1982).

Plant Factory. http://dx.doi.org/10.1016/B978-0-12-801775-3.00010-X

$$RGR = NAR \times LAR \tag{10.1}$$

RGR (relative growth rate, g g^{-1} d^{-1}), NAR (net assimilation rate, g m^{-2} d^{-1}) and LAR (leaf area ratio, m^2 g^{-1}) are further expressed with the following equations:

$$RGR = \frac{1}{W_1} \times \frac{W_2 - W_1}{T_2 - T_1} \tag{10.2}$$

$$NAR = \frac{1}{L_1} \times \frac{W_2 - W_1}{T_2 - T_1} \tag{10.3}$$

$$LAR = \frac{L_1}{W_1} \tag{10.4}$$

where W_1 and W_2 are plant dry mass (g per plant) on days T_1 and T_2 respectively and L_1 is total leaf area (m^2 per plant) on day T_1. NAR represents an average rate of leaf photosynthesis of the plant and LAR represents overall leafiness, morphological characteristics affecting light interception and therefore plant growth rate (RGR). Expressing RGR with NAR and LAR is helpful to understand the factors affecting plant growth; however, Heuvelink and Dorais (2005) noted that Equation (10.1) had some limitations in applying canopy growth with intensive mutual shading, such as a well-established tomato plant canopy. In fact, tomato canopy closure occurs very early in the vegetative stage of the entire production period. In contrast, leafy crops typically grown inside PFALs have low LAI (leaf area index) for a prolonged time before the canopy closure. Therefore, morphological enhancement to promote plant growth may be more effectively used for leafy crops than for tall canopy crops such as tomatoes. A typical growth curve of lettuce plants has been reported by many research groups and it is often a sigmoidal curve showing a rapid biomass increase towards the last 1/3 to 1/2 of the production period. Therefore, morphological enhancement to increase the light interception during the first 1/2 to 2/3 of the crop cycle may largely increase the overall growth (and thereby the yield) of a lettuce crop. Understanding crop species and cultivar specific growth curves is crucial to obtain maximum productivity in a limited production space.

ROOT GROWTH

Healthy actively growing roots should be white in color. Roots provide the active surface of water and nutrient uptake. Roots are also sink organs competing for carbohydrate allocation in plants. Root morphology is largely affected by the root zone environment. Plants develop primarily tap roots in root zone environments with relatively free access to water and nutrients, such as deep flow technique (DFT) and nutrient film technique (NFT) systems, whereas plants develop more finely branched roots with root zones having higher air porosity (and therefore higher oxygen concentration). Adequate balance of root and shoot biomass is desired, but the actual optimum ratio of root and shoot biomass is not well understood, with a few exceptions. For transplants, smaller shoot-to-root ratio (S/R ratio) is generally considered as an important criterion of greater growth capacity and quicker establishment after transplanting. For leafy crops grown in PFALs, no information is available for the optimum root-to-shoot ratios. However, research has been done in greenhouses to quantify the root growth of selected crops grown under different rhizospheric conditions. For example, Stefanelli et al. (2012) demonstrated that increasing nitrogen concentration exponentially decreased root-to-shoot ratio and linearly increased fresh biomass of lettuce grown hydroponically in a greenhouse. Nishizawa and

Saito (1998) showed restricted root and overall growth for tomato plants when root growth was physically restricted in a small container (37 cm³) compared with those in a large container (2024 or 4818 cm³).

ENVIRONMENTAL FACTORS AFFECTING PLANT GROWTH AND DEVELOPMENT

Major and well-studied environmental factors affecting plant growth include (1) temperature, (2) light intensity (daily light integral, DLI), (3) light quality, (4) humidity, or vapor pressure deficit (VPD), (5) CO_2 concentration in the air, (6) air current speed, and (7) nutrient and root-zone environments. Some of the factors are photosynthetic effects and others are morphologic or both. Understanding these key factors is crucial to maximizing the biomass production in plant factories.

TEMPERATURE AND PLANT GROWTH AND DEVELOPMENT

Plant growth and development rates are temperature dependent. A typical response curve of plant growth to temperature is shown in Figure 10.1. Note that the plant growth and developmental rates increase linearly with increasing temperature up to the maximum rate. This relationship is particularly useful to predict these rates under varied temperatures, because as long as temperatures are within the linear range, biomass and developmental stage can be well correlated with the cumulative temperature. In open-field crop production, this concept is called "growing degree days" or "heat unit" and is widely applied in predicting the phenology in crop production. However, outside temperature often exceeds the optimum temperature or may be below the minimum temperature, depending on the weather conditions and seasons. Therefore the growing degree days concept often considers these thresholds (lower and upper thresholds). Considering only the lower threshold (the minimum temperature in

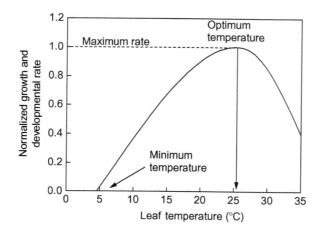

FIGURE 10.1

Plant growth and developmental rate as affected by temperature.

Figure 10.1) is a more common practice than considering both lower and upper thresholds (e.g., AZMET, http://ag.arizona.edu/azmet). In contrast, under a controlled environment, especially of PFALs, considering these thresholds is not necessary as temperature is usually maintained within the linear range. This makes the use of cumulative temperature more practical. As an example, when a leafy crop of selected cultivar reaches harvestable stage 40 days after transplanting when grown under an average temperature of 20.0°C, the cumulative temperature over the growth period is 800 degree days. Under a different average temperature of 22.0°C, the production time is expected to be 36 days (36.4 = 800/22). Average temperature could be adjusted either by altering photoperiod temperature, night period temperature or both.

DAILY LIGHT INTEGRAL

Under a given temperature condition, the plant growth is further affected by light intensity (irradiance). The response of plant growth to increase in PPF is almost linear for many cases under a controlled environment. While high PPF (greater than 1000 μmol m^{-2} s^{-1}) typically saturates leaf net photosynthetic rate (Pn) of C$_3$ plants, the whole canopy Pn at high LAI is not saturated under the same light intensity that saturates the leaf Pn. Figure 10.2 shows, as an example, the simulated influence of LAI on the linearity of canopy photosynthetic response over PPF. For this particular case, the canopy photosynthetic response curve became almost linear with LAI = 4.0 or greater. This linearity in photosynthetic light response is in fact the basis of greenhouse growers' perception of "one percent increase of light is one percent increase in yield." The linearity in the plant growth rate to daily light integral (DLI or sometimes called daily photosynthetic photon flux) has been observed for lettuce

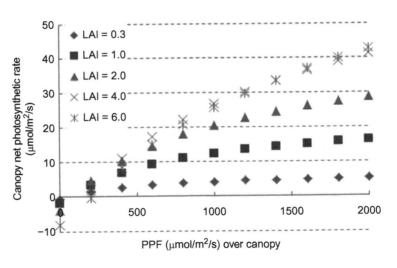

FIGURE 10.2

Canopy net photosynthetic rate as affected by PPF (incident irradience over the canopy) and canopy leaf area index (LAI). Monsi and Saeki equation (1953) was used for simulating canopy net photosynthetic rate using selected parameters ($P_{max} = 25$, $k = 0.7$, $R = 2$, $\alpha = 0.07$, where P_{max} is the maximum leaf net photosynthetic rate at saturating PPF; k is the PAR extinction coefficient of the canopy; R is the dark respiration rate and α is the quantum efficiency).

(Albright et al., 2000) for plants grown under a controlled environment. A linear relationship between cumulative yield of tomato and cumulative PAR is also reported by Cockshull et al. (1992). This linear response of plant growth over DLI and the reciprocity between the effects of PPF and photoperiod at the same DLI are useful in lighting design and deciding photoperiod and PPF to obtain target DLI. For example, the target DLI of 12 mol m^{-2} d^{-1} can be achieved by 333 μmol m^{-2} s^{-1} for a 10-h photoperiod or 185 μmol m^{-2} s^{-1} for 18-h photoperiod. Reduction of PPF reduces the number of luminaires and thereby the capital costs, while reduction of the photoperiod could take advantage of the off-peak hours of utility charges.

The minimum DLI is anecdotally recommended for various crops. Generally, target DLI is recommended as 12–17 mol m^{-2} d^{-1} for lettuce and other leafy crops (e.g., Albright et al., 2000) and 13 mol m^{-2} d^{-1} for vegetable seedlings (Fan et al., 2013). The DLI requirement for strawberries is likely to be at a similar level to these crops per our experience at the University of Arizona. Microgreens do not require high DLI but little data are available. These DLIs are relatively smaller than for tall crops. For example, optimum DLI for high-wire tomato plants is 30–35 mol m^{-2} d^{-1} (Spaargaren, 2001), more than twice the DLI for leafy crops or seedlings. The DLI requirement in fact determines which crops can practically be grown in energy intensive PFALs. Further studies are needed to identify the minimum DLI recommended for various species and cultivars grown in PFALs.

LIGHT QUALITY

Although this is currently an intensively studied area, with new information using various LEDs rapidly accumulating, generalization of light quality effects on plant growth seems to be most challenging. This may be because light quality affects both photosynthesis and morphology. The morphological response seems to be largely species and cultivar specific, while photosynthetic responses are generally similar, at least considering the quantum yield (McCree, 1972). As described in an earlier section, quantum yield is the maximum efficiency of photosynthetic photochemical reactions expressed as the yield of the reactions (CO_2 absorption) per photon absorbed. While CO_2 absorption is the result of carboxylation, the quantum yield strictly concerns the efficiency of thylakoid reactions (photosystems), and quantum yield is generally quantified under the conditions where the light is the limiting factor (at light compensation point). Also the terminology changes to quantum efficiency when irradiance (incident light) is used instead of absorbed photons (quantum efficiency = CO_2 exchange rate per photon received over the leaf). The quantum efficiency averaged for 22 plant species (McCree, 1972; Sager et al., 1988) suggests that plants potentially have higher photosynthetic efficiency in red light compared with blue and green light. As PPF increases, light use efficiency of photosynthesis declines as RuBP carboxylation becomes the limiting factor (see more information in Chapter 9). While green light has the lowest quantum yield, green light can penetrate deeper in the plant canopy due to its high transmittance and reflectance. Therefore, theoretically, quantum yield of a dense plant canopy should be more equalized over the 400–700 nm range. In fact, Paradiso et al. (2011) validated this hypothesis using a mathematical simulation considering single leaf response and canopy light penetration for rose plants, demonstrating that canopy quantum efficiency of green light is not that much lower than that of red light, compared with what McCree (1972) and Sager et al. (1988) show for leaf quantum efficiency under green vs. red light. This is quite useful information for PFALs as it encourages the scientific community and the horticultural LED R&D to focus more on plant morphological response and energy use efficiency rather than photosynthetic response to light qualities.

Another critical influence of light quality on plant photosynthesis is through stomatal conductance. Blue light is known to increase stomatal conductance and plants grown under blue rich light environments exhibit a greater leaf net photosynthetic rate than those grown under spectra limiting blue light (e.g., Bukhov et al., 1995; Hogewoning et al., 2010). However, blue light also has strong morphological influences of reducing light interception (and the overall growth) as described in subsequent sections, and the increase in stomatal conductance often does not result in increased growth.

Many studies have examined plant photomorphogenic response in the past several years. Plant growth promotion through morphological enhancement (i.e., enlarging leaf area and thereby light interception) is a unique concept and several research groups demonstrated that blue light and far-red light seem to take the major roles. Cucumber plants were grown under varied red to blue photon flux ratios of sole-source light in PFALs (Hernández and Kubota, unpublished) as well as supplemental light in a greenhouse (Hernández and Kubota, 2014) and found that increasing blue light at the same overall PPF decreased the leaf area (LAR) and thereby the growth under relatively low background solar daily light integral (mol m^{-2} d^{-1}). Supplemental far-red (700–800 nm) lighting also has significant influence on stem elongation as well as leaf expansion. Use of incandescent lamps has been practiced for many years in growth chambers for various plant research (Sager and Mc Farlane, 1997). Using far-red LEDs, Li and Kubota (2009) and Stutte (2009)) demonstrated that supplemental far-red lighting enhanced lettuce leaf area and overall growth. Both reports discuss that the lettuce growth promotion under supplemental far-red lighting was associated with an increase in leaf area and consequently improved light interception. Similar morphological influence of far-red light can be achieved by end-of-day (EOD) far-red application. Chia and Kubota (2010) and Yang et al. (2012) demonstrated that morphology of grafting rootstocks for tomato and cucurbits could be improved significantly by applying EOD far-red lighting using both stationary and movable light fixtures.

HUMIDITY (VPD)

Leaf transpiration rate (discussed further in following sections) is strongly affected by the moisture concentration (or vapor pressure deficit) of surrounding air as well as leaf temperature. Therefore, VPD indirectly affects nutrient uptake, leaf temperature, and thereby plant growth and development. Among mineral nutrition, Ca uptake is strongly affected by mass flow driven by plant transpiration, as described elsewhere. In PFALs, Ca deficiency has been recognized as a critical issue due to suboptimal airflow around the shoot tips (Goto and Takakura, 1992; Frantz et al., 2004). Increasing VPD during the photoperiod or decreasing VPD during the night period is reportedly effective to reduce the tipburn incidence of lettuce plants grown under artificial lighting (Collier and Tibbitts, 1984). VPD management can be an effective tool to control localized Ca deficiency in leafy crops, especially when creating a vertical air flow is not possible within the limited headspace inside the shelf where plants are grown at high density.

CO$_2$ CONCENTRATION

CO$_2$ enrichment is widely practiced in PFALs except in those facilities that have limited capacity of CO$_2$ control due to the highly ventilated buildings used for production. For example, in North America, use of existing buildings such as warehouses is commonly done for PFALs. According to the building standard for ventilation (ASHRAE, 2013), warehouses require a minimum of 0.3 L m^{-2} s^{-1} air exchange rate and offices require an additional ventilation (2.5 L s^{-1} per person) depending on the occupant density (number of people). For an office building of 3 m wall height, a 100 m^2 floor area that was designed to occupy 5 people has a minimum ventilation of 37.5 L s^{-1} or 0.45 h^{-1} air exchange, as a requirement. This level of ventilation rate may not economically permit increase of CO_2 concentration to the level higher than ambient because much of the CO_2 injected into the building will be lost due to the ventilation. Ohyama and Kozai (1998) demonstrated that the amount of CO_2 lost to the outside is greater than that absorbed by the plants growing in PFALs when the system has as low as 0.1 h^{-1} ventilation rate. Effect of ventilation rate on the CO_2 budget (plant uptake vs. loss to the atmosphere) is further discussed in Chapters 4 and 8 of this book. For highly vented structures, it is recommended that CO_2 be maintained at near ambient concentration (around 400 ppm) so that there will be no virtual loss of CO_2 to the outside atmosphere (Ohyama et al., 2005).

Plant responses to high CO_2 concentration have been studied intensively in the field of environmental sciences. Generally it is known that plants increase stomatal resistance under elevated CO_2 (Ainsworth and Rogers, 2007), which thereby increases the water use efficiency. Optimum temperature of photosynthesis also becomes higher under high CO_2 concentration due to the lower solubility of CO_2 at higher temperature and lower specificity of RuBisCo for CO_2, resulting in a greater effect of elevated CO_2 at higher temperatures (Long, 1991). Actual horticultural impact of CO_2 enrichment seems to be species specific as well as growth-stage specific. Kimball (1986) summarized the reported impact of CO_2 enrichment in greenhouses, where there was a notable trend that plants in the vegetative stage more pronouncedly responded to the high CO_2 concentration than did those in the reproductive stage. For example, average yield increase in tomato under CO_2 enrichment reported by 131 articles was 17%, whereas the average increase in vegetative biomass of immature crops was 52% (average of values reported by 24 articles). Similarly, CO_2 enrichment increased head lettuce marketable yield and vegetative biomass of immature crops by an average of 35% and 68%, respectively.

AIR CURRENT SPEED

Air circulation inside a plant production system is often considered for achieving better spatial uniformity of air temperature. When air circulation fans are applied, we must avoid strong air current speed around the plants that may cause mechanical stress and damage. Wind-induced shaking and the resulting mechanical stress on plants reportedly reduce internode length as well as leaf size (e.g., Biddington and Dearman, 1985). These plant responses are often observed at air current speeds greater than 1 m s^{-1}. Influence of much slower air current speed has also been studied and there are often positive influences enhancing photosynthetic or transpirational gas exchanges of leaves by reducing the boundary layer resistance. Figure 10.3 (Kitaya et al., 2000) shows the enhanced transpiration rate and net photosynthetic rate of sweetpotato leaves measured under different air current speeds.

As described in Chapter 12 on tipburn in this book, one application of this enhanced gas exchange by increasing air current speed is the prevention of tipburn in lettuce-growing greenhouses by using vertical air flow fans (Cornell Univ. Lettuce Handbook), following the well-designed studies conducted to show the proof-of-concept (Goto and Takakura, 1992). Goto and Takakura (1992) showed that localized air circulation around the shoot tip enhanced the calcium concentration of the tissue and dramatically reduced the incidence of tipburn. When applied in greenhouses, the vertical fans can be distributed to obtain a 0.5 m s^{-1} average downward air current speed over the canopy, computed from fan capacity (m^3 s^{-1}) and density (m^{-2}) (L. Albright, personal communication). As mentioned earlier,

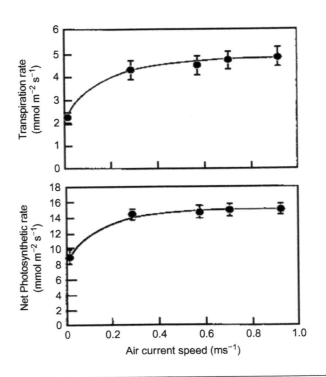

FIGURE 10.3

Effect of air current speed on the transpiration rate and net photosynthetic rates of a single leaf sweetpotato (Kitaya et al., 2000).

applying this practice in PFALs is challenging due to the limited headspace for installing vertical air-flow fans in typical vertical growing systems. Air current speed also affects the leaf energy budget and enhances the sensible and latent heat transfer between leaf and the surrounding air.

NUTRIENT AND ROOT ZONE

As discussed elsewhere in this book, plants respond to various chemical and physical properties and variables in their root zone. In fact, the interactions between these factors are complex and much research is needed to optimize the root zone environment. Major factors that we use as controlled variables are EC (electrical conductivity), pH, and DO (dissolved oxygen concentration) in PFALs as well as in greenhouse hydroponics. Keeping all factors within optimum ranges is critical in plant production. When deep flow hydroponic systems are used, plant roots are submerged in the nutrient solution and therefore maintaining a high DO is recommended. Zeroni et al. (1983) demonstrated the reduction of tomato yields with lowering DO in a deep flow hydroponic system. Design of nutrient delivery in hydroponic systems also affects the DO in the solution circulated within the system. For example, NFT systems employ different slopes to assure circulation of the nutrient solution and prevent depletion of DO in the solution. López-Pozos et al. (2011) showed that having steeper slope (4%) and gaps within the slope improved the DO in the solution and increased tomato yield grown in the NFT system.

In hydroponics, grower practices include weekly nutrient analysis to assure no critical accumulation or depletion of specific elements in the nutrient solution and adjusting the fertilizer concentration accordingly. However, because of the complexity of multiple factor interactions, optimization has not been done fully and often both researchers and commercial practitioners compromise with available fertilizer recipes and management practices. Further information can be found in Chapter 11.

DEVELOPMENT (PHOTOPERIODISM AND TEMPERATURE AFFECTING FLOWER DEVELOPMENT)

Plant flowering is generally affected by photoperiod and temperature. In general, photoperiodic flowering responses are classified into the following five types.

- Obligate short day
- Facultative short day
- Day neutral
- Obligate long day
- Facultative long day

Short day plants develop flowers when day length becomes shorter (or night length becomes longer). When the critical (threshold) length clearly distinguishes the plant's responses between longer and shorter day lengths, the photoperiodic response is classified as *obligate* type. *Facultative* response is quantitative where flowering responses are promoted by certain photoperiods (Runkle and Fisher, 2004; Erwin et al., 2004). Day neutral plants are insensitive to photoperiod. Because photoperiod is not the only factor affecting flowering responses, classification of photoperiodic flowering responses must be done in carefully designed experiments excluding interactions with other factors (such as temperature). Ideally, photoperiod should be varied by applying different durations of artificial lighting at minimum irradiance inducing the photoperiodic response so that differences between different light treatments are not largely due to DLI. Photon flux of 2 $\mu molm^{-2} s^{-1}$ seems to be adequate (e.g., Heins et al., 1997) and the photoperiod extension light should contain both red and far-red light (Craig and Runkle, 2013).

For floriculture and ornamental crops, regulation of photoperiodic response is crucial and, for that reason, foundational knowledge of species-specific photoperiod and temperature requirements for flowering is now available at least for the major ornamental species. For example, use of night interruption lighting to promote flowering of long-day plants (such as campanula) or to prevent flowering of short-day plants (such as chrysanthemum) is widely practiced. Further, more detailed classification of photoperiodic response as either facultative or obligate has been applied to many floriculture and ornamental species (Erwin et al., 2004). In contrast, limited information is available for food crop species. For tomato, there is conflicting information on its photoperiodic response. Generally it seems that tomato is believed to be day neutral, but some cultivars are regarded as quantitative short-day plants (Heuvelink, 2005; Hurd, 1973). In North America, many strawberry cultivars are classified into "day neutral" without precisely identifying their "insensitiveness" to photoperiods. Bradford et al. (2010)

Table 10.1 Typical Interactions of Temperature and Photoperiod for Flowering of Strawberry (Determined for Japanese Cultivars) (After Saga Prefecture "Strawberry Cultivation Manual")

Type	Average Temperature			
	<5°C	5–12°C	12–26°C	>26°C
June-bearing	No flower induction (dormancy)	Flower induction under all photoperiods	Short day (8–13 h) promotes flower induction	No flower induction (critical temperature depends on cultivars)
	<5°C	5–15°C	15–30°C	>30°C
Ever-bearing	No flower induction (dormancy)	Flower induction under all photoperiods	Flower induction under all photoperiods, but number of flowers increases under longer day (facultative response)	Flower induction is suppressed

challenged the traditional classification and found that the classification required correction for some of the cultivars examined. Lack of information of specific photoperiodic responses is particularly the case for crop species grown in open fields where photoperiod is an uncontrollable factor specific to the geographical location and season.

In addition to photoperiod, flower development of many species is affected by temperature. There are often interactions observed between the effects of temperature and photoperiod. Table 10.1 shows the example of Japanese strawberry flowering responses. This foundational knowledge was developed to meet the need of off-season production of the strawberry. This type of information will be needed for various flowering crops as PFALs have the capacity of controlling environments precisely to induce flowers at an ideal timing relative to the scheduled timing of harvesting.

TRANSPIRATION

Transpiration refers to evaporation from plant tissue. The process is quite passive, driven by the water vapor difference between the stomatal cavity (or intercellular space) and the surrounding air. When stomata are open, almost all transpiration occurs through the stomata, but plants also transpire through the cuticular layer, which is referred to as cuticular transpiration. Cuticular transpiration dominates in darkness as plants close stomata. As gas diffusion, the transpiration rate is affected by water vapor concentration in the surrounding air, leaf temperature, as well as boundary layer and stomatal resistances. The water diffusion process in transpiration is described using the following equation, similar to stomatal CO_2 diffusion in photosynthesis:

$$E = \frac{V_{in} - V_{air}}{R_s + R_b} \tag{10.5}$$

where E is the leaf transpiration rate (mol m^{-2} s^{-1}) through the system having water vapor concentrations in the stomatal cavity (V_{in}, mol m^{-3}) and surrounding air (V_{air}, mol m^{-3}) and stomatal and boundary layer resistances (R_s and R_b, s m^{-1}). From this equation, it is clear that increasing resistances either by closing stomata or creating a thicker boundary layer (by limiting air circulation) would reduce the leaf transpiration rate. V_{in} is usually considered to be close to saturated water vapor concentration at leaf temperature, and therefore increasing leaf temperature (by increasing shortwave radiation) would increase the transpiration rate. Lower V_{air} also contributes to increasing transpiration rate.

Another important understanding of water flux in plants is the hydraulic continuum through the root zone, roots, stems, leaves and finally surrounding air. Quantifying water potential in such a continuum is helpful to understand water flux and its driving force. Water flux occurs following the water potential gradient (from less negative to more negative). Water potential (Ψ, Pa) is defined in terms of the chemical potential of water (μ_W, J mol^{-1}) and expressed using the following equation:

$$\Psi = \frac{\mu_W - \mu_{W0}}{V_W} \tag{10.6}$$

where μ_{W0} is the chemical potential of water at a reference state consisting of pure water at the same temperature and atmospheric pressure. V_W is partial molar volume of water (m^3 mol^{-1}). Therefore Ψ equals to zero for pure water. When water content becomes less in a given system (such as plant tissue), Ψ becomes more negative. Furthermore, water potential in plants can be expressed with multiple components.

$$\Psi = \Psi_p + \Psi_o + \Psi_m \tag{10.7}$$

where Ψ_p, Ψ_o, and Ψ_m are the components associated with the pressure, osmosis, and matric pressure. Water potential of moist air can also be found using the following equation using its relative humidity (RH, %):

$$\psi = \frac{R \cdot T}{V_W} \cdot \ln \frac{RH}{100} \tag{10.8}$$

where R is the gas constant (8.31 J mol^{-1} K^{-1}) and T is the air temperature in kelvin (K). Table 10.2 has some typical values of water potentials (adopted from Hannan (1998)). The difference in water potential is the largest between air inside the stomatal cavity and the air surrounding the leaf, indicating that this is the driving force inducing the water flow in the plants. When stomata are closed at night, the difference in water potential between the root zone and the root tissue allows continuous water flux,

Table 10.2 Typical Values That may be Seen in Root Zone—Plant—Air System

Location	Water Potential (Ψ, MPa)
Hydroponic nutrient solution	−0.08[a]
Soil, 0.1 m below ground and 10 mm from root	−0.3
Plant tissue (xylem of roots and leaves)	−0.8 to −0.6
Air inside the stomatal cavity (95% RH)	−6.9
Air just outside stomata (60% RH)	−70.0
Air just outside boundary layer (50% RH)	−93.6

[a]*Value reported by Fujiwara and Kozai (1995).*
Adopted from Hannan (1998) and Nobel (1991).

maximizing turgor pressure of the plants. Excessive transpiration (due to dry air or more negative water potential of the air) could induce more negative water flux enough to lose turgor (wilting).

TRANSLOCATION

Translocation is driven by pressure flow induced by the difference in water potential between sink and source regions of phloem. Sugars synthesized in source tissues (i.e., leaves) are loaded into the sieve elements of the leaves, increasing turgor pressure of the sieve elements due to the simultaneous water flux from the xylem to the phloem. In contrast, unloading sugars into the sink tissue from the phloem increases the water potential, allowing water flux to the xylem and reducing turgor. Both loading and unloading could involve symplastic or apoplastic pathways. Apoplastic import of sugars is through the plasma membrane of the cell or the tonoplast of the vacuole in the cell, and therefore requires metabolic energy. Kitano et al. (1998) demonstrated that localized heating around tomato fruit (sink organ) enhanced sugar translocation.

Understanding translocation and the sink–source relationship is especially critical for fruiting crops such as tomato and strawberry. Sugars accumulated in leaves are known to induce negative feedback of photosynthesis by reducing hexokinase as well as RuBisCo activity, and promoting translocation of sugars from leaves to fruits can prevent this feedback inhibition. Promoting translocation of sugars is also important for leafy crops to maximize the whole plant photosynthetic rate. The sink–source relationship is well studied for tomato (e.g., Ho, 1988) whose assimilate availability (source strength) is lower than assimilate demand (sink strength), demonstrated by increased fruit size when fruit pruning is applied (Heuvelink and Dorais, 2005). The sink–source relationship may be a vague concept as it is difficult to quantify the actual strength; however, understanding the balance of sink and source is important to develop cultural practices such as pruning leaves and fruits to optimize the balance, maximizing the crop productivity.

REFERENCES

Ainsworth, E.A., Rogers, A., 2007. The response of photosynthesis and stomatal conductance to rising [CO_2]: mechanisms and environmental interactions. Plant Cell Environ. 30, 258–270.

Albright, L.D., Both, A.-J., Chiu, A.J., 2000. Controlling greenhouse light to a consistent daily integral. Trans. ASAE 43, 421–431.

ASHRAE, 2013. Ventilation for acceptable indoor air quality (ANSI approved). ASHRAE Stand. 62, 1–2013.

Biddington, N.L., Dearman, A.S., 1985. The effect of mechanically induced stress on the growth of cauliflower, lettuce, and celery seedlings. Ann. Bot. 55, 109–119.

Bradford, E., Hancock, J.F., Warner, R.M., 2010. Interactions of temperature and photoperiod determine expression of repeat flowering in strawberry. J. Am. Soc. Hortic. Sci. 135, 102–107.

Bukhov, N.G., Drozdova, I.S., Bondar, V.V., 1995. Light response curves of photosynthesis in leaves of sun-type and shade-type plants grown in blue or red light. J. Photochem. Photobiol. B Biol. 30, 39–41.

Chia, P.L., Kubota, C., 2010. End-of-day far-red light quality and dose requirements for tomato rootstock hypocotyl elongation. HortSci. 45, 1501–1506.

Cockshull, K.E., Graves, C.J., Cave, C.R.J., 1992. The influence of shading on yield of greenhouse tomatoes. J. Hortic. Sci. 67, 11–24.

Collier, G.F., Tibbitts, T.W., 1984. Effects of relative humidity and root temperature on calcium concentration and tipburn development in lettuce. J. Am. Soc. Hortic. Sci. 109, 123–131.

Craig, D.S., Runkle, E.S., 2013. A moderate to high red to far-red light ratio from light-emitting diodes controls flowering of short-day plants. J. Am. Soc. Hortic. Sci. 138, 167–172.

Erwin, J., Mattson, N., Warner, R., 2004. Light effects on annual bedding plants. In: Fisher, P., Runkle, E. (Eds.), Lighting up Profits: Understanding Greenhouse Lighting. Meister Media, Ohio, pp. 62–71.

Fan, X.X., Xu, Z.G., Liu, X.Y., Tang, C.M., Wang, L.W., Han, X.I., 2013. Effects of light intensity on the growth and leaf development of young tomato plants grown under a combination of red and blue light. Sci. Hortic. 153, 50–55.

Frantz, J.M., Ritchie, G., Cometti, N.N., Robinson, J., Bugbee, B., 2004. Exploring the limits of crop productivity: beyond the limits of tipburn in lettuce. J. Am. Soc. Hortic. Sci. 129, 331–338.

Fujiwara, K., Kozai, T., 1995. Physical microenvironment and its effects. In: Aitken-Christie, J., Kozai, T., Lila Smith, M. (Eds.), Automation and Environmental Control in Plant Tissue Culture. Kluwer Academic, Dordrecht, The Netherlands, pp. 319–369.

Goto, E., Takakura, T., 1992. Promotion of Ca accumulation in inner leaves by air supply for prevention of lettuce tipburn. Trans. ASAE 35, 641–645.

Hannan, J.J., 1998. Greenhouses. Advanced Technology for Protected Horticulture. CRC Press, Boca Raton, FL, USA.

Heins, R.D., Cameron, A.C., Carlson, W.H., Runkle, E., Whiteman, C., Yuan, M., Hamaker, C., Engle, B., Koreman, P., 1997. Controlled flowering of herbaceous perennial plants. In: Goto, E., Kurata, K., Hayashi, M., Sase, S. (Eds.), Plant Production in Closed Ecosystems. Kluwer Academic, Dordrecht, The Netherlands, pp. 15–31.

Hernández, R., Kubota, C., 2014. Growth and morphological response of cucumber seedlings to supplemental red and blue photon flux ratios under varied solar daily light integrals. Sci. Hortic. 173, 92–99.

Heuvelink, E., 2005. Developmental process. In: Heuvelink, E. (Ed.), Tomatoes. CABI, Reading, UK, pp. 53–84.

Heuvelink, E., Dorais, M., 2005. Crop growth and yield. In: Heuvelink, E. (Ed.), Tomatoes. CABI, Reading, UK, pp. 85–144.

Ho, L.C., 1988. Metabolism and compartmentation of imported sugars in sink organs in relation to sink strength. Annu. Rev. Plant Physiol. Plant Mol. Biol. 39, 355–378.

Hogewoning, S.W., Trouwborst, G., Maljaars, H., Poorter, H., van Ieperen, W., Harbinson, J., 2010. Blue light dose-responses of leaf photosynthesis, morphology, and chemical composition of *Cucumis sativus* grown under different combinations of red and blue light. J. Exp. Bot. 61, 3107–3117.

Hunt, R., 1982. Plant Growth Curves: The Functional Approach to Plant Growth Analysis. Edward Arnold, London.

Hurd, R.G., 1973. Long-day effects on growth and flower initiation of tomato plants in low light. Ann. Appl. Biol. 73, 221–228.

Kimball, B.A., 1986. Influence of elevated CO_2 on crop yield. In: Enoch, H.Z., Kimball, B.A. (Eds.), Carbon Dioxide Enrichment of Greenhouse Crops. In: Physiology, Yield and Economics, vol. 2. CRC Press, Boca Raton, Florida, pp. 105–115.

Kitano, M., Araki, T., Eguchi, H., 1998. Environmental effects on dynamics of fruit growth and photoassimilate translocation in tomato plants. I. Effects of irradiation and day/night air temperature. Environ. Control. Biol. 36, 159–167.

Kitaya, Y., Tsuruyama, J., Kawai, M., Shibuya, T., Kiyota, M., 2000. Effects of air current on transpiration and net photosynthetic rates of plants in a closed transplant production system. In: Kubota, C., Chun, C. (Eds.), Transplant Production in the 21st Century. Kluwer Academic, The Netherlands, pp. 83–90.

Li, Q., Kubota, C., 2009. Effects of supplemental light quality on growth and phytochemicals of baby leaf lettuce. Environ. Exp. Bot. 67, 59–64.

Long, S.P., 1991. Modification of the response of photosynthetic productivity to rising temperature by atmospheric CO_2 concentrations: has its importance been underestimated? Plant Cell Environ. 14, 729–739.

López-Pozos, R., Martínez-Gutiérrez, G.A., Pérez-Pacheco, R., 2011. The effects of slope and channel nutrient solution gap number on the yield of tomato crops by a nutrient film technique system under a warm climate. HortSci. 46, 727–729.

McCree, K.J., 1972. The action spectrum, absorptance and quantum yield of photosynthesis in crop plants. Agric. Meteorol. 9, 191–216.

Nishizawa, T., Saito, K., 1998. Effects of rooting volume restriction on the growth and carbohydrate concentration in tomato plants. J. Am. Soc. Hortic. Sci. 123, 581–585.

Nobel, P.S., 1991. Physicochemical and Environmental Plant Physiology. Academic Press, New York.

Ohyama, K., Kozai, T., 1998. Estimating electric energy consumption and its cost in a transplant production factory with artificial lighting: a case study. J. Soc. High Technol. Agric. 10, 96–107.

Ohyama, K., Kozai, T., Ohno, Y., Toida, H., Ochi, Y., 2005. A CO_2 control system for a greenhouse with a high ventilation rate. Acta Horticult. 691, 649–654.

Paradiso, R., Meinen, E., Snel, J.F.H., De Visser, P., van Ieperen, W., Hogewoning, S.W., Marcelis, L.F.M., 2011. Spectral dependence of photosynthesis and light absorptance in single leaves and canopy in rose. Sci. Hortic. 127, 548–554.

Runkle, E., Fisher, P., 2004. Photoperiod and flowering. In: Fisher, P., Runkle, E. (Eds.), Lighting up Profits: Understanding Greenhouse Lighting. Meister Media, Ohio, pp. 25–32.

Sager, J.C., Mc Farlane, J.C., 1997. Radiation. In: Langhans, R.W., Tibbits, R.W. (Eds.), Plant Growth Chamber Handbook. Iowa State University, Ames, Iowa, pp. 1–29.

Sager, J.C., Smith, W.O., Edwards, J.L., Cyr, K.L., 1988. Photosynthetic efficiency and phytochrome photoequilibria determination using spectral data. Trans. ASAE 31, 1882–1889.

Spaargaren, I.J.J., 2001. Supplemental Lighting for Greenhouse Crops. Hortilux Schreder, Moster, Netherlands.

Stefanelli, D., Brady, S., Winkler, S., Jones, R.B., Tomkins, B.T., 2012. Lettuce (*Lactuca sativa* L.) growth and quality response to applied nitrogen under hydroponic conditions. Acta Horticult. 927, 353–359.

Stutte, G.W., 2009. Light-Emitting Diodes for Manipulating the Phytochrome Apparatus. HortSci. 44 (2), 231–234.

Yang, Z.C., Kubota, C., Chia, P.L., Kacira, M., 2012. Effect of end-of-day far-red light from a movable LED fixture on squash rootstock hypocotyl elongation. Sci. Hortic. 136, 81–86.

Zeroni, M., Gale, J., Ben-Asher, J., 1983. Root aeration in a deep hydroponic system and its effect on growth and yield of tomato. Sci. Hortic. 19, 213–220.

NUTRITION AND NUTRIENT UPTAKE IN SOILLESS CULTURE SYSTEMS

11

Satoru Tsukagoshi[1], Yutaka Shinohara[2]

Center for Environment, Health and Field Sciences, Chiba University, Kashiwa, Japan[1] Japan Greenhouse Horticulture Association, Tokyo, Japan[2]

INTRODUCTION

Although various systems are used in soilless culture and original nutrient solutions which are suitable for each system are often proposed, the basic principles regarding the nutrient solution and its management remain the same. Taken to an extreme, the cultivation system should be chosen based on cost and ease of use. When a system is chosen, and a suitable nutrient solution for the plant species or growth stage is used, the solution must be properly managed to obtain the highest yield and quality. In other words, a proper understanding of the management of the nutrient solution is the key to successful plant cultivation even in a controlled environment.

ESSENTIAL ELEMENTS

It is known that there are 92 natural elements and about 60 of them are included in plants (Inden, 2006; Resh, 2013). However, many of these elements are not considered essential for plants. An essential element has the following characteristics: (1) if the element is absent, the plant shows symptoms and cannot complete its lifecycle, (2) various deficit symptoms appear depending on the absence of specific elements, (3) only the element can restore the symptom; other elements cannot substitute in the function, (4) the element is directly involved in plant nutrition or metabolism (Arnon and Stout, 1939).

Recent research on plant nutrition has shown that 17 elements are essential for higher plants (Epstein, 1994). These elements are divided into two main classes corresponding to the amounts required by plants. The elements required in relatively large amounts are called macronutrients (major elements), and those in small amounts are called micronutrients (minor elements, trace elements). The macronutrients include nine elements: carbon (C), oxygen (O), hydrogen (H), nitrogen (N), phosphorus (P), potassium (K), calcium (Ca), magnesium (Mg), and sulfur (S). The micronutrients include eight elements: boron (B), iron (Fe), manganese (Mn), zinc (Zn), copper (Cu), molybdenum (Mo), chlorine (Cl), and nickel (Ni). The functions, excess and deficit symptoms are summarized in Tables 11.1 and 11.2.

Plant Factory. http://dx.doi.org/10.1016/B978-0-12-801775-3.00011-1

Table 11.1 Function, Mobility, Excess and Deficiency Symptoms of Macronutrients

Nitrogen (N)

Function:	Constituent of protein, enzyme, chlorophyll, hormone, nucleic acid, etc.
Mobility:	Easy
Excess:	Nitrate nitrogen (NO_3-N): luxuriant growth, succulent growth Ammonium nitrogen (NH_4-N): chlorosis, cause of Ca deficiency
Deficit:	Yellowing of lower leaves, color fade-out of whole leaves (redden in case of strawberry), decline of plant vigor

Phosphorus (P)

Function:	Constituent of gene, sugar phosphate, cell membrane, enzyme, ATP, etc. Involved in photosynthesis and respiration
Mobility:	Easy
Excess:	Rare. White spot on cucumber leaves
Deficit:	Growth suppression, coloring of lower leaves and lower part of stem with dark green, pigmentation of anthocyanin

Potassium (K)

Function:	Involved in various metabolic and synthetic reactions through the regulation of internal pH and osmotic potential
Mobility:	Easy
Excess:	Rare. Cause of Ca, Mg deficiency
Deficit:	Outward curling of leaves, yellowing of interveinal part or edge of lower leaves, brown spot on leaves

Calcium (Ca)

Function:	Constituent of cell wall. Involved in maintenance of construction and function of cell membrane, neutralization of organic acid, transmitter of stimuli, etc.
Mobility:	Difficult
Excess:	Rare. Cause of K, Mg deficiency. Cause of micronutrient deficiency by increasing solution pH
Deficit:	Tip-burn, blossom-end rot, black heart, root rot

Magnesium (Mg)

Function:	Constituent of chlorophyll. Activation of enzymes necessary for protein synthesis, Maintenance of ribosome structure
Mobility:	Easy
Excess:	Rare. Cause of K, Ca deficiency
Deficit:	Yellowing of interveinal part or edge of lower leaves

Sulfur (S)

Function:	Constituent of sulfur-containing amino acid and vitamin. Involved in various redox reactions
Mobility:	Easy
Excess:	Rare. Cause of K, Ca deficiency
Deficit:	Color fade-out or yellowing of lower leaves

Resh (2013); Yoneyama et al. (2010).

Table 11.2 Function, Mobility, Excess, and Deficiency Symptoms of Micronutrients

Iron (Fe)

Function:	Involved in various redox reactions and electron transfer through combination with protein
Mobility:	Difficult
Excess:	Rare
Deficit:	Yellowing or chlorosis of interveinal part of new leaves, yellowing of root
Note:	Dark spot on new leaves by rapid absorption of chelate Fe if plants are transferred from soil to soilless culture. Care for precipitation by sterilization of nutrient solution by UV ray or ozone gas

Boron (B)

Function:	Involved in formation of cell wall, maturation of pollen, elongation of pollen tube, translocation of sugar
Mobility:	Difficult
Excess:	Yellowing at edge of lower leaves, browning and brown spot in lower leaves, Curling of upper leaves
Deficit:	Malformation of new leaves, poor growth of lateral root, fruit drop, and hardening of leaves and stem

Manganese (Mn)

Function:	Involved in oxygen generation in photosynthetic reaction, extinction of active oxygen, synthesis of fatty acid
Mobility:	Difficult
Excess:	Browning (chocolate brown) of veins of lower leaves, root discoloration
Deficit:	Yellowing or coloring of interveinal part of middle $\sim$ upper leaves with dirty light green, brown spot
Note:	Care for precipitation by sterilization of nutrient solution by ozone gas

Copper (Cu)

Function:	Involved in electron transfer I photosynthetic reaction, extinction of active oxygen
Mobility:	Difficult
Excess:	Yellowing of upper leaves, browning of root, and poor growth of lateral root
Deficit:	Wilting or necrosis of leaf tip, defoliation
Note:	Care for elution from piping for heating and cooling, or circulating the solution. Sufficiently contained in tap water

Zinc (Zn)

Function:	Activation of various enzymes. Involved in formation of indoleacetic acid
Mobility:	Medium $\sim$ Easy
Excess:	Yellowing of upper leaves, inhibition of root growth. Cause of Fe deficiency
Deficit:	Suppression of leaf expansion, rosette leaves, yellowing of interveinal part of leaves, yellow-brown spot
Note:	Care for elution from zinc-plating frame of greenhouse. Sufficiently contained in tap water

Molybdenum (Mo)

Function:	Constituent of nitrate reductase. Involved in N_2 fixation
Mobility:	Medium
Excess:	Yellowing of lower leaves, coloring of veins to red-purple
Deficit:	Malformation of leaves, yellowing or necrosis of interveinal part of leaves
Note	Sufficiently contained in tap water

Chlorine (Cl)	
Function:	Regulation of internal pH. Involved in oxygen generation in photosynthetic reaction
Mobility:	Easy
Excess:	Root mortality, fibrous tubers (deterioration of quality)
Deficit:	Suppression of shoot tip growth, mortality
Note:	Sufficiently supplied from nature
Nickel (Ni)	
Function:	Constituent of urease
Mobility:	Unknown
Excess:	Yellowing of leaves, mortality (details are not well known)
Deficit:	Yellowing of leaves, mortality (details are not well known)
Resh (2013); Yoneyama et al. (2010).	

Table 11.3 Mineral Concentration in Leaf (%DW)

Vegetable	N	P	K	Ca	Mg
Cucumber	4.62	0.77	3.16	3.86	0.78
Tomato	5.15	0.54	3.78	2.92	0.76
Sweet pepper	5.91	0.51	7.02	1.8.	0.82
Celery	5.74	1.44	7.00	2.06	1.42
J. hornwort (*Mitsuba*)	5.00	1.00	6.92	0.77	0.55
Lettuce	5.36	0.72	7.64	0.79	0.35
Spinach	5.79	0.67	8.38	0.72	1.43

Date (2012).

Since most fresh vegetables contain approximately 90% water, the dry matter percentage is only about 10% of the initial fresh weight. The dry matter contains 40–45% carbon (C), 40–45% oxygen (O), and about 6% hydrogen (H) (Inden, 2006). These essential elements are supplied to the plants as carbon dioxide and water, and so it is not usually necessary to consider the supply of carbon, oxygen, and hydrogen as fertilizers. Therefore, excluding these three elements, only six elements (N, P, K, Ca, Mg, S) are classified as macronutrients. The amount of macronutrients absorbed by vegetables tends to be in the order of K > N > Ca > P > Mg (Maruo, 2013), and the contents of N and K in vegetable leaves are normally higher than those of other minerals (Table 11.3), though the ratio differs depending on the vegetable species (Date, 2012). The mineral contents are important when considering the appropriate composition of the nutrient solution and the management of water and nutrient supply. In addition, vegetables have a higher demand for Ca than other plants. In either case, the essential elements must be artificially supplied to ensure they are sufficient for the plant's demand when there is little supply from nature.

Table 11.4 Beneficial Elements for Higher Plants

Element	Function
Sodium (Na)	Growth promotion for table beet
Silica (Si)	Essential for Gramineae plants, addition of disease resistance for cucumber
Cobalt (Co)	Promotion of N_2 fixation of legumes in N shortage soil
Selenium (Se)	Beneficial for broccoli, cabbage, etc. (Function is not well known)
Aluminum (Al)	Growth-promoting effect for tea plant under low soil pH condition

Yoneyama et al. (2010).

BENEFICIAL ELEMENTS

The original meaning of beneficial elements is that the elements are essential for specific plant species, or show a growth-promoting effect under specific growth conditions (Yoneyama et al., 2010). For higher plants, sodium (Na), silica (Si), cobalt (Co), selenium (Se), and aluminum (Al) are considered to be beneficial elements, and their functions, plants and relevant conditions are shown in Table 11.4.

On the other hand, "beneficial element" has another meaning. The Ministry of Health, Labour and Welfare of Japan has specified recommended values in the diet for 13 kinds of mineral, which include selenium (Se), chromium (Cr), and iodine (I). In addition, Co is defined as an essential micronutrient for humans (Ohguchi, 2012), and bromine (Br), fluorine (F) and, rubidium (Rb) are present in the human body in relatively large amounts (Yamada, 2010). In other words, it is beneficial for humans if plants contain reasonable amounts of these elements. Because it is easier to manage the amount or timing of supply of these elements in a controlled environment, products produced in plant factories may have added value.

NUTRIENT UPTAKE AND MOVEMENT

The most important physiological role of plant roots is the absorption of water and nutrients. Normally, nutrients are vigorously absorbed near the root tip and the younger part where the root grows (Johkan, 2013). The process of nutrient uptake by plants is usually described by the following two concepts (Lack and Evans, 2001; Resh, 2013). The first is the pathway through the apoplast, which is the interconnecting cell walls and intercellular spaces. Ions are transported through the apoplast (apoplastic transport) and reach the Casparian strip where some ions are selected and taken into endothelial cells by a process that requires respiration. Another is through the ion channel, which is a cellular protein that controls the inflow and outflow of specific ions to and from a cell. The force that drives the transport of ions across the channel is the difference in electrical potential and chemical potential, and thus the transport needs no energy. The nutrient ions taken into the cell are transported through the protoplasmic connection (symplastic transport). As a result of these processes, plants can take in only specific ions selectively and positively.

The nutrients move upward with water through the xylem to reach the whole plant body. Eventually the nutrients dissolved in water penetrate the cell wall, allowing the cells to absorb the nutrients. As an inevitable consequence, the distribution of the nutrients Ca and B is strongly affected by water movement (Lack and Evans, 2001; Maruo, 2013). The moving force of water is obtained mainly from transpiration flow in the daytime and from root pressure flow at night. Basically, most nutrient ions move only in xylem tissue, but certain ones are redistributed in the plant body through phloem.

NUTRIENT SOLUTION

In soilless culture, all nutrients are supplied to the plants by the nutrient solution. In short, the nutrient solution is a well-prepared liquid fertilizer. The nutrient solution should fulfill the following conditions: (1) all essential elements for the plants are contained, (2) all elements exist in ion form, (3) the concentrations of each ion and total ions are within appropriate ranges for normal plant growth, (4) harmful substances and pathogenic microorganisms are not contained, (5) the solution pH is around 5.5–6.5 and stable, (6) the solution can be made using relatively inexpensive fertilizers, 7) the concentration of each ion, ratio of ions and pH do not fluctuate greatly during the cultivation period, and (8) in the deep flow technique (DFT) and nutrient film technique (NFT), enough oxygen is dissolved for root respiration.

The composition of the nutrient solution indicates the concentration of each nutrient ion in the solution. Many studies on this issue have been conducted and various compositions (formulas) have been designed. The typical formulas widely used in Japan are summarized in Table 11.5. A formula should be chosen depending on the kind of plant, growth stage, growing season, open or closed system, whether a substrate is used or not, kind of substrate, target quality of products, and so forth. Nevertheless, the only difference is the proportion of each ion and the concentration. In soilless culture, a slight change of nutrient management can greatly influence plant growth and the quality of the product. Therefore, adequate nutrient management can be achieved by not only considering the composition and concentration of ions but also other conditions such as the concentration of dissolved oxygen, temperature, and flora of microorganisms.

Table 11.5 Typical Formula and Composition of Nutrient Solution

Formula		Concentration (mM)					
		NO_3-N	NH_4-N	PO_4-P	K	Ca	Mg
Enshi (Nat. Hort. Exp. Sta.)		16	1.3	1.3	8	4	2
Yamasaki	Melon	13	1.3	1.3	6	3.5	1.5
	Cucumber	13	1	1	6	3.5	2
	Tomato	7	0.7	0.7	4	1.5	1
	Strawberry	5	0.5	0.5	3	1	0.5
	Sweet Pepper	9	0.8	0.8	6	1.5	0.8
	Lettuce	6	0.5	0.5	4	1	0.5
	Eggplant	10	1	1	7	1.5	1
Hoagland & Arnon (1938)		14	1	3	6	8	4
Holland (Rockwool, Tomato)		10.5	0.5	1.5	7	3.8	1
Date (2012).							

SOLUTION pH AND NUTRIENT UPTAKE

The solution pH (potential hydrogen) means the concentration of H^+ ion in the nutrient solution. The solution is neutral if the pH is 7, acidic if lower than 7 and alkaline if higher than 7. The solution pH affects the effectiveness of the nutrients through the change of ion form and/or solubility. For example, plants mainly absorb P in $H_2PO_4^-$ form, and more than 80% of P exists as $H_2PO_4^-$ when the solution pH is between 5 and 6. However, P in $H_2PO_4^-$ form decreases to 30% when the pH is 7. In addition, Ca easily combines with P to form a precipitate, while the stability of chelate Fe, such as Fe-EDTA, decreases at high pH. The pH is typically controlled between 5.5 and 6.5, but the desirable solution pH is 4.5–5.5 for leaf onion and some other species to increase Ca and P efficiency.

Fluctuations of solution pH are mainly caused by the unbalanced absorption of cation and anion by plants. A plant releases H^+ from the root when it absorbs cations, and OH^- or HCO_3^- when it absorbs anions. Therefore, the solution pH tends to increase when anion uptake is higher than cation uptake, and vice versa. In addition, in substrate culture the pH of the substrate affects the solution pH. Generally, the solution pH tends to increase with inorganic substrates such as rockwool and rice husk charcoal, and to decrease with organic ones such as peat and coir.

NITROGEN FORM

Although the main nitrogen form in nutrient solutions is NO_3^-, the NO_3^- absorbed by plants is reduced to NH_4^+ by nitrate reductase and then assimilated to amino acids. Since enzyme activity is dependent on light intensity, the supply of NH_4^+ tends to be less than that demanded by the plant under low light intensity if NO_3^- is the only nitrogen source. Therefore, the use of a small amount of NH_4^+ with NO_3^- usually promotes plant growth under low light intensity. However, if the NH_4^+ absorption rate is higher than the NH_4^+ assimilation rate, NH_4^+ accumulates in plant cells and becomes harmful because of abnormal N metabolism in the plant. NH_4^+ also inhibits Ca uptake because of antagonism, so the plant easily exhibits Ca deficit. In addition, the characteristics for which the nitrogen form is preferentially absorbed by the plant depend on the plant species, solution pH, solution temperature, and so forth. Consequently, NH_4^+ will be effective as a growth promoter for more plant species in soilless culture when: (1) the solution pH is kept constant, (2) the NH_4^+ concentration in the solution is kept lower, (3) the NH_4^+ is supplied in proper quantities in accordance with the rate of nitrogen assimilation by the plant.

NEW CONCEPT: QUANTITATIVE MANAGEMENT

As mentioned previously, many kinds of formulas for nutrient solutions have been proposed. However, such formulas have been basically designed for controlling the electrical conductivity (EC) or ion concentration in the solution (concentration management). Under this condition, the actual quantity of nutrient supplied to the plants is influenced by the actual solution volume in the cultivation system. In addition, some ions such as NO_3^-, $H_2PO_4^-$, and K^+ tend to be absorbed rapidly by plants when the concentration is managed, resulting in the absorption of a large quantity of such ions by plants.

However, when the absorption exceeds a certain amount (luxury absorption), the yield and quality of the product no longer increase with a further increase of absorption of this kind of ion. Moreover, the vegetative growth of fruit vegetables tends to be over-luxuriant rather than reproductive growth, and the unbalanced growth has an adverse effect on the fruit set and fruit productivity of the plant. Quantitative management of nutrients is a new concept of fertilizer application so as not to allow luxury absorption by plants (Date, 2012; Terabayashi et al., 2004). When the application of fertilizer is regulated quantitatively, the fertilizer is added to the nutrient solution tank once every week or two weeks, and the weight of the fertilizer is determined by preliminary measurements of the amount of nutrient uptake by the plant. For example, if the plant absorbs 1 g of NO_3^- a day, then N fertilizer containing 7 g of NO_3^- is added to the solution tank once a week. Even if the concentration of nutrient fluctuates by this management, ions can be vigorously absorbed from the solution under a wide range of solution concentrations (Maruo et al., 2004). The quantitative management of nutrients has additional advantages, such as avoiding excessive fertilization and reducing the cost of fertilizers, as well as minimizing pollution caused by draining the nutrient solution into the environment since all of the applied fertilizer is consumed by the plants. Application of this method is expected to increase with the accumulation of data on plant nutrient uptake and the development of automatic systems for controlling fertilizer usage.

REFERENCES

Arnon, D.I., Stout, P.R., 1939. The essentiality of certain elements in minute quantity for plant with special reference to copper. Plant Physiol. 14, 371–375. http://www.plantphysiol.org/content/14/2/371.full.pdf+html.

Date, S., 2012. Preparation and management of nutrient solution (written in Japanese: Baiyoueki no chousei, kanri). In: Japan Greenhouse Horticulture Association, , Hydroponic Society of Japan, (Eds.), All About Hydroponics. Seibundo Shinkosha, Tokyo, pp. 64–101.

Epstein, E., 1994. The anomaly of silicon in plant biology. Proc. Natl. Acad. Sci. U. S. A. 91, 11–17.

Inden, H., 2006. Nutrient uptake (written in Japanese: Youbun kyushu). In: Kozai, T., Goto, E., Fujiwara, K. (Eds.), Saishin shisetsu engeigaku. Asakura Publishing, Tokyo, pp. 35–40.

Johkan, M., 2013. Growth and development of plants (written in Japanese: Seicho to hatsuiku). In: Shinohara, Y. (Ed.), Yasai engeigaku no kiso. Rural Culture Association Japan, Tokyo, pp. 15–28.

Lack, A.J., Evans, D.E., 2001. Instant Note in Plant Biology. Taylor and Francis, UK, pp. 129–151.

Maruo, T., 2013. Response to environmental conditions and metabolism of plants (written in Japanese: Kankyo hannou to taisha). In: Shinohara, Y. (Ed.), Yasai engeigaku no kiso. Rural Culture Association Japan, Tokyo, pp. 45–47.

Maruo, T., Takagaki, M., Shinohara, Y., 2004. Critical nutrient concentrations for absorption of some vegetables. Acta Horticult. 644, 493–499.

Ohguchi, K., 2012. Mineral nutrition (written in Japanese: Mukishitsu no eiyou). In: Taji, Y. (Ed.), Kiso eiyougaku. Yodosha, Tokyo, pp. 128–136.

Resh, H.M., 2013. Hydroponic Food Production, seventh ed. CRC Press, Florida.

Terabayashi, S., Asaka, T., Tomatsuri, A., Date, A., Fujime, Y., 2004. Effect of the limited supply of nitrate and phosphate on nutrient uptake and fruit production of tomato (*Lycopersicon esculentum* Mill.) in hydroponic culture. Hortic. Res. 3, 195–200 (in Japanese with English abstract).

Yamada, T., 2010. Mineral nutrition (written in Japanese: Mukishitsu no eiyou). In: Gomyo, T., Watanabe, S., Yamada, T. (Eds.), Basic Nutrition Science. Asakura Publishing, Tokyo, pp. 95–105.

Yoneyama, T., Hasegawa, I., Sekimoto, H., Makino, A., Mato, T., Kawai, N., Morita, A., 2010. New Plant Nutrition (written in Japanese: Shin-shokubutsu eiyou, hiryou-gaku). Asakura Publishing, Tokyo.

TIPBURN

Toru Maruo, Masahumi Johkan
Graduate School of Horticulture, Chiba University, Matsudo, Chiba, Japan

INTRODUCTION

Tipburn is a physiological disorder during rapid plant growth involving necrosis at the leaf apex of young developing leaves (Figure 12.1). This physiological disorder commonly affects many vegetables such as lettuce (*Lactuca sativa* L.), strawberry (*Fragaria × ananassa*), cabbage (*Brassica oleracea* L. var. *capitata*), and Chinese cabbage (*B. pekinensis* Rupr.), and causes major economic losses (Saure, 1998). Tipburn is a serious physiological disorder in the production of lettuce plants in plant factories with artificial lighting (PFALs) as well as greenhouses and open fields.

The major cause of tipburn is often considered to be Ca^{2+} deficiency. Ca^{2+} is a component of cell walls and maintains cell function as a messenger signal, so Ca^{2+} deficiency causes cell death or necrosis because of abnormal cell formation. Internal browning of cabbage and blossom end rot of tomato fruit are also due to Ca^{2+} deficiency in the same way as tipburn. This chapter explains the physiological disorder of Ca^{2+} deficiency, using tipburn of lettuce plants as an example.

CAUSE OF TIPBURN

Ca^{2+} deficiency is caused by an imbalance between the Ca^{2+} required by plant cells and the amount supplied by the root zone. In an optimum environment in which plant growth rate increases, Ca^{2+} deficiency tends to occur because the amount supplied cannot keep pace with the increasing amount of Ca^{2+} required by the plant cells. However, tipburn is caused not only by Ca^{2+} deficiency in the plant, but also by the interaction of various factors.

INHIBITION OF CA^{2+} ABSORPTION IN ROOT

The inhibition of Ca^{2+} absorption involves two factors: the low concentration of Ca^{2+} in the root zone and the low absorption ability of Ca^{2+} in the roots. Tipburn often occurs when the plant root cannot absorb Ca^{2+} sufficiently: the plant cells become Ca^{2+} deficient even though the root zone is rich in Ca^{2+}.

Factors preventing the absorption of Ca^{2+} include a drastic pH change in the root zone which damages the roots, unsuitable temperature for root growth, and water and salinity stresses in the root zone. Furthermore, NH_4^+ and K^+ act antagonistically to the Ca^{2+} absorption by root. Under NH_4^+ and K^+ rich

FIGURE 12.1

Tipburn symptoms in hydroponic butter head lettuce.

Table 12.1 Chiba University's Formula for Nutrient Solution for Butter Head Lettuce (me/L) (Maruo et al., 1992)

Formula	NO_3^-	NH_4^+	PO_4^{3-}	K^+	Ca^{2+}	Mg^{2+}	SO_4^{2-}
Starter fertilizer	12	0	4	8.3	5	2	2
Adding fertilizer	12	1.3	4	8	4	2	2

conditions, the root can absorb more than it needs, so Ca^{2+} absorption is relatively suppressed. A formula for nutrient solution for suppressing lettuce tipburn in hydroponics was devised (Table 12.1).

INHIBITION OF Ca^{2+} TRANSFER FROM ROOT TO SHOOT

Ca^{2+} absorbed by the roots is transferred to shoots by the transpiration stream along with other mineral nutrients. The transpiration stream in xylem is under very strong tension, which pulls the water upward through the root and stem to the leaves. Since the transpiration rate per leaf of immature leaves is less than that of mature leaves, tipburn tends to occur in immature leaves because the supply of Ca^{2+} is insufficient compared with the plant cell requirement. Similarly, suppression of transpiration by a high-humidity environment and a hot, dry environment in which the stomata close is a major factor preventing the transfer of Ca^{2+} from root to shoot.

The stream of root pressure, which is osmotic pressure of root cells, also transfers the water upward from root to shoot. The tension of the root pressure stream is weaker than that of the transpiration stream, but it plays an important role in transferring water at night. Therefore, low activity of the root due to water and salinity stresses, low temperature and low oxygen in the root zone also causes Ca^{2+} deficiency.

COMPETITION FOR Ca^{2+} DISTRIBUTION

Ca^{2+} is an essential component of plant cell walls and is not mobile once it is fixed to the cell wall. Therefore, Ca^{2+} does not move from tissues with sufficient Ca^{2+} to tissues with insufficient Ca^{2+}, and each tissue usually competes for the distribution of Ca^{2+} supplied from the root. One factor affecting the Ca^{2+} distribution is leaf area, and the calcium content in a leaf is strongly dependent on the cumulative transpiration rate. The pulling force of Ca^{2+} from the vascular tissue in immature leaves is weak, because the leaf area and transpiration rate of immature leaves are less than those of mature leaves. If the proportion of mature leaves with large leaf area increases in a plant, the amount of Ca^{2+} distribution in immature leaves relatively decreases.

COUNTERMEASURE

Ca^{2+} deficiency is caused by an imbalance between the Ca^{2+} required by plant cells and the amount supplied, and causes physiological disorders such as tipburn (Table 12.2). It is effective to use tipburn-resistant cultivars because the varietal characteristic of Ca^{2+} deficiency varies drastically. Although foliar application of Ca^{2+} is an effective preventive measure, direct application to immature leaves is important, because mature leaves often have enough calcium content. It is also important to slow down the production to some extent because tipburn often occurs under the optimal environment for plant growth. In management, therefore, it is important to find the marginal conditions under which tipburn of only a few percent occurs.

Table 12.2 Factors Causing Tipburn in Hydroponic Lettuce

Factors	Incidence Conditions	Related Factors
Shoot	• Higher growth rate • Transpiration suppression	Light period and intensity Humidity (vapor pressure deficit) Wind speed Temperature CO_2 concentration Sudden increase in light intensity
Plant	• Resistant varieties • Imbalance proportion of mature and immature leaves	Cultivar Leaf number and area Growth rate
Root	• Inhibition of Ca^{2+} absorption • NH_4^+ and K^+ rich in root zone	Nutrient solution Changes with time of pH

REFERENCES

Maruo, T., Ito, T., Ishii, S., 1992. Studies on the feasible management of nutrient solution in hydroponically growth lettuce (*Lactuca sativa* L.). Tech. Bull. Fac. Chiba Univ. 46, 235–240.

Saure, M.C., 1998. Cause of the tipburn disorder in leaves of vegetables. Sci. Hortic. 76, 131–147.

FUNCTIONAL COMPONENTS IN LEAFY VEGETABLES

13

Keiko Ohashi-Kaneko

Research Institute, Tamagawa University, Tokyo, Japan

INTRODUCTION

Plant factories with artificial lighting (PFALs) enable close control of environmental factors such as light, temperature, humidity, and composition of nutrient solutions. When these factors fluctuate, plants have some ability to acclimate, adjusting their morphological characteristics such as weight, leaf area, stem length, and leaf color, as well as their composition of minerals, pigments, proteins, organic acids, and secondary metabolites. Therefore, environmental control in plant factories enables the production of vegetables with characteristics that better meet customer demands. This chapter introduces the production of functional vegetables by controlling the composition of the hydroponic nutrient solution and the quality of the lighting.

LOW-POTASSIUM VEGETABLES

Patients suffering from kidney disease or those undergoing dialysis face restrictions on their potassium intake. However, they can eat fresh salads containing lettuce with a low potassium content. Aizufujikako Co., Ltd., a semiconductor manufacturer since 1967, succeeded in mass-producing low-potassium lettuce for the first time in Japan (Suzuki, 2014). One of their factory's dust-free clean rooms, which receives no solar radiation, was diverted to the cultivation of functional vegetables, including low-potassium lettuce, with almost no renovation required.

The method of cultivating vegetables with a low potassium concentration was established for spinach by Ogawa et al. (2007). The potassium concentration of low-potassium spinach was determined to be about one-fifth (170 mg potassium per 100 g) of that in spinach supplied with the conventional nutrient solution. Starting about 2 weeks before harvest, the supply of potassium was reduced by replacing the KNO_3 in the nutrient solution with HNO_3, while maintaining the pH at 6.5 with 0.1 M NaOH. Eliminating potassium 2 weeks before harvest did not cause any significant change in fresh weight. However, this restriction of potassium uptake resulted in an increase in sodium concentration in the leaves; the sodium concentration of low-potassium spinach was 13 times higher than that of spinach in the conventional nutrient solution. It is likely that sodium ions were used for maintaining the intracellular osmotic potential instead of potassium ions. This increase in sodium is undesirable for patients suffering from kidney disease or undergoing dialysis.

A modified cultivation method, which became an open patent in 2012, enabled the production of lettuce with low potassium concentration without increasing the sodium concentration (Ogawa et al., 2012).

Plant Factory. http://dx.doi.org/10.1016/B978-0-12-801775-3.00013-5

In this method, leaf lettuce was cultivated in a nutrient solution containing neither potassium nor sodium ions from 11 to 17 days before harvesting. The potassium concentration of this lettuce was less than 100 mg per 100 g. According to the standard tables of food composition in Japan, compiled by the Ministry of Education, Culture, Sports, Science and Technology (MEXT), the potassium concentration of regular lettuce is 490 mg per 100 g (MEXT, 2002). Low-potassium lettuce has little bitterness and is expected to be more appetizing to children. In this modified method, the lower limit of the magnesium concentration was up to 5% (w/w) 11–17 days before harvest; sufficient absorption of magnesium by low-potassium lettuce was effective for preventing disease in leaves (Ogawa et al., 2012). Later, Aizufujikako modified other cultivation methods and acquired a patent in 2014 (Matsunaga, 2014). In this method, the concentration of the nutrient solution is managed by carefully controlling the EC value throughout the cultivation period, and the ratio of phosphate to nitrogen concentration in a potassium-containing nutrient solution used in the initial period of cultivation is altered by adding potassium-free nutrient solution in the subsequent stage of cultivation. The potassium concentration of leaf lettuce cultured by this modified method is stable at about one-fourth of that of leaf lettuce cultured by conventional nutrient solution even in the case of large-scale hydroponics (Matsunaga, 2014).

Aizufujikako is reportedly developing low-potassium melon, tomato, and strawberry (Aizufujikako Co., Ltd., http://drvegetable.jp/about/); patients suffering from kidney disease may soon be able to enjoy a fruit salad.

LOW-NITRATE VEGETABLES

Plants accumulate the excess nitrogen not used for growth as nitrate in vacuoles. This is considered to be a reserve pool that is utilized during nitrogen starvation. However, although not very toxic itself, nitrate is harmful to human health, and it is hazardous when ingested because it is converted to nitrite in the gastrointestinal tract. Nitrite leads to the formation of nitrosoamines, which are potent carcinogens (Sohár and Domoki, 1980). The nitrate concentration in leafy vegetables can be reduced by eliminating NO_3-N in the nutrient solution or replacing it with NH_4-N a few weeks before harvest. Further, by promoting nitrate reductase (NR) activity, the nitrate in vacuoles is used for assimilation and the nitrate content in the plants is reduced.

RESTRICTION OF FEEDING NITRATE FERTILIZER TO PLANTS

More than 90% of the nitrogen-containing nutrients in most nutrient solutions, including Hoagland's solution and Otsuka-A nutrient solution (Otsuka AgriTechno Co., Ltd., Japan), which is used widely in Japan, are in the form of NO_3-N. In order to prevent an increase in the nitrate concentration in vegetables, they should be fed a suitable amount of nitrogen fertilizer that can be exhausted during growth. The complete removal of nitrate from the nutrient solution 1 week before harvest results in a 56% decrease in nitrate concentration of celery leaves compared with those grown in regular nutrient solution (Martignon et al., 1994). When plants are deprived of N, they are able to reuse nitrate accumulated in vacuoles for growth, and as a result, the accumulated nitrate decreases (Santamaria et al., 1998). Cutting N completely 7–10 days before harvest seems to decrease the fresh mass of lettuce leaves at harvest (Santamaria et al., 1998).

Replacing some of the NO_3-N in the nutrient solution with NH_4-N can decrease the nitrate concentration of vegetables without decreasing yield (Saigusa and Kumazaki, 2014). In general, vegetables cultivated in the field, such as lettuce, spinach, rocket, chicory, and endive, prefer NO_3-N over NH_4-N, and feeding NH_4-N is considered to cause stunting. However, some of these plant species that absorb NO_3-N preferentially, including lettuce, can grow vigorously in nutrient solution in which the main source of nitrogen is NH_4-N, provided the pH is kept strictly at 5.5. Their biomass in lettuce plants is almost the same as that of lettuce plants cultured in conventional nutrient solution (Moritugu et al., 1980; Moritugu and Kawasaki, 1982, 1983). When plants take up ammonium ions, they excrete protons into the nutrient solution in order to maintain their internal electric balance. This reduces the pH of the nutrient solution (Tsukagoshi, 2002), and plants sensitive to acidic conditions are injured (Troelstra et al., 1990). However, spinach and Chinese cabbage cultured in a solution in which the main source of nitrogen is NH_4-N are stunted even when the pH is held constant, and they seem to be sensitive to both ammonium and pH (Moritugu and Kawasaki, 1983).

In leaf lettuce, nitrate content was remarkably low without a decrease in yield when available N in the nutrient solution was cut by one-fifth (1375 ppm vs. 6450 ppm for plants fed with the conventional nutrient solution), with the NO_3-N/NH_4-N molar ratio kept at 1 from transplanting to harvest (Saigusa and Kumazaki, 2014). When chicory plants were transferred from 4 mM NO_3-N to 1 mM NO_3-N plus 3 mM NH_4-N 6 days before harvesting, nitrate content in leaves decreased by 58% without a decrease in fresh mass compared with growth with no change in nutrient solution (Santamaria et al., 1998).

REDUCTION IN ACCUMULATED NITRATE BY ASSIMILATION OF NITRATE

The nitrate content in tobacco leaves changes diurnally, and the nitrate content at the end of the light period is lower than at the beginning (Matt et al., 1998; Geiger et al., 1998). This diurnal fluctuation in nitrate content seems to be related to NR activity. NR content and activity are low at the end of the night, increase to a maximum at 2–3 h into the photoperiod, and decline during the second part of the photoperiod (Geiger et al., 1998). It seems that nitrate content declines several hours after NR activity is maximal, so it may be desirable to harvest in the afternoon. NR activity in plants increases with an increase in light intensity (Carrasco and Burrage, 1992). Since the light intensity in PFAL is very low compared with sunlight, it is easy for plants to accumulate nitrate; thus, the nitrogen concentration of the nutrient solution should be kept low.

NR activity is affected by light quality (Jones and Sheard, 1977; Ohashi-Kaneko et al., 2007). NR activity in maize leaves under blue light is higher than under red light, but leaf nitrate content is not affected by light quality (Jones and Sheard, 1977). Nitrate accumulation in plants is also related to nitrate uptake. Blue light promotes stomatal opening (Sharkey and Raschke, 1981; Kinoshita et al., 2001), which may promote nitrate uptake due to an increase in transpiration. Therefore, it may be difficult to regulate nitrate content by only controlling light quality.

IMPROVING THE QUALITY OF LEAFY VEGETABLES BY CONTROLLING LIGHT QUALITY

In general, in PFAL, the minimum light intensity and light period for growing plants with commercial-grade fresh weight are set so as to minimize the electricity cost. Therefore, it is desirable to achieve high productivity and high vegetable quality under limited light intensity. For example, cultivation techniques that enhance antioxidant concentration by controlling the light environment have been developed. This section introduces data on the functional ingredient content of leafy vegetables and herbs irradiated with UV-B, UV-A, or blue light.

LEAFY VEGETABLES

Figure 13.1 shows chlorophyll, carotenoid, L-ascorbic acid, soluble sugar, and nitrate content in the leaves of leaf lettuce, spinach, and komatsuna grown under white, red, and blue light from fluorescent lamps (FLs) and a mixture of red and blue light (the ratio of number of red FL to number of blue FL was 1) (Ohashi-Kaneko et al., 2007). The effects of light quality on the concentration of each component and biomass production (data not shown) were quite different depending on the plant species. Irradiation from blue FL alone or a combination of red FL and blue FL can produce high-quality leaf lettuce rich in L-ascorbic acid, but no such blue light effect was observed in spinach or komatsuna. Ooshima

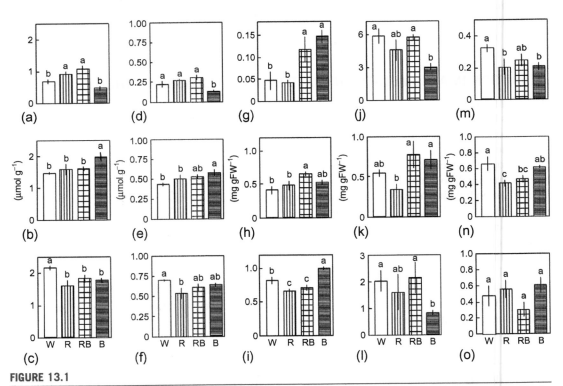

FIGURE 13.1

Chlorophyll (a, b, c), total carotenoid (d, e, f), L-ascorbic acid (g, h, i), soluble sugar (j, k, l), and nitrate (m, n, o) in the leaves of leaf lettuce (a, d, g, j, m), spinach (b, e, h, k, n), and komatsuna (c, f, i, l, o) grown under white light (W), red light (R), blue light (B) from FL or a mixture of R and B (RB) at 37d after germination for leaf lettuce and spinach and 32d after germination for komatsuna (Ohashi-Kaneko et al., 2007). The vertical bars indicate the SE (n = 4). Means with different letters within each panel are significantly different at the 5% level by LSD.

et al. (2013) cultivated red leaf lettuce in an environment with a varying ratio of red to blue light from a red light-emitting diode (LED) and a blue LED and observed that both the anthocyanin and ascorbic acid content increased with an increase in blue light ratio. However, they also noted that when the blue light ratio exceeded 20%, the fresh weight of above-ground tissue at harvest was lower, less than 80 g per plant, which is the commercial-grade fresh weight. Further investigation is required to determine the optimum ratio of blue light and red light that improves the ascorbic acid content in plants that reach commercial-grade fresh weight.

When red leaf lettuce plants are cultivated under irradiation with mixtures of red light and blue light, the leaf anthocyanin content and the expression of genes encoding several enzymes related to the anthocyanin synthesis pathway increase as the blue light ratio increases (Shoji et al., 2010). Anthocyanin synthesis in leaves of red leaf lettuce is induced by UV-B (Park et al., 2007) and UV-A light (Voipio and Autio, 1995). Supplemental irradiation with UV-B light alone or UV-A light alone during the dark period increased the anthocyanin content in the leaves of red leaf lettuce compared with those receiving no supplemental irradiation, and the anthocyanin content was higher under UV-B irradiation than UV-A irradiation (Ebisawa et al., 2008); furthermore, the anthocyanin content in leaves grown under supplemental irradiation of UV-B plus 60 μmol m^{-2} s^{-1} of blue light during the dark period was seven times higher than that of plants grown with no supplemental irradiation. Since high-quality leaf lettuce rich in anthocyanin content can be shipped if harvested in the first half of a light period, this is a useful technology.

HERBS

The functional ingredients in herbs are mainly products of secondary metabolism, such as essential oils, polyphenols, carotenoids, anthocyanins, and alkaloids. The secondary metabolite ingredients of plants function as defensive chemical substances. For example, when plants are exposed to biotic stress, such as being eaten by insects or animals, synthesis of essential oils is promoted (Koseki, 2004). Production of secondary metabolites is also promoted by abiotic stress, such as high temperature, low temperature, dryness, and UV light irradiation. In addition, light quality triggers the synthesis of secondary metabolites (Koseki, 2004).

UV irradiation raises the essential oil concentration in leaves of herbs. Hikosaka et al. (2010) observed that the limonene and *l*-menthol concentrations of the upper leaves of Japanese mint plants grown under white light from FL supplemented with UV-B light were, respectively, 1.5 and 2 times higher than in plants grown under white light with no supplemental UV-B light. When sweet basil grown in a glasshouse into which sunlight enters is irradiated with supplementary UV-B in the early morning, levels of several essential oils including eugenol and linalool were enhanced compared with no UV-B supplementation (Johnson et al., 1999).

We have also investigated the effects of visible light quality on biomass production and essential oil concentration of perilla, coriander, and rocket. Figure 13.2 shows the main essential oil content of perilla and coriander plants grown under different lighting conditions (Ohashi-Kaneko et al., 2013). Red and blue LEDs and white FL (control) were used as the light sources for cultivation. The biomass production of perilla plants grown under blue light was much smaller than that under red light. The concentration of perillaldehyde, which makes up 50% of the essential oil, in the leaves of the perilla plants grown under blue light was 1.8 times higher than that of the plants grown under red light. The biomass production and (*E*)-2-decenal and (*E*)-2-dodecenal concentrations were highest in coriander grown

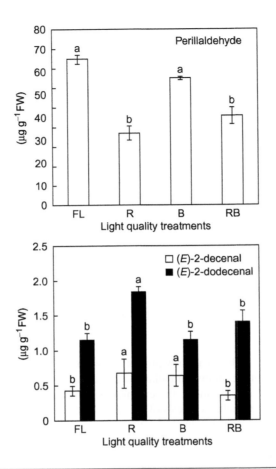

FIGURE 13.2

Perillaldehyde content in the leaves of perilla plants at 35d after transplanting (upper panel) and (*E*)-2-decenal and (*E*)-2-dodecenal contents in the shoots of coriander plants at 21d after transplanting (lower panel) (Ohashi-Kaneko, 2013; Ohashi-Kaneko et al., 2013). These plants were grown under white light from FL (FL), red light (R), blue light (B) from LED, or a mixture of R and B (RB). The vertical bars indicate the SE ($n = 2$–3 for perilla plants, $n = 3$ for coriander plants). Means with different letters within a panel are significantly different at the 5% level by the Tukey-Kramer honest significant difference (HSD) test.

under red light (Ohashi-Kaneko et al., 2013). To produce high-quality coriander, the concentration of heptane, (*E*)-2-hexenal, and octanal, which give it a fresh grassy fragrance, should also be raised (Kohara et al., 2006). Therefore, we plan to examine the effects of light quality on the composition of these essential oils. The smell of salad rocket resembles that of sesame seeds, and is considered to originate in essential oils such as benzaldehyde with an almond-like flavor, furfural with flavors resembling almond and fresh baked bread (Jirovetz et al., 2002), and anethol (Awano et al., 2006). The composition of its essential oil differs depending on the light quality of the environment in which the rocket plants are cultivated (Ohashi-Kaneko et al., 2013). By detailed comparison of the composition of the essential oil and the flavor and taste of the leaves under different lighting conditions, control of the light quality will enable the quality of rocket to be adjusted. It may soon be possible to produce herbs matched with their purpose, such as cooking materials, raw materials for fragrances, and medicinal materials, in an enclosed artificially lit plant factory.

CONCLUSION

Since methods for producing low-potassium and low-nitrate vegetables are reasonably well established, the next step is stable mass production. In addition, controlling the light quality will be useful for increasing biomass productivity and also the concentration of functional ingredients of vegetables and fresh herbs, although the most effective light treatment will depend on the species. The optimum light quality conditions for growing plants with commercial-grade fresh weight and rich in functional ingredients differ, and include variations in light intensity and light period. We plan to analyze the growth of plants cultivated under various lighting conditions.

REFERENCES

Awano, K., et al., 2006. Rocket (in Japanese). In: Inabata, K. (Ed.), Encyclopedia of Food Scents (Written in Japanese: Tabemono kaori hyakka jiten). Asakura Publishing Co., Ltd., Tokyo, pp. 628–629

Carrasco, G.A., Burrage, S.W., 1992. Diurnal fluctuations in nitrate accumulation and reductase activity in lettuce (*Lactuca sativa* L.) grown using nutrient film technique. Acta Hortic. 323, 51–56.

Ebisawa, M., Shoji, K., Kato, M., Shimomura, K., Goto, F., Yoshihara, T., 2008. Effect of supplementary lighting of UV-B, UV-A, and blue light during night on growth and coloring in red-leaf lettuce (in Japanese). J. Sci. High Technol. Agric. 20, 158–164.

Geiger, M., Walch-Liu, P., Engels, C., Harnecker, J., Schulze, E.D., Ludewig, F., Sonnewald, U., Scheible, W.R., Stitt, M., 1998. Enhanced carbon dioxide leads to a modified diurnal rhythm of nitrate reductase activity in older plants, and a large stimulation of nitrate reductase activity and higher levels of amino acids in young tobacco plants. Plant Cell Environ. 21, 253–268.

Hikosaka, S., Ito, K., Goto, E., 2010. Effects of ultraviolet light on growth, essential oil concentration, and total antioxidant capacity of Japanese mint. Environ. Control Biol. 48, 185–190.

Jirovetz, L., Smith, D., Buchbauer, G., 2002. Aroma compound analysis of *Eruca sativa* (*Brassicaceae*) SPME headspace leaf samples using GC, GC-MS, and olfactometry. J. Agric. Food Chem. 50, 4643–4646.

Johnson, C.B., Kirby, J., Naxakis, G., Pearson, S., 1999. Substantial UV-B mediated induction of essential oils in sweet basil (*Ocimum basilicum* L.). Phytochemistry 51, 507–510.

Jones, R.W., Sheard, R.W., 1977. Effects of blue and red light on nitrate reductase level in leaves of maize and pea seedlings. Plant Sci. Lett. 8, 305–311.

Kinoshita, T., Doi, M., Suetsugu, N., Kagawa, T., Wada, M., Shimazaki, K., 2001. phot1 and phot2 mediate blue light regulation of stomatal opening. Nature 414, 656–660.

Kohara, K., Sakamoto, Y., Hasegawa, H., Kozuka, H., Sakamoto, K., Hyata, Y., 2006. Fluctuations in volatile compounds in leaves, stems, and fruits of growing coriander (*Coriandrum sativum* L.) plants. J. Jpn. Soc. Hortic. Sci. 75, 267–269.

Koseki, Y., 2004. Secondary metabolites and plant defense (in Japanese). In: Nishitani, K., Shimazaki, K. (Eds.), Plant Physiology (Written in Japanese: Shokubutsu seirigaku). Baifukan, Tokyo, pp. 282–312.

Martignon, G., Casarotti, D., Venezia, A., Sciavi, M., Malorgio, F., 1994. Nitrate accumulation in celery as affected by growing system and N content in the nutrient solution. Acta Hortic. 361, 583–589.

Matsunaga, S. 2014. Japan patent 5628458.

Matt, P., Shurr, U., Klein, D., Krapp, A., Stitt, M., 1998. Growth of tobacco in short-day conditions leads to high starch, low sugars, altered diurnal changes in the *Nia* transcript and low nitrate reductase activity, and inhibition of amino acid synthesis. Planta 207, 27–41.

Ministry of Education, Culture, Sports, Science and Technology (MEXT), 2002. Standard tables of food composition in Japan, fifth revised and enlarged edition. Kagawa Nutrition University Publishing Division, Tokyo, pp. 100.

Moritugu, M., Kawasaki, T., 1982. Effect of solution pH on growth and mineral uptake in plants under constant pH condition. Ber. Ohara Inst. Landwirtsch. Forsch., Okayama Univ. 18, 77–92.

Moritugu, M., Kawasaki, T., 1983. Effect of nitrogen source on growth and mineral uptake in plants under nitrogen restricted culture condition. Ber. Ohara Inst. Landwirtsch. Forsch., Okayama Univ. 18, 145–158.

Moritugu, M., Suzuki, T., Kawasaki, T., 1980. Effect of nitrogen sources upon plant growth and mineral uptake. 1. Comparison between constant pH and conventional culture method. JSSSPN 51, 447–456.

Ogawa, A., Taguti, S., Kawashima, C., 2007. A cultivation method of spinach with a low potassium content for patients on dialysis. Jpn. J. Crop Sci. 76, 232–237.

Ogawa, A., Udzuka, K., Toyofuku, K., Ikeda, T., 2012. Japan patent P2012-183062A.

Ohashi-Kaneko, K., 2013. Nutrient ingredients and functional components in herb plants cultured at an enclosed artificial lighting plant factory (in Japanese). In: Takatsuji, M., Kozai, T. (Eds.), Important Problem and Measure Against Plant Factory Management (Written in Japanese: Shokubutsu kojo keiei no juuyo kadai to taisaku). Johokiko Co., Ltd., Tokyo.

Ohashi-Kaneko, K., Takase, M., Naoya, K., Fujiwara, K., Kurata, K., 2007. Effect of light quality on growth and vegetable quality in leaf lettuce, spinach and komatsuna. Environ. Control Biol. 45, 189–198.

Ohashi-Kaneko, K., Ogawa, E., Ono, E., Watanabe, H., 2013. Growth and essential oil content of perilla, rocket and coriander plants grown under different light quality environments (in Japanese). J. Sci. High Technol. Agric. 25, 132–141.

Ooshima, T., Hagiya, K., Yamaguchi, T., Endo, T., 2013. Commercialization of LED used plant factory. Nishimatsu Construction Report 36, 1-6.

Park, J.S., Choung, M.G., Kim, J.B., Hahn, B.S., Kim, J.B., Bae, S.C., Roh, K.H., Kim, Y.H., Cheon, C.I., Sung, M.K., Cho, K.J., 2007. Genes up-regulated during red coloration in UV-B irradiated lettuce leaves. Plant Cell Rep. 26, 507–516.

Saigusa, M., Kumazaki, T., 2014. Nitrate concentration and trial of decreasing nitrate concentration in factory vegetables (in Japanese). In: Takatsuji, M., Kozai, T. (Eds.), Important Problem and Measure Against Plant Factory Management (Written in Japanese: Shokubutsu kojo keiei no juuyo kadai to taisaku). Johokiko Co., Ltd., Tokyo, pp. 208–214.

Santamaria, P., Elia, A., Parente, A., Serio, F., 1998. Fertilization strategies for lowering nitrate content in leafy vegetables: chicory and rocket salad cases. J. Plant Nutr. 21, 1791–1803.

Sharkey, T.D., Raschke, K., 1981. Effect of light quality on stomatal opening in leaves of *Xanthium strumarium* L. Plant Physiol. 68, 1170–1174.

Shoji, K., Goto, E., Hashida, S., Goto, F., Yoshihara, T., 2010. Effect of red light and blue light on the anthocyanin accumulation and expression of anthocyanin biosynthesis gene in red-leaf lettuce (in Japanese). J. Sci. High Technol. Agric. 22, 107–113.

Sohár, J., Domoki, J., 1980. Nitrite and nitrate in human nutrition. Bibl. Nutr. Dieta 29, 65–74.

Suzuki, H., 2014. Development of low potassium vegetables and exploitation of a market (in Japanese). SHITA report 31, 1–7.

Troelstra, S.R., Wagenaar, R., Smant, W., 1990. Growth responses of plantago to ammonium nutrition with and without pH control: comparison of plants precultivated on nitrate or ammonium. In: van Beusichem, M.L. (Ed.), Plant Nutrition - Physiology and Applications, Developments in Plant and Soil Sciences 41. Springer, Netherlands, pp. 39–43.

Tsukagoshi, S., 2002. pH (in Japanese). In: Japan Greenhouse Horticulture Association, New Manual of Hydroponics (Written in Japanese: Yoeki Saibai no shin-manuaru). Seibundo-Shinkosha, Tokyo, pp. 180-183.

Voipio, I., Autio, J., 1995. Responses of red-leaved lettuce to light intensity, UV-A radiation and root zone temperature. Acta Hortic. 399, 183–187.

MEDICINAL COMPONENTS

Sma Zobayed

SHIRFA Biotech, Pitt Meadows, British Columbia, Canada
JRT Research and Development, Aldergrove, British Columbia, Canada

INTRODUCTION

Medicinal herbs were the primary health care agents for many centuries before the advances of modern medicine and are still used worldwide as a significant part of the health care system. In a recent report of the World Health Organization, the percentage of the population which has used plant-based medicine at least once is 48% in Australia, 70% in Canada, and 75% in France (WHO report: Traditional Plants, 2003). Currently, up to 80% of the population in Africa and 40% in China use traditional medicines to meet their health care needs (WHO report: Traditional Plants, 2003). As a result, international trade in medicinal plants has become a major part of the global economy and demand is increasing in both developing and industrialized nations. According to a recent report by Global Industry Analysts, the global herbal supplements market will reach $107 billion by 2017 (Global Industry Analysts Report, 2014). However, this explosive growth of the consumption of plant-based medicines has been accompanied by issues of quality and consistency, compromising the safety and efficacy and leading to serious health issues (Zobayed et al., 2005a). These issues include: (i) adulteration and contamination with misidentified plant species; (ii) contamination with plant and soil-borne microbial and environmental pollutants leading to wide variation of the concentration of medicinal metabolites in the plant tissues; (iii) genetic variability of seed-grown medicinal plants leading to variations in medicinal contents; and (iv) medicinal metabolites from wild-collected plants or from field cultivation can be influenced by seasonal variations, as well as variations in growing conditions such as temperature, photoperiod, sunlight (UV and other light qualities), light intensity, water content, soil pH, soil nutrient condition, etc. However, there is virtually no cultivation of medicinal plants on any significant scale worldwide and the current procedure for preparing medicinal plant agents involves mainly harvesting wild plants. In a report, Williams (1996) estimated that about 99% of the 400–550 species currently sold for use in traditional medicine in Africa originate from wild sources. Therefore, it is necessary to develop new technologies to ensure the efficacy and safety of medicines based on plants and to maximize the biomass and metabolite contents in plants.

Plant Factory. http://dx.doi.org/10.1016/B978-0-12-801775-3.00014-7

GROWING MEDICINAL PLANTS UNDER CONTROLLED ENVIRONMENTS: MEDICINAL COMPONENTS AND ENVIRONMENTAL FACTORS

Environmental factors such as temperature, relative humidity, light intensity, light quality, water content, minerals, and CO_2 concentration in the air influence the growth of a plant and production of medicinal metabolites. Recent research indicates that growing medicinal plants under controlled environments with artificial light can ensure the efficacy and safety of the medicinal plant products, ensure year-round harvesting of products, and maximize biomass production by optimizing nutrient uptake and environmental factors such as temperature and CO_2 concentration. A number of studies have been conducted in recent years to manipulate environmental factors to optimize medicinal components while growing under controlled environments.

CO_2 CONCENTRATION AND PHOTOSYNTHETIC RATES

Among the environmental factors, CO_2 concentration is one of the most important affecting medicinal components in plants under controlled environments. A number of studies have been conducted on *Hypericum perforatum* plants growing under a controlled environment with CO_2 enrichment; those studies showed that CO_2 can enhance biomass and secondary metabolite production by increasing net photosynthetic rates. For example, Mosaleeyanona et al. (2005) showed that medicinal metabolite contents (mg/plant) in the leaf tissues were significantly higher in plants grown under controlled environments than those in the field (control). They found that hypericin and pseudohypericin contents in plants grown under controlled environments were 30 and 41 times greater, respectively, than those of the field-grown plants. They also showed that hypericin and pseudohypericin contents in the leaf tissues increased with increasing photosynthetic rate under elevated levels of CO_2. In another study, Zobayed and Saxena (2004) reported that growing *H. perforatum* plants under a controlled environment with CO_2 enrichment increased the hypericin, pseudohypericin, and hyperformine concentrations. A positive relationship between photosynthetic rate and medicinal composition of *H. perforatum* has also been described (Zobayed et al., 2006). These studies showed that in plants grown under elevated CO_2 concentration, the leaf photosynthetic rates, and the contents of medicinal components increased. In *Scutellaria baicalensis*, a CO_2 enrichment system showed higher concentrations of wogonin, baicalein, and baicalin in plant tissues compared with a non-enriched system (Zobayed et al., 2004).

TEMPERATURE STRESS

Temperature stress is known to cause many physiological, biochemical, and molecular changes in plant metabolism and alter the production of secondary metabolites in plants. Elevated temperatures reduce stomatal conductance and thus reduce photosynthesis and growth of many plant species (Berry and Bjorkman, 1980). The photochemical efficiency of photosystem II (PSII) also decreases at elevated temperatures, indicating increased stress (Gamon and Pearcy, 1989; Maxwell and Johnson, 2000). When plants are stressed, secondary metabolite production may increase because growth is often inhibited more than photosynthesis, and the carbon that is fixed but not allocated to growth is instead allocated to secondary metabolites (Mooney et al., 1991). Several studies have examined the effects of increased temperatures on the production of secondary metabolites in plants, but most of these studies yielded contradictory results. Some report that secondary metabolites increase in response to elevated temperatures (Litvak et al., 2002), while others report that they decrease (Snow et al., 2003). In the case of *H. perforatum*, plants grown under controlled environments for 70 days followed by low-temperature (15°C) treatment for 15 days showed a reduction of photosynthetic efficiency and the maximum quantum efficiency of PSII photochemistry of the dark-adapted leaves (ϕ^P_{max}). High-temperature (35°C) treatment decreased net photosynthesis and maximum quantum efficiency of PSII as well, but increased the leaf total peroxidase activity and also increased the hypericin, pseudohypericin, and hyperforin concentrations in the shoot tissues (Zobayed et al., 2005b). These results provide the first indication that temperature is an important environmental factor to optimize secondary metabolite production in *H. perforatum* and controlled environment technology can allow the precise application of such specific stresses.

WATER STRESS

Water stress is known to increase the concentrations of secondary metabolites in plant tissues and a severe water stress condition may cause oxidative stress due to the formation of reactive oxygen species and photoinhibitory damage. A detailed study was conducted by Zobayed et al. (2007) to evaluate the changes in physiological status, especially the photosynthetic efficiency and the medicinal component profile of the leaf tissues of *H. perforatum* plants exposed to water stress. In the leaf tissues of plants grown under a water stress condition, both hypericin and pseudohypericin concentrations reduced with time and the concentration was significantly lower than that of the control; in contrast, the hyperforin concentration increased significantly. The net photosynthetic rates of the leaves of plants grown under a water stress condition were significantly lower than those of the control. The maximum quantum efficiency of PSII photochemistry (ϕ^P_{max}) of the dark-adapted leaves was similar for both wilted and recovered plants although these values were significantly lower than those of the control. The leaf tissues of the plants under a water stress condition had a significantly higher capacity to detoxify oxygen radicals with an approximately 2.5-fold increase over the antioxidant potential of the leaves of nontreated (control) or recovered plants. In another study, Abreu and Mazzafera (2005) showed that water stress increased the levels of medicinal compounds such as betulinic acid, quercetin, rutin, 1,5-dihydroxyxanthone, and isouliginosin *B* in *Hypericum brasiliense*, an important medicinal plant. They concluded that there is a reallocation of carbon, with water-stressed plants showing a reduction in growth while the levels of the compounds increased.

SPECTRAL QUALITY AND UV RADIATION

Glycyrrhizin, the major bioactive component of *Glycyrrhiza uralensis*, is widely used as a natural sweetener. In a recent study, Afreen et al. (2006) evaluated the effects of different spectral qualities of light including red, blue, white, and UV-B radiation on the production of glycyrrhizin in a controlled environment. Plants were grown under artificial lights with elevated CO_2 concentration and assigned to red, blue, and white light treatments. The concentrations of glycyrrhizin quantified in the root tissues were highest in the plants grown under red light compared with white or blue light. In plants exposed to UV-B radiation (3 days at 1.13 W m^{-2}; wavelength: 280–315 nm) while growing under a controlled environment, the concentration of glycyrrhizin increased significantly in the root tissues of 3-month-old plants. These results indicate another advantage of growing plants in a controlled environment, i.e. a high concentration of secondary metabolite production can be achieved within a short production period. Exposure of plants to UV radiation results in multiple responses including increased synthesis of UV-screening pigments, as well as reinforcement of the antioxidant system and other defense mechanisms. Many plants are capable of avoiding UV radiation by accumulating UV-filtering flavonoids and other secondary metabolites; for instance, epidermal flavonoids are enhanced in response to increased UV-B (Bornman et al., 1997). The synthesis of the basic skeletons for active secondary metabolites is dependent on the carbon assimilated during photosynthesis and the rate of photosynthesis is known to be influenced by light intensity, spectrum of the available light, duration of light, etc. Plant responses to red, blue, or UV radiation induce different expression via signal transduction pathways using signal transducing photoreceptors, which are a switch for controlling the expression of specific genes involved with growth, developmental processes, and secondary metabolism in plants (Jenkins et al., 1995). Cosgrove (1981) mentioned that blue light reduces cell expansion and thus inhibits leaf growth and stem elongation. This growth inhibition is mediated by blue light photoreceptors and is considered distinct from inhibitory phytochrome effects.

Melatonin (*N*-acetyl-5-methoxytryptamine) is known to be synthesized and secreted by the pineal gland in vertebrates. Evidence for the occurrence of melatonin in the roots of *G. uralensis* plants and the response of this plant to the spectral quality of light including red, blue, and white light (control) and UV-B radiation (280–315 nm) under controlled environment systems for the synthesis of melatonin

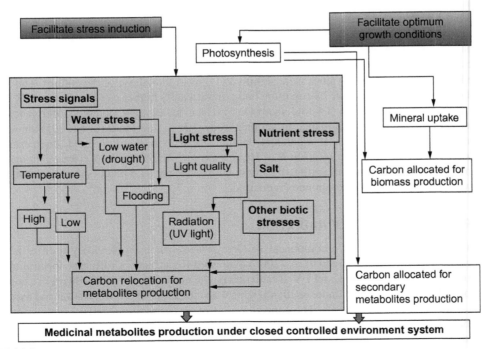

FIGURE 14.1

Pathways facilitating the production of medicinal metabolites under a closed controlled environment system.

were investigated in a recent study (Afreen et al., 2007). Melatonin concentrations were highest in plants exposed to red light and varied depending on the wavelength of the light spectrum in the following order: red ≥ blue ≥ white light. Interestingly, in a more mature plant (6 months) melatonin concentration was increased considerably. The concentration of melatonin quantified in the root tissues was highest in the plants exposed to high-intensity UV-B radiation.

CONCLUSION

The production of medicinal plants under controlled environments described in this article represents significant progress for phytopharmaceutical production. The production of medicinal plants in a closed controlled environment system has several advantages: (a) plants can be grown in sterile, standardized conditions and are free from biotic and abiotic contaminations, (b) uniform plant growth with consistent plant material can be achieved, and (c) consistent biochemical profiles can be achieved with uniform production of metabolites. Closed systems can ensure optimum environmental conditions to maximize biomass production (Figure 14.1) and facilitate the induction of stresses, thus inducing the reallocation of carbon to the production of medicinal metabolites (Figure 14.1). The adaptation of the growing system described here may facilitate the development of safe, consistent, and high-quality phytopharmaceutical products of medicinal plants.

REFERENCES

Abreu, I.N., Mazzafera, P., 2005. Effect of water and temperature stress on the content of active constituents of *Hypericum brasiliense* Choisy. Plant Physiol. Biochem. 43, 241–248.

Afreen, F., Zobayed, S., Kozai, T., 2006. Spectral quality and UV-B stress stimulate glycyrrhizin concentration of *Glycyrrhiza uralensis*. Plant Physiol. Biochem. 43, 1074–1081.

Afreen, F., Zobayed, S., Kozai, T., 2007. Occurrence of melatonin, a neuro-hormone in the underground part of *Glycyrrhiza uralensis*. J. Pineal Res. 41, 108–115.

Berry, J., Bjorkman, O., 1980. Photosynthetic response and adaptation to temperature in higher plants. Annu. Rev. Plant Physiol. 31, 491–543.

Bornman, J.F., Reuber, S., Cen, Y.P., Weissenböck, G., 1997. Ultraviolet radiation as a stress factor and the role of protective pigments. In: Lusden, P. (Ed.), Plants and UV-B. Responses to Environmental Change. Society for Experimental Biology Seminar Series 64. Cambridge University Press, Cambridge, pp. 157–168.

Cosgrove, D.J., 1981. Rapid suppression of growth by blue light. Plant Physiol. 67, 584–590.

Gamon, J.A., Pearcy, R.W., 1989. Leaf movement, stress avoidance and photosynthesis in *Vitis californica*. Oecologia 79, 475–481.

Herbal supplements and remedies: a global strategic business report. Global Industry Analysts Inc., 2014 (http://www.strategyr.com/Herbal_Supplements_and_Remedies_Market_Report.asp; visited 7th December, 2014).

Jenkins, G.I., Christie, J.M., Fuglevanc, G., Long, J.C., Jackson, J.A., 1995. Plant response to UV and blue light: biochemical and genetic approaches. Plant Sci. 112, 117–138.

Litvak, M.E., Constable, J.V.H., Monson, R.K., 2002. Supply and demand processes as controls over needle monoterpene synthesis and concentration in Douglas fir [*Pseudotsuga menziesii* (Mirb.) Franco]. Oecologia 132, 382–391.

Maxwell, K., Johnson, G.N., 2000. Chlorophyll fluorescence: a practical guide. J. Exp. Bot. 51, 659–668.

Mooney, H.A., Winner, W.E., Pell, E.J., 1991. Response of plants to multiple stresses. Academic Press, San Diego, California, USA.

Mosaleeyanona, K., Zobayed, S., Afreen, F., Kozai, T., 2005. Relationships between net photosynthetic rate and secondary metabolite contents in St. John's wort. Plant Sci. 169, 523–531.

Snow, M.D., Bard, R.R., Olszyk, D.M., Minster, L.M., Hager, A.N., Tingey, D.T., 2003. Monoterpene levels in needles of Douglas fir exposed to elevated CO_2 and temperature. Physiol. Plant. 117, 352–358.

Williams, V.L., 1996. The Witwatersrand multi trade. Veld Flora 82, 12–14.

World Health Organization, Fifty-sixth World Health Assembly, Provisional agenda item 14.10; Traditional medicine, Report by the Secretariat, 2003.

Zobayed, S., Saxena, P.K., 2004. Production of St. John's wort plants under controlled environment for maximizing biomass and secondary metabolites. In Vitro Cell. Dev. Biol. 40, 108–114.

Zobayed, S., Murch, S.J., Rupasinghe, H.P.V., de Boer, J.G., Glickman, B.W., Saxena, P.K., 2004. Optimized system for biomass production, chemical characterization and evaluation of chemo-preventive properties of *Scutellaria baicalensis* Georgi. Plant Sci. 167, 439–446.

Zobayed, S., Afreen, F., Kozai, T., 2005a. Necessity and production of medicinal plants under controlled environments. Environ. Control Biol. 43, 243–252.

Zobayed, S., Afreen, F., Kozai, T., 2005b. Temperature stress can alter the photosynthetic efficiency and secondary metabolite concentrations in St. John's wort. Plant Physiol. Biochem. 43, 977–984.

Zobayed, S., Afreen, F., Goto, E., Kozai, T., 2006. Plant-environment interactions: accumulation of hypericin in dark glands of *Hypericum perforatum*. Ann. Bot. 98, 793–804.

Zobayed, S., Afreen, F., Kozai, T., 2007. Phytochemical and physiological changes in the leaves of St. John's wort plants under a water stress condition. Environ. Exp. Bot. 59, 109–116.

PRODUCTION OF PHARMACEUTICALS IN A SPECIALLY DESIGNED PLANT FACTORY

Eiji Goto

Graduate School of Horticulture, Chiba University, Matsudo, Chiba, Japan

INTRODUCTION

Plant factories with artificial lighting (PFALs), where leafy vegetables are cultivated until harvest in closed plant production systems with artificial lighting (CPPSs), were proposed, developed, and implemented in Japan during the 1980s. During the 1990s, the products from these factories received high evaluations by the food service industry, to which they primarily catered. During the 2000s, commercial production of nursery plants of fruits and vegetables, flowers and ornamental plants was initiated in CPPSs or PFALs (see Chapter 19). PFALs also provide good cultivation systems for the production of plants for pharmaceutical use.

Recently, genetically modified (GM) plants have attracted much attention because of their use in the production of valuable materials, such as pharmaceutical proteins, for humans or livestock. Pharmaceutical products that have been or are being developed include oral vaccines for humans or livestock, agents that prevent lifestyle-related diseases, substances with antimicrobial or anti-inflammatory effects, and functional proteins and secondary metabolites.

The production of pharmaceutical materials by using GM plants (plant-made pharmaceuticals, PMPs) has advantages over conventional production methods using animals or microorganisms (Daniell et al., 2001; Ma et al., 2003), such as (1) low risk of contracting infectious diseases from animals or microorganisms used, (2) elimination of the need to maintain a cold chain, (3) low production cost, (4) ease of large-scale production, and (5) utilization as functional food.

Food crops can be utilized for the production of valuable materials because their edible components have a low risk of transmitting infectious diseases. In Japan, research groups have introduced functional protein-encoding genes into rice, potato, soybean, lettuce, tomato, and strawberry plants (Table 15.1) (for oral vaccine see Nochi et al. (2007); for miraculin see Sugaya et al. (2008) and Kim et al. (2010)). Expression of the inserted DNA and accumulation of the required protein in the target plant organ are significantly influenced by plant and climatic conditions (Stevens et al., 2000). Therefore, a highly controlled environmental system is required to achieve an annual production of a valuable material with consistent quality.

Plant Factory. http://dx.doi.org/10.1016/B978-0-12-801775-3.00015-9

Table 15.1 Pharmaceutical Materials Developed or Being Developed in Japan

Type	Target	Material	Host Plant
Oral vaccine	Human	Vaccine for cholera	Rice
	Human	Influenza vaccine	Rice
	Livestock	Vaccine for swine edema disease	Lettuce
	Livestock	Bird influenza vaccine	Potato
Pharmaceutical material	Human	Human thioredoxin	Lettuce
	Pet animal	Canine interferon for periodontal disease	Strawberry
Functional protein	Human	Lactoferrin	Strawberry
	Human	Miraculin	Tomato

The materials were selected from research projects subsidized by the METI as "Development of fundamental technologies for production of high-value materials using transgenic plants" (2006-2010).

A CPPS is ideal for the production of medicinal substances that have been approved by pharmaceutical law. Advantages of the CPPS over open fields include: (1) stable plant production under complete environmental control, (2) efficient use of water and CO_2 gas, (3) fulfillment of the Good Manufacturing Practice (GMP) specifications, and (4) low risk of gene diffusion. Because responses of GM plants to environmental conditions have not been investigated, it is necessary to research the physiological characteristics of the plants and establish suitable cultivation methods for each plant under artificial conditions.

The author's research group developed a CPPS for stable accumulation of high concentrations of desired functional proteins in the edible components of strawberry and rice plants. The GM strawberry plant accumulates a functional protein in the achenes and the fruit, which is known to prevent lifestyle-related diseases. The GM rice plant accumulates an oral vaccine in the protein body of the seed. This chapter summarizes the production system and appropriate environmental conditions for plant production of valuable materials in large quantities.

CANDIDATE CROPS FOR PMPs

Considering the selection of plant species, established genetic engineering technology for the plants is primarily important, and the mechanism by which a plant accumulates the target material should also be investigated. For commercialization of a large-scale production system, reduction in labor in terms of cultivation management and easy installation of mechanization and automation are also of importance.

We have tested a variety of food crops in the CPPS and calculated the production efficiency of photoassimilates per unit light energy for rice, strawberry, lettuce, and tomato from our experimental data (Table 15.2). Leafy vegetables and fruits are cultivated in commercial PFALs and greenhouses. However, from the viewpoint of productivity, the production efficiency of photoassimilates per unit light energy does not differ significantly among food crops. Therefore, besides vegetables, crops such as cereals, pulses, and potatoes are candidate crops. For example, cereal and pulse crops are suitable for the production of valuable materials derived from carbohydrates and proteins, and cereal crops and

Table 15.2 Comparison of Candidate Crops for PMPs Based on Experimental Data

		Rice	Strawberry	Lettuce	Tomato
PPF	μmol m^{-2} s^{-1}	800	300	200	600
Light period	h day^{-1}	12	16	16	12
Daily PPF (DLI)	mol m^{-2} day^{-1}	34.6	17.3	11.5	25.9
Integrated PPF	mol m^{-2} year^{-1}	12,440	6220	4150	9330
Cultivation period	Days	100	90	30	90
Edible biomass per cultivation[a]	gFW m^{-2}	800	3200	2500	7500
Harvest index		0.5	0.5	0.9	0.5
Dry matter ratio	%	85	10	5	6
Total dry biomass per cultivation	gDW m^{-2}	1360	640	139	900
Number of cultivations		3.6	4	12	4
Total dry biomass per year	gDW m^{-2} year^{-1}	4896	2560	1667	3600
Production efficiency of photoassimilates per unit light energy	g mol^{-1}	0.39	0.41	0.40	0.39

[a]*Highest production was calculated for optimal cultivation management at a light condition based on our experimental data.*

potatoes can be stored for long periods at normal temperatures. Leafy vegetables have a high harvest index and a short cultivation period, and we do not need to focus on the diffusion of pollen because they are harvested before flowering. Fruits originally have functional compounds such as polyphenols; therefore, GM fruit plants may have multiple positive effects if plant modification is designed to accumulate a valuable material that can be used for the prevention of lifestyle diseases.

CONSTRUCTION OF GM PLANT FACTORIES

We constructed two CPPSs with artificial lighting for stable accumulation of high concentrations of desired functional proteins in the edible components of strawberry and rice plants. Figure 15.1 shows the facility design and air and water flow of a closed rice production system. The production system is an advanced model of a commercial PFAL using only artificial light. However, the design is unique from the viewpoint of transgenic crop management. The system must have facilities to prevent diffusion of plant parts, such as pollen, and to inactivate the unused parts of the crop. If the system specifications meet the standards, an ideal environment for GM plant production can be provided.

With regard to environmental control, spatial uniformity of light, temperature, humidity, and gas in a plant canopy is crucial to the production of a material bearing consistent quality. The CPPS can be used to manipulate not only the aerial conditions but also the root zone conditions. By using hydroponics, the nutrient solution can be optimized for each growth stage. Therefore, the system can provide the plant with a variety of artificial stimuli and physiological stresses to maximize accumulation of the target material. Because the system provides appropriate growth conditions and an agrochemical-free environment by blocking pest invasion, agronomic plant traits to overcome natural environment-related problems are not required.

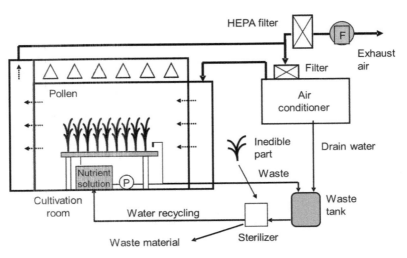

FIGURE 15.1

Facility design and air and water flow in the typical cultivation room of a CPPS.

We developed a prototype of a strawberry cultivation room (20 m²) equipped with multilayer shelves with white fluorescent lamps (FHF32-EX-N-H; Panasonic Electric Works Co., Ltd.) and a commercial NFT hydroponic system for GM strawberry production (Figure 15.2). We used non-GM everbearing strawberry (*Fragaria × ananassa* Duch. cv. "HS138") plants for the cultivation experiment. The seedlings were derived from the micropropagation of stem meristems (Hikosaka et al., 2009; Miyazawa et al., 2009) and were then cultivated in the room. Air temperature, relative humidity, and CO_2 concentration in the room could be regulated to suitable levels. In the flowering stage, pollination was carried out manually every 2–3 days.

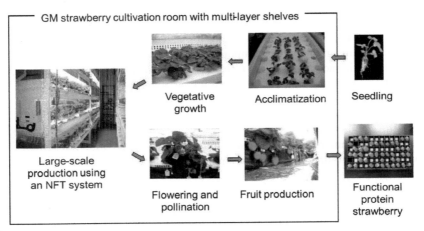

FIGURE 15.2

Cultivation of GM strawberry plants in a CPPS.

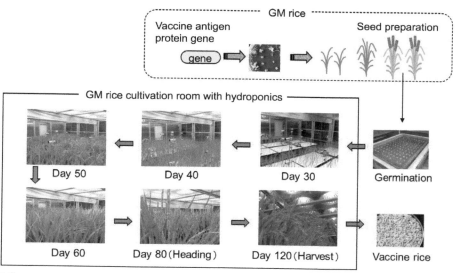

FIGURE 15.3

Cultivation of GM rice plants in a CPPS. The day is an example of a standard cultivation condition.

We developed a prototype of a rice cultivation room with a lighting system by using a high-intensity light source for GM rice production (Figure 15.3). This prototype has a hydroponic cultivation area of about 15 m^2, a height of about 2 m, and is equipped with 56 ceramic metal halide lamps (M400FCEH-W/BUD/H0; Iwasaki Electric Co., Ltd.). The maximum light intensity is about 1200 µmol m^{-2} s^{-1} of photosynthetic photon flux (PPF) at 1 m above the floor surface. It is possible to vary the light intensity to the required level within a range of 65–100% by using a light control system. Furthermore, the thermal conditions of the cultivation area can be controlled by separating the lamp and cultivation spaces with glass and circulating air exhausted from the cultivation space to the lamp space. Non-GM rice (*Oryza sativa* L., Nipponbare) plants were used for material plants in the cultivation experiment.

OPTIMIZATION OF ENVIRONMENT CONDITIONS FOR PLANT GROWTH

STRAWBERRY

We adopted the same approach as that of the rice plants to establish optimal light conditions for the strawberry plants in the CPPS. According to the results of the growth experiment using various air temperatures and CO_2 concentrations, we set the average daily temperature at 23°C and the CO_2 concentration at 1000 µmol mol^{-1}. Figure 15.4 shows fruit yield of an individual plant for 35 days from when the first flower was fertilized. A DLI (daily light integral) of 19.4 mol m^{-2} day^{-1} produced 380 g of fruits. With three cultivations per year and a planting density of 5 plants m^{-2}, the annual yield would be 42% greater than the average yield of greenhouses. The results indicate that suitable cultivation conditions in the CPPS enhance photosynthesis of strawberry plants and subsequently facilitate maximum fruit production.

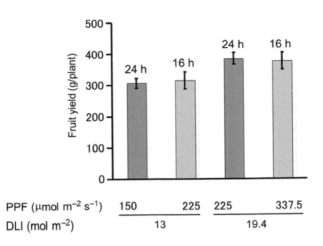

FIGURE 15.4

Fruit yield of strawberry plants grown in the CPPS under different PPF, light period, and DLI conditions.

We studied the effects of light quality on plant growth and concentration of human adiponectin (hAdi) in transgenic everbearing strawberry (*Fragaria × ananassa* Duch. "HS 138") (Hikosaka et al., 2013). The hAdi plants in the flowering stage were exposed to three different light qualities (white, blue, and red) for a 16-h light period under fluorescent lamps until the fruits were harvested. The hAdi concentration on a fresh-weight basis in plants grown under red light condition was significantly highest among all treatments.

TOMATO

Kato et al. (2010) created a new variety of dwarf tomato suitable for high productivity of fruits that contain the target pharmaceutical material in a closed cultivation system under stable environmental conditions. We investigated the effects of light intensity on tomato fruit yield and the accumulation of recombinant miraculin, a taste-modifying glycoprotein, in transgenic tomato fruits (Kato et al. 2011). Plants were cultivated at various PPFs of 100–400 $\mu mol\ m^{-2}\ s^{-1}$ under white fluorescent lamps in the CPPS. Miraculin production per unit of energy used was the highest at 100 $\mu mol\ m^{-2}\ s^{-1}$; however, miraculin production per unit area was the highest at 300 $\mu mol\ m^{-2}\ s^{-1}$. The commercial productivity of recombinant miraculin in transgenic tomato fruits largely depended on light conditions in the PFAL.

RICE

To determine the optimal light conditions for rice plant production, we analyzed light conditions of the conventional rice cultivation method in paddy fields and established a new lighting strategy. In Hikone City, Shiga Prefecture, located in the center of Japan's main island, the DLI exceeds 65 $mol\ m^{-2}\ day^{-1}$ on a clear and sunny day in mid-June. However, the average DLI in paddy fields for approximately 5 months from transplanting to harvest is 43 $mol\ m^{-2}\ day^{-1}$ (Figure 15.5). In the CPPS, it is not necessary to recreate a typical sunlight curve of 43 $mol\ m^{-2}\ day^{-1}$ with a maximum intensity of 1400 $\mu mol\ m^{-2}\ s^{-1}$ because initial costs of lamps and ballasts are high and air temperature and humidity control becomes relatively complicated with large changes in light intensity. Consequently, we set light intensity at

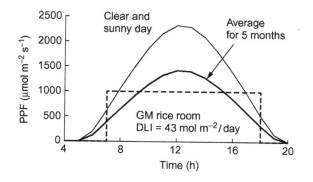

FIGURE 15.5

Light conditions of conventional cultivation of rice in paddy fields and a lighting strategy specialized for a CPPS.

1000 μmol m^{-2} s^{-1} PPF and light period at 12 h day^{-1} so the DLI reached 43 mol m^{-2} day^{-1}. The air temperature in the light period was controlled at 27°C to maximize both photosynthetic and growth rates. The air temperature in the dark period was set at 23°C, so the average daily air temperature was 25°C. The relative humidity was set at 70%, and the CO_2 concentration was set at 400 μmol mol^{-1}.

Rice plants were cultivated under various planting densities and DLI conditions. The results showed that the yield of rice grown in the room was 40–60% greater than the average yield of the cultivar grown in paddy fields in Japan (Figure 15.6). The main reason was probably that there were no cloudy or rainy days in the CPPS so that the plants photosynthesized continuously under optimal conditions of light

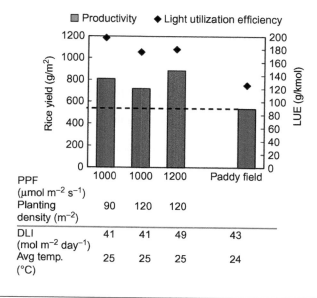

FIGURE 15.6

Comparison of productivity and light utilization efficiency of rice plants grown in the CPPS and in paddy fields.

intensity and air temperature. This was supported by higher light utilization efficiencies of the three treatments tested in the room. Further, the efficiency at 1000 μmol m^{-2} s^{-1} PPF was higher at a planting density of 90 plants m^{-2} than at a density of 120 plants m^{-2}. This means that an optimal planting density for rice production exists under certain environmental conditions, as shown in paddy fields.

CONCLUDING REMARKS

Manipulation of aerial environmental factors such as light and air temperature is a key technique to promote accumulation of desired functional proteins. For maximally utilizing the CPPS for accumulating a valuable material, plants that can be grown under continuous lighting and have the potential to induce the target material in response to aerial environmental stresses are very promising. Plant production using this technology is not a part of agriculture; it is a new plant-based industry that combines molecular biology and environmental engineering.

REFERENCES

Daniell, H., Streatfield, S.J., Wycoffc, K., 2001. Medical molecular farming: production of antibodies, biopharmaceuticals and edible vaccines in plants. Trends Plant Sci. 6, 219–226.

Hikosaka, S., Sasaki, K., Aoki, T., Goto, E., 2009. Effects of in vitro culture methods during the rooting stage and light quality during the seedling stage on the growth of hydroponic everbearing strawberries. Acta Horticult. 842, 1011–1014.

Hikosaka, S., Yoshida, H., Goto, E., Tabayashi, N., Matsumura, T., 2013. Effects of light quality on the concentration of human adiponectin in transgenic everbearing strawberry. Environ. Control Biol. 51, 31–33.

Kato, K., Yoshida, R., Kikuzaki, A., Hirai, T., Kuroda, H., Hiwasa-Tanase, K., Takane, K., Ezura, H., Mizoguchi, T., 2010. Molecular breeding of tomato lines for mass production of miraculin in a plant factory. J. Agric. Food Chem. 58, 9505–9510.

Kato, K., Maruyama, S., Hirai, T., Hiwasa-Tanase, K., Mizoguchi, T., Goto, E., Ezura, H., 2011. A trial of production of the plant-derived high-value protein in a plant factory: photosynthetic photon fluxes affect the accumulation of recombinant miraculin in transgenic tomato fruits. Plant Signal. Behav. 6, 1172–1179.

Kim, Y.-W., Kato, K., Hirai, T., Hiwasa-Tanase, K., Ezura, H., 2010. Spatial and developmental profiling of miraculin accumulation in transgenic tomato fruits expressing the miraculin gene constitutively. J. Agric. Food Chem. 58, 282–286.

Ma, J.K.-C., Drake, P.M.W., Christou, P., 2003. Genetic modification: the production of recombinant pharmaceutical proteins in plants. Nat. Rev. Genet. 4, 794–805.

Miyazawa, Y., Hikosaka, S., Aoki, T., Goto, E., 2009. Effects of light conditions and air temperature on the growth of everbearing strawberry during the vegetative stage. Acta Horticult. 842, 817–820.

Nochi, T., Takagi, H., Yuki, Y., Yang, L., Masumura, T., Mejima, M., Nakanishi, U., Matsumura, A., Uozumi, A., Hiro, T., Morita, S., Tanaka, K., Takaiwa, F., Kiyono, H., 2007. Rice-based mucosal vaccine as a global strategy for cold-chain- and needle-free vaccination. PNAS 104, 10986–10991.

Stevens, L.H., Stoopen, G.M., Elbers, I.J.W., Molthoff, J.W., Bakker, H.A.C., Lommen, A., Bosch, D.W.J., 2000. Effect of climate conditions and plant developmental stage on the stability of antibodies expressed in transgenic tobacco. Plant Physiol. 124, 173–182.

Sugaya, T., Yano, M., Sun, H.-J., Hirai, T., Ezura, H., 2008. Transgenic strawberry expressing the taste-modifying protein miraculin. Plant Biotechnol. 25, 329–333.

SYSTEM DESIGN, CONSTRUCTION, CULTIVATION AND MANAGEMENT

PLANT PRODUCTION PROCESS, FLOOR PLAN, AND LAYOUT OF PFAL

16

Toyoki Kozai

Japan Plant Factory Association, c/o Center for Environment, Health and Fields Sciences, Chiba University, Kashiwa, Chiba, Japan

INTRODUCTION

The floor plan of a plant factory with artificial lighting (PFAL) and layouts of equipment and culture (or cultivation) beds are designed to achieve efficient operations by workers and smooth flows of materials such as the plants and supplies in the plant production system. In addition, it is essential that the PFAL be designed and operated to maintain a high level of sanitation for food safety.

However, PFALs are relatively new production systems, so the optimum production technologies have not yet been established (Schueller, 2014) and there is much room for improvement in the plant production process. This chapter describes the state of the art regarding the plant production process, the floor plan of a PFAL, and sanitation issues. Sanitation is discussed in more detail in Chapter 21 from a different viewpoint. The procedures of seeding, seedling production, and transplanting are described in detail in Chapter 18. Data collection, analyses, visualization, diagnosis, and improvement methods for PFALs are described in Chapter 22.

MOTION ECONOMY AND PDCA CYCLE

To continuously improve operations in the plant production process, it is often useful to introduce the principles of motion economy and the PDCA (plan-do-check-act) cycle.

PRINCIPLES OF MOTION ECONOMY

For an existing PFAL, the manual operations in the production process need to be improved weekly or monthly to eliminate "muri" (overburden), "mura" (inconsistency), and "muda" (waste). This improvement process was invented and is being used by Toyota Motor Corporation, and is called "kaizen" (Kanawaty, 1992; Meyers and Stewart, 2002). In kaizen, motion economy and the PDCA cycle are an important concept and methodology.

The principles of motion economy form a set of rules and suggestions to improve the manual work in manufacturing and reduce fatigue and unnecessary movements by the worker, which can lead to a reduction in work-related trauma (http://en.wikipedia.org/wiki/Principles_of_motion_economy). The

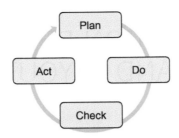

FIGURE 16.1

Scheme showing the PDCA (plan-do-check-act) cycle.

principles of motion economy can be classified into three groups: (1) principles related to the use of the *human body*, (2) principles related to the arrangement of the *workplace*, and (3) principles related to the *design of tools and equipment*. These three principles are further divided into five to ten steps (Wikipedia).

PDCA CYCLE

The PDCA cycle is a four-step model for carrying out change in any action process (Figure 16.1). The PDCA cycle is repeated to continuously improve the process. In Step 1 (plan), an opportunity is recognized and a change is planned. In Step 2 (do), the change is tested and a small-scale study is carried out. In Step 3 (check), the test is reviewed; the results are analyzed and what was learned is identified. In Step 4 (act), action is taken based on what was learned in Step 3. If the change did not work, the PDCA cycle is repeated with a different plan. If the change was successful, what was learned from the test is incorporated into wider changes and used to plan new improvements, thus beginning the PDCA cycle again.

PLANT PRODUCTION PROCESS

A typical flow of operations in the plant production process focusing on the movement of plants is shown in Figure 16.2. Panel, floor, and culture bed cleaning, etc. are not shown. Each step of the process is further broken down. Figure 16.3 shows an example of the breakdown of Step 1 (seeding for germination).

Currently, most handling operations in Figure 16.2 are conducted manually if the daily production capacity of leaf greens is about 5000 heads or less. If production capacity is over 10,000 heads/day, the following handling operations are semiautomated or automated in most cases: seeding, uploading, unloading, and transporting of culture panels to and from the tiers (see Chapter 23), weighing of leaf greens, packing the harvested produce in plastic bags, labeling the plastic bags containing leaf greens, packing the plastic bags into container boxes, labeling the container boxes, and wrapping the boxes before cooling. In the near future, the transplanting and transporting of boxes in the PFAL will be automated in large-scale PFALs. Operations that are difficult to fully automate include trimming damaged leaves and packing plastic bags into boxes.

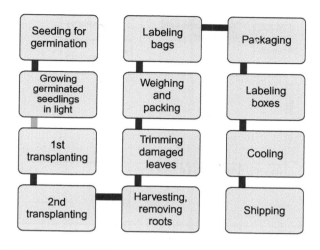

FIGURE 16.2

A typical flow of operations in the plant production process focusing on the movements of plants. Panel, floor, and culture bed cleaning, etc. are not shown.

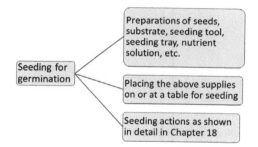

FIGURE 16.3

Breakdown of the first step (seeding for germination) of the plant production process shown in Figure 16.2.

On the other hand, most manual operations in a PFAL are relatively simple, safe, and light work under a comfortable working environment (air temperature of 20–22°C), and so PFALs can create manual job opportunities suited to elderly and handicapped persons.

LAYOUT

FLOOR PLAN

A rough floor plan of a PFAL is shown in Figure 16.4. The PFAL consists mainly of an operation room and a culture room. The outer units of air conditioners are installed close to the walls outside (Figure 16.5) and/or on the rooftop. Exposing an outer unit to direct solar light raises the temperature of its heat exchanger and increases the electricity cost for cooling, and thus should be avoided by

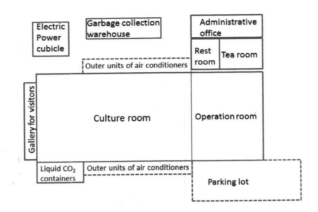

FIGURE 16.4

Rough floor plan of a PFAL.

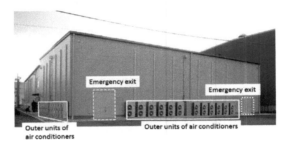

FIGURE 16.5

Outer units of air conditioners installed close to the building walls and emergency exits. The outer units can also be installed on the rooftop.

(photo courtesy of Mirai Co., Ltd.)

shading. The distance between the inner and outer units of air conditioners should be minimized to achieve a high coefficient of performance (COP) for cooling or electric energy efficiency of air conditioners for cooling. Accumulation of snow at the air inlet and outlet of outer units must be strictly avoided. When a PFAL is built in a cold region, air conditioners with special specifications must be chosen, which provide a high COP for cooling when the outside air temperature is 0°C or lower.

The position of the emergency exit is shown in Figure 16.5. The door of the emergency exit needs to be thermally insulated to avoid water condensation on its inner surface and the surroundings in cool and cold seasons. Invasion by small insects and dust through air gaps around the door must be strictly prevented. The warehouse for garbage collection is placed at least 20 m away from the culture and operation rooms to discourage the movement of insects and small animals between the rooms and the warehouse.

The rest room, tea room, and administrative office are physically separated from the culture and operation rooms to maintain a high level of sanitation of these rooms, even though this is slightly inconvenient for workers. Since visitors cannot enter the operation and culture rooms, a corridor with glass windows through which they can look into the culture room may be provided.

OPERATION ROOM

Figure 16.6 shows a typical floor plan of the operation room. The installation of a hot water shower or air shower depends on the risk management policy of the PFAL operator. Hot water showers will lower the risk of biological contamination of the culture room, but take much longer time to use than an air shower. Those intending to enter the operation room need to take either a hot water shower or an air shower before putting on clean clothing. Two separate hot water/air showers and changing rooms are provided for use by women and men, respectively. Those intending to enter the culture room through the operation room need to put on clean (disinfected) overalls, a head cap with a mask, a pair of gloves, and boots (Figure 16.7), and then take an air shower again for 15–30 seconds before entering the culture room.

Supplies to be used in the culture room are transported through the pass box from the operation box. The air shower room and pass box in the operation room have doors at both sides to prevent the exchange of air between the operation and culture rooms. There is a glass window between the culture room and operation room, and video cameras can also be installed in both the operation and culture rooms. Work tables are movable. Supplies transported into the culture room through the operation room are sterilized in the UV sterilization box in the operation room.

Both the culture and operation rooms are clean, but the air in the culture room is cleaner than that in the operation room in terms of the population density of microorganisms (bacteria, fungi, and viruses). Thus, weighing, trimming of plants, packing of plants into plastic bags, and sealing of the plastic bags are conducted in the culture room, not in the operation room. After the plastic bags, have been sealed, they are packed into the container boxes in the operation room, so the harvested produce packed and sealed in the culture room is cleaner (CFU/g is lower) than that packed and sealed in the operation room. However, it should be noted that the population density of microorganisms in the culture beds can become very high if the beds are not sterilized often.

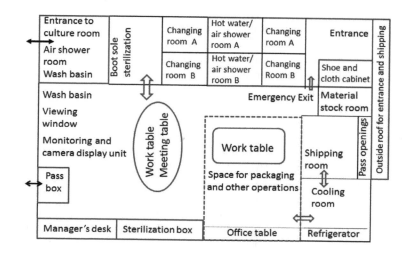

FIGURE 16.6

A typical floor layout of the operation room at a PFAL with a daily production capacity of 5000–10,000 lettuce heads. The culture room is located on the left-hand side. Floor areas of rooms and spaces are not shown to scale.

FIGURE 16.7

A worker wearing clean overalls. The cap, mask, and gloves are mostly disposable. The overalls and boots are usually used again after washing and sterilizing.

(photo courtesy of Mirai Co., Ltd.)

CULTURE ROOM

The following steps are taken by workers entering the culture room: (1) enter the changing/locker room in the operation room, (2) take off all clothing, (3) take a hot water shower or air shower for the whole body including hair/head washing, (4) put on clean underwear, overalls, cap, mask, and gloves, (5) wash the hands while wearing the gloves using a disinfectant for sterilization, (6) put on clean boots and sterilize the boot soles, (7) take an air shower again, and (8) enter the culture room. This procedure is similar to that for entering the micropropagation room for pathogen-free plants (Chun and Kozai, 2000).

Figure 16.8 shows a typical layout of tiers with culture beds and equipment in the culture room. The culture area ratios of seeding, seedling production, and cultivation are roughly 1:12:50. Each shelf of the tiers is 1–2 m wide with an average of about 1.5 m. The tier length depends on the size of the culture room. The vertical distance between tiers is 30–100 cm with an average of around 50 cm. The walkway is 1 m wide on average but it largely depends on the method of transplanting and harvesting: (1) transplanting and harvesting are conducted at both ends of tiers and the walkway is used for maintenance, cleaning of culture beds, etc. only, or (2) transplanting and harvesting are conducted along the walkway. The walkway is narrower in the former case than the latter.

The directions of air outflow from the air conditioners and air circulation fans must be the same as those of the tiers. It is also important to move the room air vertically to obtain uniform vertical distributions of air temperature, VPD, air current speed, CO_2 concentration, and so forth.

Table 16.1 lists the equipment and sensors installed at fixed places in the culture room. In addition, there are movable carts with or without batteries for carrying culture panels, harvested produce, plant

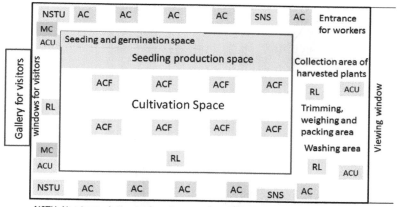

FIGURE 16.8

A typical floor layout of the culture room at a PFAL. Floor areas of rooms and spaces are not shown to scale.

Table 16.1 Equipment and Environmental Sensors Typically Installed at Fixed Places in the Culture Room

Category	Equipment and Environmental Sensors
Electricity supply	Power distribution box, breakers, and relays
Air conditioning	Inner units of air conditioners with plumbing for refrigerant
	Plumbing for recycling use of drained water
	Air circulation fans
	Actual/set point temperature display unit
	Air cleaners with filters and ozone (O_3) gas generator
Nutrient solution supply	Culture beds with circulation pumps
	Plumbing with strainers and valves
	Sterilization unit (filters, UV (ultraviolet) lamp, and O_3 gas generator) (Figure 16.9)
	Tank with a floating switch, and stock solution tanks
	Plumbing for civil or clean water supply, plumbing for drainage in emergency discharge
Lighting	Light source with reflectors
	Power stabilizer, inverters, and AC-DC converters
CO_2 supply	Control unit with distribution tubes
Sanitation control	Washing/cleaning machine of culture panels (Figure 16.10)
	Cleaning tools for floor and culture beds
Stock room	Supplies for plant production, sanitation, etc.
Sensors for environmental control	Air: Temperature, relative humidity (VPD), CO_2 concentration, and CO_2 supply rate
	Nutrient solution: pH, EC (electric conductivity), temperature, water supply rate, and circulating nutrient solution flow rate
	Electrical energy: Watt meter, watt-hour meter

FIGURE 16.9

Sterilization unit with filters, UV lamp, and ozone generator for nutrient solution.

FIGURE 16.10

Culture panel cleaning machine.

(photo courtesy of Sashinami Seisakujo Corporation)

FIGURE 16.11

A lift with batteries for uploading and unloading the culture panels.

(photo courtesy of Hokuetsu Industries Co., Ltd.)

residue, packing boxes, and so forth. Also, there are movable lifts with or without batteries for uploading and unloading the culture panels with plants (Figure 16.11).

SANITATION CONTROL
BIOLOGICAL CLEANNESS

The culture and operation rooms must be clean biologically as well as physically and chemically. In many industries, the cleanness of room air is expressed as the number of particles (physical objects) with size of 0.1 (or 0.5) μm or larger per unit volume of air. Chemical cleanness of food is often expressed as the concentrations of heavy metals, agrochemicals, and other toxic chemical substances.

In a PFAL, biological cleanness of the room air, harvested produce, or nutrient solution is expressed as CFU (colony forming unit) (see Chapter 21), which is a measure of the population density of live microorganisms. It is generally considered that fresh vegetables can be eaten without washing if the CFU of the harvest is less than 300 per gram, while the CFU of field-grown lettuce heads ranges from 10,000 to 100,000 if not washed.

Microorganisms can be classified into four groups: (A) potentially harmful to humans but not to plants, (B) potentially harmful to plants but not to humans, (C) potentially harmful neither to humans nor to plants, and (D) potentially harmful to both humans and plants. Thus, in addition to CFU tests, weekly or monthly tests are necessary for measuring the population densities of harmful microorganisms such as the *Coliform* group and *Staphylococcus aureus* which may cause human disease

(Chapter 21). These pathogens are often brought in by workers having symptoms of diarrhea, so these workers should not be allowed to enter the operation and culture rooms.

In order to avoid the dispersion of pathogens that may cause diseases in plants, knives and scissors used for harvesting must be disinfected every time the workers harvest new culture beds or tiers. The propagation of microorganisms (fungi) such as *Pythium* spp. *Fusarium oxysporum* Schlecht. f.sp. occurs mostly in the nutrient solution in the culture beds and wet surfaces where there is food for microorganisms.

In the nutrient solution in culture beds, there are many fine roots and algae (some dead and some alive) favored by microorganisms as food. Some microorganisms grow very rapidly under such favorable conditions, so the culture panels and culture beds must be cleaned by removing the fine roots, algae, and microorganisms every 2-4 weeks.

Continuous sterilization of recirculating nutrient solution is essential, although many microorganisms stay and propagate in the culture beds and do not move with the recirculating nutrient solution, while some pathogens in the nutrient solution move into the roots, propagate in the plants and cause disease.

Typically, the population density of microorganisms is much higher in the nutrient solution than in the room air, so the following actions should be avoided: (1) dripping the nutrient solution onto the aerial part of plants, (2) dipping the aerial part of plants into the nutrient solution, (3) harvesting or touching the plants with gloves after touching the roots with gloves, and (4) leaving the roots on the floor, culture beds, and other places in the culture room.

ISO22000 AND HACCP FOR FOOD SAFETY

There are two international standards for food safety control: ISO22000 and hazard analysis and critical control point (HACCP). The ISO22000 family of International Standards addresses food safety management. The ISO's food safety management standards help organizations identify and control food safety hazards. The ISO22000 family contains a number of standards, each focusing on different aspects of food safety management (http://en.wikipedia.org/wiki/ISO_22000).

HACCP is a system that helps food business operators assess how they handle food and introduces procedures to make sure the food produced is safe to eat. As of 2014, there are few PFALs producing vegetables under the ISO22000 and HACCP standards.

In Japan, the Third Party Accreditation Committee on Products and Product System (TPAC-PPS) was established in 2013 for evaluating products produced in PFALs in terms of pathogenic microorganisms, nutrient components for human health (vitamins A, B, C, etc.), toxic heavy metals, agrochemicals, and nutrient components for plants (NO_3^-, etc.).

REFERENCES

Chun, C., Kozai, T., 2000. Closed transplant production system at Chiba University. In: Kubota, C., Chun, C. (Eds.), Transplant Production in the 21st Century. Kluwer Academic Publishers, Dordrecht, The Netherlands, pp. 20–27.

Kanawaty, G., 1992. Introduction to Work Study. International Labour Office, Geneva. ISBN 978-92-2-107108-2.

Meyers, F.E., Stewart, J.R., 2002. Motion and Time Study: For Lean Manufacturing. Prentice Hall, New Jersey. ISBN 978-0-13-031670-7.

Schueller, J.K., 2014. Implications for plant factories from manufacturing and agricultural equipment technologies. In: Proceedings of Plant Factory Conference, Kyoto, 5.

HYDROPONIC SYSTEMS

17

Jung Eek Son[1], Hak Jin Kim[2], Tae In Ahn[1]
Department of Plant Science, Seoul National University, Seoul, South Korea[1] Department of Biosystems Engineering, Seoul National University, Seoul, South Korea[2]

INTRODUCTION

Hydroponic systems are essential tools for plant production in indoor farming such as plant factories with artificial lighting (PFALs). Among various hydroponic systems, the deep flow technique (DFT), nutrient film technique (NFT), and aeroponic systems have been commercially used with recirculated nutrient solutions. Because ion concentrations in the nutrient solutions change with time causing a nutrient imbalance (Son and Takakura, 1987; Zekki et al., 1996; Cloutier et al., 1997), nutrient control systems with real-time measurements of all nutrients are required, but such systems are not yet available on a commercial basis. Instead, ion electrical conductivity (EC)-based hydroponic systems have been the second-best choice but suffer from nutrient imbalance (Savvas and Manos, 1999; Ahn and Son, 2011). For improving the nutrient balance, periodic analysis of nutrient solutions and adjustment of nutrient ratios were conducted (Ko et al., 2013). As an advanced method, ion-selective electrodes (ISEs) and artificial neural networks (ANNs) were used to estimate the concentration of each ion (Dorneanu et al., 2005; Gutierrez et al., 2007; Kim et al., 2013). To protect the plants in plant factories from disease, disinfection systems, such as ultraviolet (UV) systems, are required. The light intensity and exposure time of UV radiation are related to the disinfection ratio of pathogens (e.g., Runia, 1995). This chapter describes the hydroponic systems, sensors and controllers, nutrient management systems, ion-specific nutrient management, and nutrient sterilization systems required for plant production in plant factories.

HYDROPONIC SYSTEM

A hydroponic system, or hydroponics, is a method of growing plants using mineral nutrient solutions in water without soil. For growing leafy vegetables in general, the main hydroponic systems used are the DFT and NFT systems. In the DFT system, nutrient solutions are supplied to the plants whenever the water level in the culture bed becomes lower than the set value, and are recirculated and supplied to the bare roots of plants, at constant time intervals, in the culture bed with a 1/100 slope. NFT systems and modified DFT systems similar to an ebb-and-flow system have been widely used in plant factories (Figure 17.1).

In recirculation systems, nutrient solutions that are not absorbed by plants return to the nutrient tank. Therefore, water and nutrient absorption by plants can be easily estimated by measuring the loss of the

Plant Factory. http://dx.doi.org/10.1016/B978-0-12-801775-3.00017-2

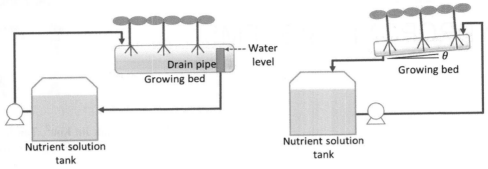

FIGURE 17.1

Schematic diagrams of DFT (left) and NFT (right) hydroponic systems.

nutrient solutions in the tank. In addition, aeroponic systems, which directly spray the nutrient solutions to the roots of the plants, are used in plant factories.

SENSORS AND CONTROLLERS

Root-zone environmental factors such as nutrient concentration, pH, dissolved oxygen, and temperature directly affect the growth of hydroponically grown plants. For real-time measurement of these factors, corresponding sensors are required. By assuming that the electric current increases with an increase of ionized nutrients in the nutrient solutions, EC sensors are used to measure nutrient concentrations. Nutrient solutions absorbed by transpiration can be measured by water-level sensors in large nutrient solution tanks and by load cells in relatively small tanks. Ultrasonic wave or laser sensors can also be conveniently used for noncontact measurement of the water level in the tanks. The process of controlling the water and nutrient supplies is as follows: determine the amounts of stock nutrient solutions and water, inject them into the nutrient solution tank and mix them, and supply the mixed nutrient solutions to the plants. Nutrient control systems consisting of sensors and controllers are used commercially in hydroponic farms.

NUTRIENT MANAGEMENT SYSTEMS
OPEN AND CLOSED HYDROPONIC SYSTEMS

Hydroponic nutrient solutions are composed of 13 essential elements. Each nutrient has a suitable concentration and relative ratios for the normal growth of a plant, and these are the target values of a nutrient control system. However, ionic concentration in the nutrient solutions changes with time, and subsequently a nutrient imbalance occurs in the closed hydroponic system (Figure 17.2).

Thus, in order to achieve optimal control, all nutrients should be measured in real time. However, such a system has both economic and technical limitations. High-precision instrument analysis is relatively expensive, and ion sensors are still at the research stage for their durability and stability. To date,

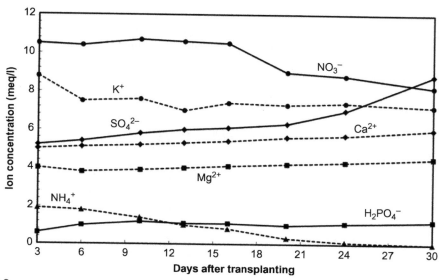

FIGURE 17.2

Changes in ionic concentration of the recirculated nutrient solutions controlled at a fixed EC (an example for lettuce).

the field application of real-time measurement systems for individual nutrients is difficult; instead, an EC and water level sensor system for controlling total ion concentration is widely used. Closed systems that monitor the results of the control process are preferable for nutrient solution management (Figure 17.3).

However, scaling-up the system requires a modular structure of the culture bed and drainage tank, and increasing the number of modules increases the installation cost. In contrast, open-loop control systems can have a relatively simple structure, even in the case of large-scale systems. But because they lack feedback, such systems might be inappropriate for plants that have higher fluctuations in uptake concentrations.

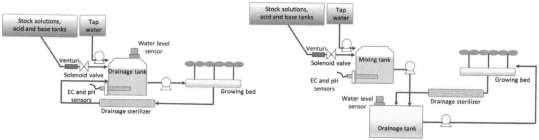

FIGURE 17.3

Schematic diagrams of closed (left) and open (right) hydroponic systems.

The concept behind EC-based hydroponic systems involves controlling total ion concentrations while minimizing nutrient imbalance through the injection of a stock solution. In order to operate the system normally, a theoretical understanding of the nutrient solution mixing procedure is required. The mixing process of nutrient solutions is discontinuously conducted by measuring changes in the amounts of nutrients and water in the drainage tank. This can be expressed as:

$$V_t EC_t = V_c EC_c + aU \quad U = \frac{V_t EC_t - V_c EC_c}{a} \tag{17.1}$$

where V_t is the target volume of the nutrient solution stored in the drainage tank; EC_t is the target EC value (dS m^{-1}); V_c and EC_c represent the current volume of nutrient solution and EC in the drainage tank, respectively; U is the total amount of nutrients absorbed by the plants in milliequivalents; and a is the empirical coefficient for conversion of total salt concentration and EC. For hydroponic solutions, $a = 9.819$ was suggested as the value of the coefficient by Savvas and Adamidis (1999), which can be in the range of 0.8–4.0 dS m^{-1}. Based on Equation (17.1), the required injection amount of stock solution can be calculated by:

$$V_t EC_t = V_c EC_c + V_w EC_w + V_{stk} EC_{stk} \quad V_{stk} = \frac{V_t EC_t - V_c EC_c - V_w EC_w}{EC_{stk}} \tag{17.2}$$

where V_w is the required injection amount of tap water; EC_w is EC of tap water; V_{stk} is the required injection amount of stock solution; and EC_{stk} is the conversion of the milliequivalent concentration of stock solution to EC. The sum of V_c, V_w, and V_{stk} should be equal to V_t. Therefore, from this relationship, the following equation can be derived:

$$V_{stk} = \frac{V_t EC_t - V_c EC_c - EC_w(V_t - V_c)}{EC_{stk} - EC_w} \tag{17.3}$$

Once V_{stk} and V_w are calculated, these values are then submitted to the controller as reference input values, and actuators such as pumps and solenoid valves are activated.

CHANGES IN NUTRIENT BALANCE UNDER EC-BASED HYDROPONIC SYSTEMS

The transport of nutrients and water in the hydroponic system in Figure 17.3 (left) can be expressed as a differential equation. Silberbush and Ben-Asher (2001) constructed this model, and the above-mentioned EC-based nutrient solution mixing process is applicable in the model. The following equations are simplified models of nutrient and water transport in EC-based hydroponic systems.

$$V_b \frac{dC_b^i}{dt} = Q_{ir} C_{drg}^i - \frac{V_{max} C_b^i}{K_m + C_b^i} - Q dr C_b^i \tag{17.4}$$

$$V_{drg} \frac{dC_{drg}^i}{dt} = Q dr C_b^i + Q_{inj} C_{inj}^i + Q_{wtr} C_{wtr}^i - Q_{ir} C_{drg}^i \tag{17.4}$$

$$\frac{dV_b}{dt} = Q_{ir} - Q_{trs} - Q dr \tag{17.5}$$

$$\frac{dV_{drg}}{dt} = Q dr + Q_{inj} + Q_{wtr} - Q_{ir} \tag{17.6}$$

where V_b is the volume of the nutrient solution in the culture bed; V_{drg} is the stored nutrient solution in the drainage tank; C represents the concentration of individual nutrients, superscripts indicate the names of

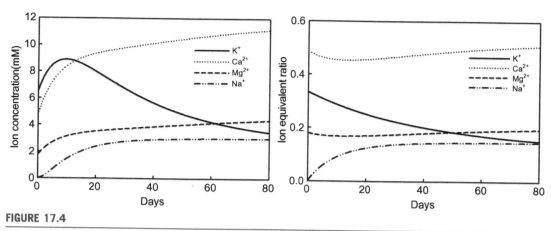

FIGURE 17.4

Simulated changes in the ion concentration (left) and the ion equivalent ratio (right) in the culture bed under an EC-based nutrient management system.

nutrients, and subscripts indicate the position of the nutrients; Q_{ir} is the irrigation rate; Q_{trs} is the transpiration rate; $Q dr$ is the drainage rate; Q_{inj} is the stock solution injection rate; and Q_{wtr} is the tap water injection rate. Q_{inj} and Q_{wtr} are calculated by the preceding equation, and V_{max} and K_m are coefficients for Michaelis–Menten kinetics.

Based on this model, changes in nutrient concentration and balance under EC-based hydroponic systems can be simulated. Figure 17.4 shows the simulated results of the changes in the concentrations and ratios of K^+, Ca^{2+}, Mg^{2+}, and Na^+. In practice, the absorption ratios between nutrients are not equal to the ratios in the nutrient solution, and therefore changes in the nutrient balance are observed during cultivation periods. The simulated results adequately show these tendencies. Nutrients that have relatively lower absorption rates such as Na^+ gradually increase in the nutrient solution. In EC-based hydroponic systems, the total ion concentration in the drainage tank is fixed at a target value; thus, it limits the injection amount of the stock solution. Therefore, the nutrient ratio of Na^+ in the nutrient solution rapidly increases, and this can result in a decrease of injection of other nutrients from the stock solution and lead to the development of salinity stress. However, if the tap water Na^+ concentration is in a certain range, the dynamic equilibrium can then be observed (Savvas et al., 2008). A major source of Na^+ inflow is tap water; therefore, the concentration at equilibrium is decided by the concentration in the tap water, the transpiration rate, and the absorption rate. If the predicted equilibrium concentration is above the threshold of a cultivated plant, then the application of a desalinator needs to be considered. Furthermore, ratios of other essential nutrients are important for normal plant growth; thus it is necessary to periodically analyze the nutrient solutions and adjust the nutrient ratios in the stock solution before deficiency or toxicity develops (Ko et al., 2013).

ION-SPECIFIC NUTRIENT MANAGEMENT

Hydroponic solutions contain various nutrients essential for crop growth. These nutrients are generally taken into plants in various ionic forms, such as NO_3^-, $H_2PO_4^-$ or HPO_4^{2-}, and K^+ through a combination of root interception and diffusion. Since the reserves of nutrients for the plants are limited in hydroponic greenhouse cultivation, an inaccurately balanced nutrient solution may result in an unbalanced nutrient composition in the root environment.

In closed hydroponic systems that reuse the drainage solution, the build-up of salts is managed by dramatically reducing the fertilizers that are dissolved in the water, which are added to replenish the drainage solution. However, in closed systems such as NFT, DFT, and ebb-flow systems, it is important to determine the nutrient concentrations in the reused solution to regenerate a nutrient solution of optimal composition because prolonged reuse of drainage water may lead to an accumulation of some nutrients, resulting in alterations in the nutrient ratios (Gutierrez et al., 2007) (Figure 17.5).

To overcome such limitations regarding the use of closed systems on a practical scale, a feedback control loop that uses an ion analyzer is needed to dilute liquid fertilizers on the basis of on-line measurements of nutrient concentrations (Gieling et al., 2005). In addition, an algorithm implemented in a computer program is used to generate the time duration values needed to activate the valves for the addition of liquid fertilizer by the injection unit (Savvas and Manos, 1999). Current practices for managing hydroponic nutrients in closed systems are usually based on automatic control of EC in the nutrient solution. A main problem with this practice is that EC measurements provide no information on the concentrations of individual ions and therefore do not allow individual real-time corrections to be made to each nutrient in response to the demand of the crop (Cloutier et al., 1997). Improved efficiency of fertilizer utilization may be possible through accurate measurement and control of the individual nutrients in the solution in real time.

The need for such fast, continuous monitoring has led to the application of ISE technology to measure hydroponic nutrients (Cloutier et al., 1997; Gutierrez et al., 2007). The advantages over standard

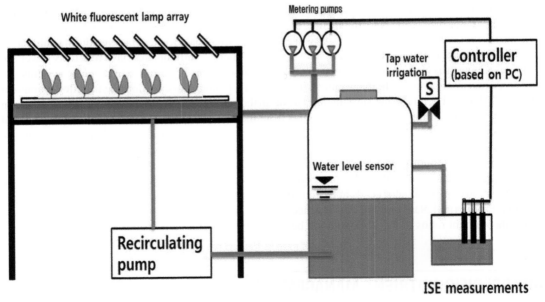

FIGURE 17.5

A schematic diagram of a closed hydroponic system based on nutrient feedback control.

analytical methods (spectroscopic techniques) include simple methodology, direct measurement, sensitivity over a wide concentration range, low cost, and portability (Heinen and Harmanny, 1992). The key component of an ISE is the ion-selective membrane that responds selectively to one ion in the presence of other ions in a solution. There are currently ion-selective membranes available for most of the important hydroponic nutrients, including NO_3^-, K^+, Ca^{2+}, and Mg^{2+}. However, there are several potential disadvantages of ISEs as compared to standard analytical methods. One is chemical interference by other ions because ISEs are not truly specific but rather respond more or less to a variety of interfering ions. To overcome interference issues, various data processing methods such as multivariate calibration and ANN methods can be used. The multivariate calibration method is useful for allowing the cross responses arising from primary and interfering ions for accurately determining individual ion concentrations (Forster et al., 1991). An ANN method that uses data obtained with an array of multiple electrodes can be used for the simultaneous determination of various ions (Gutierrez et al., 2007). For example, as shown in Figure 17.6, in a previous study that determined NH_4^+, K^+, Na^+, and NO_3^- ions in a hydroponic solution, the ANN was able to predict the concentrations of the tested ions with an acceptable level for almost three days without the need to remove the interfering effects.

Another disadvantage is reduced accuracy due to electrode response drift and biofilm accumulation caused by the presence of organic materials in hydroponic solutions (Carey and Riggan, 1994; Cloutier et al., 1997). In particular, stability and repeatability of sensor response may be a major concern when considering an in-line management system that includes continuous immersion of ISEs in hydroponic solution because the accuracy of the measurement may be limited by the drift in electrode potential over time. The use of a computer-based automatic measurement system would improve the accuracy and precision of determining nutrient concentrations because consistent control of sample preparation, sensor calibration, and data collection can reduce variability among multiple electrodes during replicate measurements (Dorneanu et al., 2005; Kim et al., 2013). Ideally, an automated sensing system for

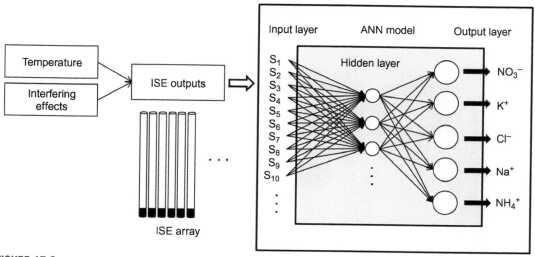

FIGURE 17.6

A schematic representation of the ANN method using an array of ion-selective electrodes.

hydroponic nutrients would be able to periodically calibrate and rinse the electrodes and continuously measure the nutrients in the hydroponic solution while automatically introducing solutions for calibration and rinsing as well as measurement.

STERILIZATION SYSTEM

In general, contamination of crops by pathogens is lower in hydroponic cultures than in soil cultures; however, the pathogens can quickly spread to neighboring plants through the nutrient solutions when even one plant is contaminated. In order to minimize this risk, disinfection systems using filters, heat, ozone, and UV radiation are used in hydroponics.

Filtering systems eliminate the pathogens and other soluble solids in the nutrient solution, and their capacity is determined by pore size. Heating systems sterilize the pathogens by heating the nutrient solutions. Because the temperature for sterilization differs depending on the pathogen species, adequate ranges of temperatures should be set. After the heating treatment, a cooling process is required for reuse of the nutrient solutions on crops. Ozone (O_3) systems sterilize the nutrient solutions by using the oxidizing capacity of ozone gas. UV systems have the advantage of fast sterilization of pathogens in the nutrient solutions passing through the tube of the sterilizer. A light intensity of at least 250 mJ cm^{-2} is required to disinfect the pathogens perfectly (Runia, 1995). Even though the light intensity is within an adequate range, the sterilizing capacity decreases due to a lower transmittance of UV radiation if the turbidity of the nutrient solutions increases. To prevent this, filters should be installed in front of the UV sterilizer (Figure 17.7). A disadvantage of UV systems is that the Fe-EDTA (chelating agent, ethylenediaminetetraacetic acid) in the nutrient solutions is precipitated and cannot be used. Therefore, addition of Fe-EDTA is required to compensate for the loss of Fe in hydroponic systems.

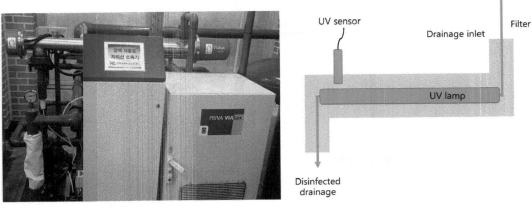

FIGURE 17.7

Commercial UV sterilizer (left, installed at ATEC in Korea) and schematic diagram of the structure (right).

REFERENCES

Ahn, T.I., Son, J.E., 2011. Changes in ion balance and individual ionic concentrations to EC reading at different renewal intervals of nutrient solution under EC-based nutrient control in closed-loop soilless culture for sweet peppers (*Capsicum annuum* L. 'Fiesta'). Korean J. Hortic. Sci. Technol. 29, 29–35.

Carey, C.M., Riggan, W.B., 1994. Cyclic polyamine ionophores for use in a dibasic-phosphate-selective electrode. Anal. Chem. 66, 3587–3591.

Cloutier, G.R., Dixon, M.A., Arnold, K.E., 1997. Evaluation of sensor technologies for automated control of nutrient solutions in life support systems using higher plants. In: Proceedings of the Sixth European Symposium on Space Environmental Control Systems. Noordwijk, The Netherlands.

Dorneanu, S.A., Coman, V., Popescu, I.C., Fabry, P., 2005. Computer-controlled system for ISEs automatic calibration. Sensors Actuators B 105, 521–531.

Forster, R.J., Regan, F., Diamond, D., 1991. Modeling of potentiometric electrode arrays for multicomponent analysis. Anal. Chem. 63, 876–882.

Gieling, T.H., Van Straten, G., Janssen, H., Wouters, H., 2005. ISE and Chemfet sensors in greenhouse cultivation. Sensors Actuators B Chem. 105, 74–80.

Gutierrez, M., Alegret, S., Caceres, R., Casadesus, J., Marfa, O., del Valle, M., 2007. Application of a potentiometric electronic tongue to fertigation strategy in greenhouse cultivation. Comput. Electron. Agric. 57, 12–22.

Heinen, M., Harmanny, K., 1992. Evaluation of the performance of ion-selective electrodes in an automated NFT system. Acta Horticult. 304, 273–280.

Kim, H.-J., Kim, W.-K., Roh, M.-Y., Kang, C.-I., Park, J.-M., Sudduth, K.A., 2013. Automated sensing of hydroponic macronutrients using a computer-controlled system with an array of ion-selective electrodes. Comput. Electron. Agric. 93, 46–54.

Ko, M.T., Ahn, T.I., Cho, Y.Y., Son, J.E., 2013. Uptakes of nutrients and water of paprika (*Capsicum annuum* L.) as affected by renewal period of recycled nutrient solution in closed soilless culture. Hortic. Environ. Biotechnol. 54, 412–421.

Runia, W.T.H., 1995. A review of possibilities for disinfection of recirculation water from soilless cultures. Acta Horticult. 382, 221–229.

Savvas, D., Adamidis, K., 1999. Automated management of nutrient solutions based on target electrical conductivity, pH, and nutrient concentration ratios. J. Plant Nutr. 22, 1415–1432.

Savvas, D., Manos, G., 1999. Automated composition control of nutrient solution in closed soilless culture systems. J. Agric. Eng. Res. 73, 29–33.

Savvas, D., Chatzieustratiou, E., Pervolaraki, G., Gizas, G., Sigrimis, N., 2008. Modelling Na and Cl concentrations in the recycling nutrient solution of a closed-cycle pepper cultivation. Biosys. Eng. 99, 282–291.

Silberbush, M., Ben-Asher, J., 2001. Simulation study of nutrient uptake by plants from soilless cultures as affected by salinity buildup and transpiration. Plant Soil 233, 59–69.

Son, J.E., Takakura, T., 1987. A study on automatic control of nutrient solutions in hydroponics. J. Agric. Met. 43 (2), 147–151.

Zekki, H., Gauthier, L., Gosselin, A., 1996. Growth, productivity, and mineral composition of hydroponically cultivated greenhouse tomatoes, with or without nutrient solution recycling. J. Am. Soc. Hortic. Sci. 121, 1082–1088.

SEEDING, SEEDLING PRODUCTION AND TRANSPLANTING

Osamu Nunomura, Toyoki Kozai, Kimiko Shinozaki, Takahiro Oshio

Japan Plant Factory Association, c/o Center for Environment, Health and Field Sciences, Chiba University, Kashiwa, Chiba, Japan

INTRODUCTION

A standard procedure for producing seedlings in a plant factory with artificial lighting (PFAL), consisting of preparation, seeding, seedling production and transplanting, is described. The procedure described here needs to be modified depending on the plant species, culture system, daily production capacity of the PFAL, etc. The numerical values given in the description are merely examples and should be modified to suit the particular situation. Leaf lettuce (*Lactuca sativa* L. vari. crispa) is chosen as the plant material in this chapter.

The objective of this chapter is to provide the standard procedure for achieving percent seed germination of 98% or higher (the percentage achieved by beginners would be around 90%), and percent transplantable seedlings from germinated seeds exceeding 98%. Accordingly, the percent transplantable seedlings from seeds sown exceeds 96% ($=98 \times 98/100$) or higher. The percent salable harvests of seeds sown and of transplanted seedlings should exceed, respectively, 95% or higher and 99%.

PREPARATION

(1) Choose either untreated or treated seeds (Figure 18.1). Treated seeds include naked (shell-less) and coated seeds. Treated seeds for enhancing rapid and uniform germination are preferable to untreated seeds.

(2) Caution is necessary regarding whether or not to treat the seeds with a fungicide. The choice depends on the purpose of seedling production and the biological characteristics of the seeds.

(3) Eliminate low-quality seeds based on their size, shape, color and (specific) weight through visual checking or by using an automatic grading/sorting machine.

(4) Prepare a sponge-like or foamed urethane seeding mat (called "mat" hereafter) ($28 \times 58 \times 2.8$ cm) (Figure 18.2) consisting of 300 cubes or cuboids ($2.3 \times 2.3 \times 2.8$ cm) each with a small hole (7–10 mm in diameter, 5–10 mm in depth) on the upper surface of the cube. Each cube can be separated easily by hand from the mat (Figure 18.3).

Plant Factory. http://dx.doi.org/10.1016/B978-0-12-801775-3.00018-4

FIGURE 18.1

Lettuce seeds untreated (left) and coated (right). Treated seeds are classified into naked (shell-less) seeds, coated seeds, and coated naked seeds.

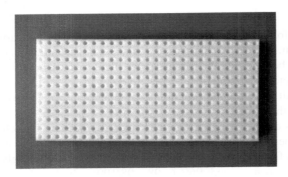

FIGURE 18.2

White-colored seeding mat (sponge-like foamed urethane mat) (28 cm wide, 58 cm long and 2.8 cm high) consisting of 300 cuboids (2.3 × 2.3 × 2.8 cm).

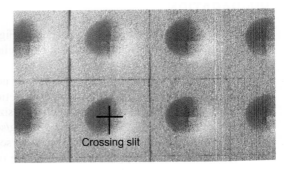

FIGURE 18.3

The cuboids each with a hollow at the center of the upper surface. There is a crossing slit at the center of each hollow to encourage the radicle of the germinated seed to grow downward smoothly.

(5) At the center of each hole, there is a crossing slit (10 mm long each) reaching the bottom of the cube to encourage the radicle (the youngest primary root) of a germinated seed to grow downward easily.

(6) Prepare a foamed plastic tray (outside dimensions: $30 \times 60 \times 4.0$ cm) to hold the mat (Figure 18.4). Measure the weight of the empty tray prior to putting the mat in the tray.

(7) Prepare a predetermined volume (3 liters per tray) of nutrient solution the strength of which is 1/4–1/8 that of the nutrient solution to be used after the second transplanting.

(8) Press down the mat surface uniformly using a flat plate (30×60 cm) with many small holes or a press machine (Figure 18.5) with a stacking and automatic feeding unit to drive out all the air from the mat. This step can be conducted by hand if only a few mats are to be pressed.

(9) Concurrently, soak the mat in nutrient solution so that all the capillary tubes and/or pores in the mat are filled with nutrient solution, resulting in no air bubbles in the mat.

(10) Weigh the tray containing the mat and nutrient solution, and add or remove a small amount (0.2 l) of nutrient solution to obtain the predetermined target weight of 3280 g (e.g., 215 g for tray, 3000 g for nutrient solution, and 65 g for the mat) common to all the trays.

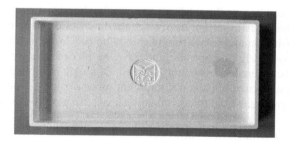

FIGURE 18.4

Formed, foamed polystyrene tray (relatively hard) to store the seeding mat (outer dimensions: 30 cm wide, 60 cm long, and 4 cm high; inner dimensions: 28 cm wide, 58 cm long, and 2.8 cm high).

FIGURE 18.5

A press machine for foamed mats to drive out all the air from the mat and fill the pores with water. The machine can be connected to a stacking and automatic feeding unit.

Sasinami Seisakujo Corporation, Aichi Prefecture, Japan.

(11) Check that the surface of the horizontal mat is wet evenly and that the level of free nutrient solution is just at the bottom of each hole of the mat. An evenly wet mat surface and free nutrient solution at the bottom of holes are important to achieve uniform seed germination over the mat. If the wetness at the mat surface is uneven, often due to insufficient capillary suction within the mat, the mat needs to be replaced with a new one.

SEEDING

(12) Place one seed in the wet hole of each cube (Figure 18.6) using tweezers, a seeding plate, a semi-automatic seeder (Figure 18.7), or an automatic seeding machine (Figure 18.8). Confirm that the seed touches the wet center of the hole.

(13) Cover the mat surface with thin plastic film (0.02 mm thick) to keep its surface always moist during the germination stage (Figure 18.9).

(14) Move the trays to a germination space. Ensure that the trays are always horizontal to prevent the seeds and nutrient solution from moving in the tray during transportation.

(15) The trays are stacked in case the seeds are negatively photoblastic (dim light is unnecessary for germination of photoblastic seeds). Apply a slight downward pressure on the mat surface of the uppermost tray using an additional thin plastic plate (0.5–1.0 mm thick).

(16) If the seeds are positively photoblastic, the trays are placed in a germination rack with the vertical distance between the tiers being 7–10 cm. Dim light is applied to the surface of each mat from the side using vertically placed fluorescent tubes or string-type LED lamps.

(17) Microorganisms grow easily in the germination space, so periodic cleaning and/or sterilization are required.

(18) The set point of air temperature varies between 15°C and 30°C with the plant species and cultivar (15–22°C for lettuce plants).

(19) Two to three days after sowing, remove the plastic film from the tray.

(20) At seed germination, a radicle comes out first (Figure 18.10). Then, the hypocotyl with unfolded cotyledon leaves covered with seed coat appears (Figure 18.11).

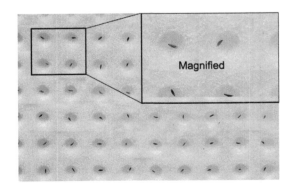

FIGURE 18.6

Lettuce seeds sown on the mat. The seeds touch the wet center of the hole to germinate smoothly.

FIGURE 18.7

Tweezers (upper left), seeding plate (lower left) and semi-automatic seeding tool (right: Minoru Industrial Co., Ltd., Okayama Prefecture, Japan). The seeding plate has two transparent plastic plates with small holes in grid-like fashion. A handle is connected to the right-hand side of the upper plate. When one seed has been placed in each hole of the upper plate, the seeding plate is placed above the mat and the handle is pulled to the right-hand side. Then, each seed falls down at each center of the hole. The seeds are stacked in the upper container of the semi-automatic seeding tool. The machine is moved to the left-hand side manually. Then, one seed drops down from a hole at the bottom of the rotary container. The distance between the holes can be adjusted manually to suit different mats.

FIGURE 18.8

Automatic seeding machine for sponge-like seeding mat.

Nansei-Kobashi Co., Ltd., Mie Prefecture, Japan.

FIGURE 18.9

Formed polystyrene tray with seeds covered with thin plastic film (0.02 mm thick) (left) and covered with an empty foamed polystyrene tray (right) to keep the mat surface wet.

FIGURE 18.10

Germinated lettuce seed with radicle. Hypocotyl and cotyledon are still in the seed coat.

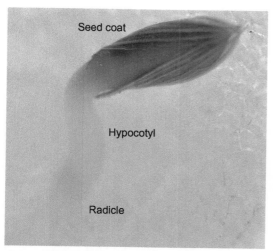
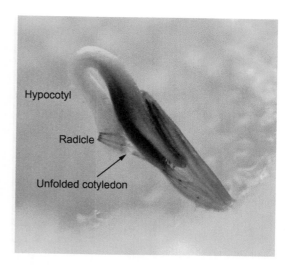

FIGURE 18.11

Lettuce (*Lactuca sativa* L. var. crispa) seed 22 h after seeding (left) and 42 h after seeding (right). Cotyledon is still folded and partly in the seed coat.

(21) Confirm that all the radicles are growing downward into the mat. By applying slight downward pressure to the germinating seeds, the radicles grow downward smoothly through the crossing slit into the bottom of the cube. If the downward pressure is too low, the radicle will not penetrate the mat (Figure 18.12).

(22) At this stage, percent germination is expected to be 98% or higher. If not, it is necessary to analyze the cause to improve the percent germination.

(23) Move the trays from the germination space to a seedling production space. Hereafter, the seedlings grow by photosynthesis.

(24) Grow the germinated seeds to seedlings with unfolded green cotyledons at photosynthetic photon flux (PPF) of about 50–100 μmol m^{-2} s^{-1}. Green cotyledons usually expand fully in 4 to 7 days after seeding (4 days in the case of lettuce seeds) (Figures 18.13 and 18.14).

(25) Algae grow quickly on the wet mat surface under light (Figure 18.15). To prevent this, the level of free nutrient solution in the tray needs to be lowered by 10 mm to keep the mat surface dry, when the radicle with fine roots or rootlets comes out into the nutrient solution (Figure 18.16).

(26) A black-colored sponge-like, foamed urethane seeding mat is effective to suppress algae growth, although PPFD is reduced due to its low light reflectivity (Figure 18.17).

SEEDLING PRODUCTION AND TRANSPLANTING

(27) Three to seven days after seeding, cotyledons are unfolded and green in color. In the case of lettuce seeds, cotyledons are fully unfolded 72 h after seeding. Then, small true leaves can be observed by 5 days after seeding. Seven days after seeding, two true leaves are unfolded (Figure 18.18).

FIGURE 18.12

Germinated lettuce seed with radicle which did not penetrate into the substrate. The radicle was lifted up from the substrate, (a) 2 days after seeding, and (b) 7 days after seeding.

FIGURE 18.13

Lettuce (*Lactuca sativa* L. var. crispa) seeds with cotyledons 72 h after seeding. The cotyledon is nearly unfolded and starts photosynthetic activity at this growth stage.

(28) The seedlings start growing quickly 10–12 days after seeding. The seedlings 14–16 days after seeding (Figure 18.19) are suitable for the first transplanting to a culture panel ($30 \times 60 \times 1$ cm) having 24–30 holes (Figure 18.20)

(29) Transplant each seedling with the cube to the culture panel for further growth after the radicle with rootlets has come out from the bottom surface of the cube and the radicle is submerged about 30 mm beneath the bottom surface. Transplanting can be done manually or by using a transplanting machine.

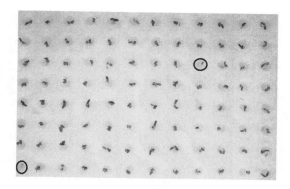

FIGURE 18.14

Ninety-six germinated lettuce seeds on the foamed urethane mat, 5 days after seeding. For commercial plant production, the germination percentage needs to be 98% or higher. In the photograph, the two seeds in the circle germinated 2 days later than the other seeds.

FIGURE 18.15

Algae grow quickly in the presence of light and nutrient solution on the wet seeding mat surface. (a) Spotted algae growth on the mat; (b) spotted algae growth on the mat; (c) algae growth on the surface and within the mat.

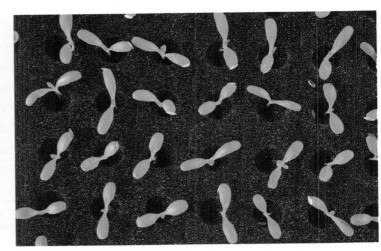

FIGURE 18.16

Seedling with roots which penetrated the cuboid. At this timing, the level of nutrient solution is lowered by 10 mm to allow the seeding mat surface to dry out and thus inhibit rapid growth of algae.

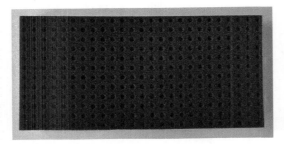

FIGURE 18.17

Black-colored seeding mat (sponge-like foamed urethane mat) which suppresses the growth of algae.

(30) Twelve (=300/25) culture panels (formed, foamed urethane board) each with 26 holes (Figure 18.20) or ten (=300/30) culture panels with 30 holes are necessary for each mat having 300 cubes.

(31) The culture panel with 25–30 seedlings is placed in a hydroponic system with a culture bed and nutrient solution circulation unit to grow the seedlings further at a PPF of 100–150 μmol m^{-2} s^{-1} for about 10–15 days to obtain larger seedlings ready for the second transplanting (Figure 18.21), when the ratio of projected leaf area to the culture panel area exceeds 0.9.

FIGURE 18.18

Lettuce seedlings with unfolded cotyledon 3 days (upper left), 5 days (upper right), and 7 days (bottom) after seeding.

FIGURE 18.19

Lettuce (*Lactuca sativa* L. vari. crispa) seedlings 14 days after seeding, ready for the first transplanting. Black-colored foamed urethane cubes are used as substrate.

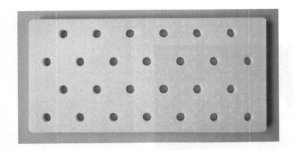

FIGURE 18.20

Formed culture panel (relatively hard board) with 26 holes (2.2 cm in diameter) for the first transplanting (29.8 cm wide, 59.6 cm long, 1.4 cm thick).

FIGURE 18.21

Lettuce seedlings (*Lactuca sativa* L. vari. crispa) 24 days after seeding, ready for the second transplanting. Fresh weight: 10–12 g/plant, Root fresh weight ratio: 0.40–0.45.

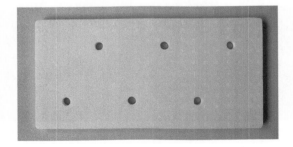

FIGURE 18.22

Formed culture panel (relatively hard board) with 6 holes (2.2 cm in diameter) for the second transplanting (29.8 cm wide, 59.6 cm long, 1.4 cm thick).

(32) The second transplanting is conducted using a culture panel (formed, foamed urethane board) with 6–8 holes (Figure 18.22). The larger the seedlings, the harder it is to pick them up and transplant them. However, the total cultivation area required for seeding, seedling production, and cultivation after the second transplanting can be reduced if larger seedlings are used for the second transplanting.

(33) If the hypocotyl of a transplant is too elongated, the transplant tends to lie down on the culture panel and its growth is delayed for a few days.

(34) The growth after the second transplanting is enhanced when the CO_2 concentration is increased up to around 1000 ppm.

TRANSPLANT PRODUCTION IN CLOSED SYSTEMS

INTRODUCTION

The closed transplant production system (CTPS) is a type of plant factory with artificial lighting (PFAL) or closed plant production system (CPPS) that is designed and operated specifically for the production of transplants (seedlings including grafted seedlings from seeds, and plantlets from cuttings/explants) (Kozai et al., 2004). The CTPS is used also for the acclimatization of micropropagated plantlets and for the vegetative propagation (multiplication) of plants (Chun and Kozai, 2000).

The CTPS is useful for producing transplants with desirable ecophysiological characteristics by environmental control (Kozai, 2007). Examples include: (1) tomato seedlings with flower buds at the eighth node or lower which enable early first harvesting and thus higher annual yield, (2) seedlings with thick and short hypocotyl (stem), which are resistant to environmental stresses including high winds and heavy rain, (3) seedlings with soft and elongated hypocotyl, which are necessary for easy cutting of hypocotyl by hand or a cutting machine to produce scions or root stocks for grafted seedlings, and (4) transplants that are physiologically and morphologically uniform all year round regardless of the weather.

The CTPS is also useful for: (1) the production of insect- and pesticide-free transplants, (2) shortening the transplant production period by 30–40% compared with that using a greenhouse by optimally controlling the environment including CO_2 enrichment, (3) increasing the annual transplant productivity per unit land area, and (4) improving the traceability of the transplant production process (Kozai, 2006).

In this chapter, firstly, the main components and functions of the CTPS are described, as well as the annual productivity and electricity cost per transplant. Secondly, the ecophysiology of transplant production is discussed. Thirdly, production schemes for several medicinal and horticultural transplants are described. Finally, the propagation and production of blueberry and strawberry transplants are introduced.

MAIN COMPONENTS AND THEIR FUNCTIONS

Toyoki Kozai

Japan Plant Factory Association, c/o Center for Environment, Health and Field Sciences, Chiba University, Kashiwa, Chiba, Japan

Plant Factory. http://dx.doi.org/10.1016/B978-0-12-801775-3.00019-6

MAIN COMPONENTS

The CTPS has been commercially used since 2004 and is now used at more than 300 locations in Japan as of 2014. The main components of the CTPS are the same as those of the PFAL, as shown in Figure 19.1. However, there are some unique characteristics of the CTPS which are somewhat different from those of the PFAL in order to produce the high-quality transplants mentioned in the introduction of this chapter.

The standard size of the basic CTPS unit being used in Japan is 4.5 m wide, 3.6 m long, and 2.4 m high (floor area: 16.2 m^2) (see Figure 19.2a) with two rows each having four tiers. This CTPS can hold 96 plug trays (each 30 cm wide × 60 cm long). By using plug trays each having 100 or 200 cells, 9600 or 19,200 transplants can be produced at one time. In the case of tomato seedling production, for example, it takes about 20 days after seeding or 17 days after germination to get the seedlings ready for transplanting. Therefore, the annual production capacity of one basic unit of CTPS is about 200,000 seedlings (=9600 × 365/17) for 100-cell plug trays and 400,000 seedlings for 200-cell plug trays. CO$_2$ concentration during the photoperiod is usually kept at around 1200 ppm to enhance the photosynthesis and thus growth. Liquid CO$_2$ in a pressurized container is used as the CO$_2$ source.

The basic units can be connected with each other as shown in Figure 19.2b. In this case, extra facilities such as seeding machines, germination boxes, germination rooms, and insect net rooms are often installed next to the CTPS (Figure 19.3).

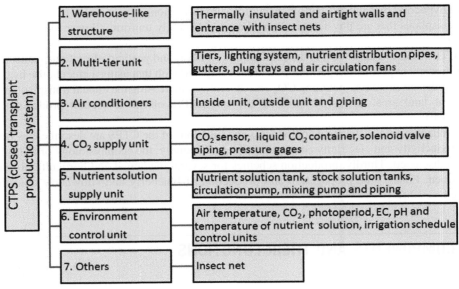

FIGURE 19.1

Main components of CTPS.

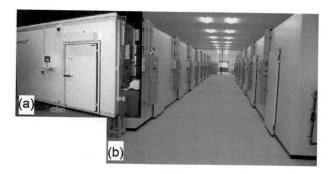

FIGURE 19.2

Exterior of CTPS. (a) Basic unit of CTPS (floor area: 16 m²) for a small-scale grower (left). (b) CTPS consisting of 21 basic units (right).

Photo courtesy of Mitsubishi Chemicals Inc.

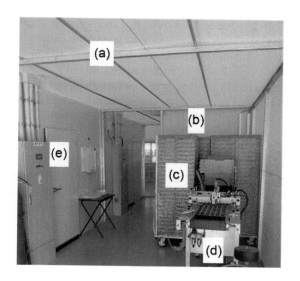

FIGURE 19.3

Extra facilities often used together with multiple CTPSs. (a) Insect net room to prevent insects from entering the CTPS; (b) germination room; (c) germination box to be placed in the germination room after seeding; (d) seeding machine, (e) CTPS.

LIGHT SOURCE, AIR CONDITIONERS, AND SMALL FANS

Fluorescent lamps (FLs) have been used as light sources until 2014 but they will be replaced by LED lamps in the near future. Five or six FLs (1.2 m long, 32 W) on each shelf provide photosynthetic photon flux density (PPFD) of around 300 or 350 μmol m^{-2} s^{-1} at the plug tray surface (Figure 19.4). However, it should be noted that the PPFD decreases with time by about 30% after being used for 10,000 h.

FIGURE 19.4

Inside view of the plug tray shelf. Five to six FL (fluorescent) tubes (32 W each, 300–350 µmol m^{-2} s^{-1} and four small sucking fans (2–3 W each) are installed for one shelf unit holding four plug trays.

Photo courtesy of Mitsubishi Chemicals Inc.

Four household air conditioners are installed above the tiers in the basic unit of CTPS. One small air circulation fan (electricity consumption: 6 W, air flow rate: 2.9 m^3 min^{-1}) is installed for each plug tray, or four fans for four plug trays (Figure 19.4). This arrangement of air conditioners and fans generates the air flow pattern shown in Figure 19.5. With this arrangement, relatively laminar horizontal air current is generated over the plug trays and through the transplant canopy in each shelf (Figure 19.6). By using fans with an inverter for controlling the fan rotation speed, the air current speed over the transplant canopy is continuously controlled in a range between 20 cm s^{-1} and 100 cm s^{-1}, depending on the transplant growth stage, planting density, transplant morphology, etc. The horizontal air current within the transplant canopy with a height of about 10–15 cm keeps the relative humidity in the transplant canopy lower than 85%, and hence elongation of hypocotyl (stem) can be avoided even at a planting density of 1.5–2.0 times that in a greenhouse (Figure 19.7).

In most CTPS, no humidification unit is installed because the transpiration by plants and evaporation from the substrate humidify the room air. For a CTPS full of plug trays and with an air temperature of around 25°C, the relative humidity over the plug trays is kept around 75–80% by the balance between the evapotranspiration of liquid water and the condensation of water vapor by air conditioners. In contrast, the relative humidity tends to be as low as 50% when the CTPS is almost empty.

ELECTRICITY COSTS

The electricity cost for lighting per transplant can be estimated easily as a function of electricity consumption per lamp (W), number of lamps per shelf (N_L), photoperiod hours per day (P), days required to produce the transplants after germination (D), and the number of cells per plug tray (N_P). Assuming that five fluorescent tubes ($N = 5$) are used for four plug trays (30 cm × 4 = 120 cm) and that $W = 0.035$ kW, $P = 16, D = 17$, and $N_P = 1280$ (these assumptions correspond to tomato seedling production), then the electricity consumption for lighting per transplant (E) is expressed by:

$$E = \frac{W \times N_L \times P \times D}{4 \times N_P} = \frac{0.035 \times 5 \times 16 \times 17}{4 \times 128} = 0.094 \text{kWh} \qquad [19.1]$$

If the price of electricity per kWh is US$0.10 and 0.20, the electricity cost per transplant is US$0.0094 (=0.12 × 0.10) and 0.019, respectively.

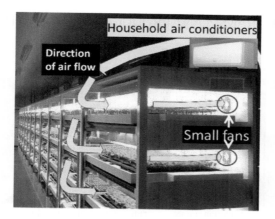

FIGURE 19.5

Air flow pattern in the CTPS. Uniform air currents over the plug trays are generated by using small suction fans inside the shelves and household air conditioners.

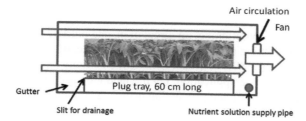

FIGURE 19.6

Schematic diagram of longitudinal section of the culture shelf. Arrows indicate the direction of air flow. Air current speed is uniform over the trays and within the transplant canopy, and can be controlled by a fan with inverter (see also Figure 19.8).

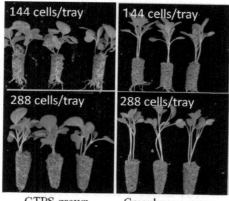

FIGURE 19.7

A horizontal air flow of constant speed is generated inside the plant canopy, so the relative humidity inside the canopy is kept at about 85% or less, enabling high-density transplanting without elongation of hypocotyl of transplants.

Nutrient solution supply
system

Ebb and flow nutrient supply
unit

FIGURE 19.8

Left: Buried nutrient solution tank (a), two stock solution tanks (b), and the control unit (c). Right: Gutter for nutrient solution return (d), and the slit for adjusting the outflow rate and the level of nutrient solution in the culture bed (e). Nutrient solution which is not absorbed by the substrate is returned to the nutrient solution tank to be recycled.

The electricity cost for air conditioning, air circulation, and nutrient solution circulation is roughly one quarter of the cost for lighting (Kozai, 2007), so the total electricity cost per transplant is around US\$0.012 = 0.094 × ((80 + 20)/(80)) or 0.026. Although electricity costs are incurred, (1) no labor or labor cost is required for pesticide application, (2) the land area used for transplant production is reduced by 90% compared with transplant production using a greenhouse, which greatly reduces the labor cost for handling the plug trays (Kozai, 2006, 2007), and (3) no heating cost is required during the photoperiod, and the heating cost is minimized even during the dark period because the CTPS is thermally well insulated and almost airtight.

NUTRIENT SOLUTION SUPPLY

Nutrient solution is typically supplied only once or twice using the ebb and flow system (Figure 19.8) to restrict the water uptake by transplants and thus to produce transplants with thick and short stems. Drained nutrient solution is returned to the nutrient solution tank after filtering to be recycled. EC, pH, and temperature of nutrient solution can be controlled by installing controllers.

The CTPS is always cooled by air conditioners during the photoperiod to remove the heat energy generated by lamps. The water condensed at the cooling panel of air conditioners is also returned after filtering to the nutrient solution tank. Therefore, no drainage pipes are required in the CTPS. By recycling the nutrient solution and condensed water, the water consumption of the CTPS is about 90% less than that of a greenhouse (Kozai, 2007).

ECOPHYSIOLOGY OF TRANSPLANT PRODUCTION

Toshio Shibuya

Graduate School of Life and Environmental Sciences, Osaka Prefecture University, Osaka, Japan

INTRODUCTION

The optimization of plant growth is one of the most important objectives of plant production in controlled environments. Most studies have focused on quantitative growth. Recent studies have also focused on the improvement of qualitative growth through specific nutrient components by controlling physical environmental factors. In transplant production, plant quality should be evaluated from different aspects than for fruit and vegetable production: as greenhouse growers demand transplants which have high productivity after transplantation, the quality of transplants should be evaluated in terms of their growth potential (measured as photosynthetic performance and biotic and abiotic stress resistance) after transplantation. Growth potential is strongly affected by ecophysiological responses to physical environmental factors; plants sense factors such as temperature, humidity, and light, and acclimatize accordingly. In CTPS in which these factors can be controlled (Kozai et al., 2006; Kozai, 2007), controlling the factors might therefore improve the growth potential of transplants.

In addition to plant–environment interactions, plant–plant interactions are important for improving the quality of densely grown transplants, because the micrometeorology within the plant canopy depends on the structure of and gas exchange within the plant canopy, which depend on interactions between neighboring plants. Thus, it is important to understand these interactions in order to optimally control the environmental factors.

This section describes the plant–environment interactions and plant–plant interactions in transplant production from an ecophysiological viewpoint.

EFFECTS OF LIGHT QUALITY ON PHOTOSYNTHETIC PERFORMANCE IN TRANSPLANTS

Light is one of the most important factors for the ecophysiological acclimatization of plants to their environment. The light environment determines photosynthetic performance (Lichtenthaler et al., 1981). Leaves acclimatized to bright light (sun leaves) have a greater maximum photosynthetic rate, whereas those acclimatized to dim light (shade leaves) have a lower light compensation point (LCP). In addition to light quantity, light quality, which is indicated by the ratio of red to far-red light (R:FR), determines photosynthetic acclimatization. In natural ecosystems, the absorption of red light by neighboring vegetation reduces the R:FR, which increases shoot elongation and leaf expansion and decreases leaf thickness (Smith and Whitelam, 1997). This "shade avoidance" response allows plants to tolerate or avoid shading (Franklin, 2008).

The R:FR ratio also affects physiological and morphological properties when it is greater than that of natural light. The typical commercial FLs used in CTPS emit little FR light, making their R:FR generally much greater than that of natural sunlight. Leaves acclimatized to a higher R:FR are thicker (Figure 19.9) and have more chlorophyll, thereby improving their photosynthetic performance. Shibuya et al. (2010a) compared photosynthetic performance between leaves acclimatized to illumination from metal halide lamps (MHL) with spectra similar to that of natural light (R:FR = 1.2) and that from FLs with a high R:FR (=11). The maximum photosynthetic rate of FL-acclimatized leaves was 1.4 times that of MHL-acclimatized leaves at a range of light intensities (Figure 19.10). The photosynthetic and morphological characteristics of high-R:FR-acclimatized leaves were similar to those of sun leaves (Lichtenthaler et al., 1981). The response curve of FL-acclimatized leaves at the acclimatization photosynthetic photon flux (PPF) of $100 \, \mu mol \, m^{-2} \, s^{-1}$ was almost the same as that of MHL-acclimatized leaves at $300 \, \mu mol \, m^{-2} \, s^{-1}$. This indicates that a photosynthetic performance equivalent to that of plants acclimatized to natural light can be obtained by FL

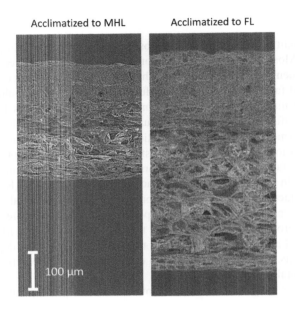

FIGURE 19.9

Cross sections of cucumber cotyledons acclimatized to illumination from metal halide lamps (MHL) with spectra similar to that of natural light (R:FR = 1.2) or fluorescent lamps (FL) with high R:FR (=11).

Modified from Shibuya et al. (2011).

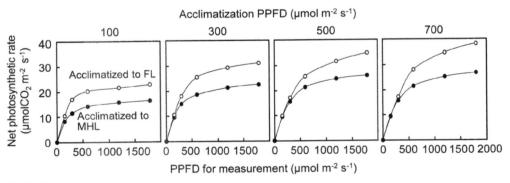

FIGURE 19.10

Light response curves of cucumber seedlings acclimatized to different light qualities and quantities. MHL: metal halide lamps with spectra similar to that of natural light (R:FR = 1.2); FL: fluorescent lamps with high R:FR (=11).

Modified from Shibuya et al. (2012).

acclimatization with one third the quantity of light. This greater photosynthetic performance results from a greater biomass for photosynthesis per unit leaf area and from a higher stomatal density and greater stomatal size (Shibuya et al., 2015). These results mostly agree with the report that an increased proportion of active phytochrome, which is correlated with R:FR, enhances stomatal development (Boccalandro et al., 2009). However, the resultant increased stomatal conductance of high-R:FR-acclimatized leaves reduces the water-use efficiency (=net photosynthetic rate/transpiration rate) below that of normal-R:FR-acclimatized leaves. This lower water-use efficiency may be disadvantageous when plants must be moved to water-deficit conditions, such as during transplantation for horticultural production. Thus, it may be necessary to monitor and prevent water stress when transplanting high-R:FR-acclimatized plants.

The heightened photosynthetic performance of high-R:FR-acclimatized leaves reflects the opposite physiological and morphological responses typical of shade avoidance. This acclimatization is probably advantageous when plants are to be transplanted to high-PPF conditions. On the other hand, high-R:FR illumination inhibits leaf expansion and consequent plant growth during the acclimatization period. This is probably due to greater allocation of biomass to leaf thickening. Trade-offs between quantitative growth and plant defense against environmental stress cause changes in the distribution of biomass (Bazzaz et al., 1987). The acclimatization to higher light intensity that results from a high R:FR may be a defense against environmental stress. In this case, there may be a trade-off between faster growth and improved photosynthetic performance, which is beneficial under higher light intensity. When selecting artificial light for transplant production, we may need to consider which of the two advantages is more important.

EFFECTS OF THE PHYSICAL ENVIRONMENT ON BIOTIC STRESS RESISTANCE IN TRANSPLANTS

Plants acclimatized under shade have less resistance to herbivores and pathogens (McGuire and Agrawal, 2005; Ballaré, 2009; Moreno et al., 2009; Cerrudo et al., 2012). Conversely, plants acclimatized under high-R:FR light have improved resistance to biotic stress. High-R:FR light improved powdery mildew resistance in cucumber seedlings (Shibuya et al., 2011). High-R:FR-acclimatized leaves had thicker epidermis, palisade, and spongy mesophyll (Figure 19.9). As the morphological properties of host plants affect the severity of pathogenic fungal infection (Szwacka et al., 2009), the improved powdery mildew resistance in high-R:FR-acclimatized leaves might be due in part to morphological changes induced by the improved acclimatization to light described above. There are many reports that plant resistance to diseases can be improved by illumination at particular wavelengths (e.g., Wang et al., 2010). Acclimatization to light quality also affects plant–herbivore interactions. Shibuya et al. (2010b) evaluated the preference of adult sweet potato whitefly for cucumber seedlings that had been acclimatized to FL or MHL; there were significantly fewer whitefly adults on FL-acclimatized seedlings, which had a deeper color and thicker leaves, 24 h after release. Environmental factors often have indirect effects on herbivore behavior through their effects on the morphological and physiological characteristics of plants (Berlinger, 1986; Waring and Cobb, 1992; McAuslane, 1996; Chu et al., 1999). The difference in preference of whitefly adults between FL- and MHL-acclimatized leaves may be explained by the morphological characteristics which were induced by light quality. Low-R:FR light strongly reduced plant sensitivity to jasmonates, which are key regulators of plant immunity (Moreno et al., 2009). Therefore, it is likely that the powdery mildew resistance of high-R:FR leaves was improved not only through morphological changes, but also through systemic defense mechanisms.

A similar phenomenon can be observed when humidity is changed. Humidity influences the properties of the leaf surface, and consequently its resistance to water loss (Koch et al., 2006). The structural properties that protect against water loss through the epidermis also have an important function in a plant's interaction with fungal pathogens. In inoculation tests, infection by powdery mildew fungus was inhibited on leaves that had been acclimatized to a high vapor-pressure deficit (VPD) compared with leaves that had been acclimatized to low VPD (Itagaki et al., 2014). The high-VPD-acclimatized leaves were thicker and had greater dry mass per area. The inhibition was caused mainly by changes in leaf morphological properties for protecting against water loss. Humidity also affects plant–herbivore interactions through host-plant responses. Adult sweet potato whitefly preferred low-VPD-acclimatized leaves to high-VPD-acclimatized leaves, which are thicker and darker and have more trichomes (Shibuya et al., 2009a).

Protection from environmental stresses is essential for plant survival. Typical defense responses include the improvement of systemic resistance through antioxidant systems and hardening of epidermal ultrastructure. The improved biotic resistance described in the preceding paragraphs probably results from defense against higher light intensity in high-R:FR-acclimatized plants and against excess water loss in high-VPD-acclimatized plants. But the biomass allocation required for improvement of defense can affect other plant properties. For example, leaf structure is hardened at the expense of a lower leaf expansion rate. This suggests that plant defense takes priority over leaf expansion in the allocation of biomass. Therefore, there would be trade-offs between the growth rate during transplant production and the potential growth ability after transplantation. Transplant growers may have to optimize the environmental conditions to balance these two conflicting benefits.

EFFECTS OF PLANT–PLANT INTERACTIONS ON GAS EXCHANGE WITHIN TRANSPLANT CANOPY

In transplant production, seedlings are often grown densely. Gas exchange between plants and the atmosphere occurs through the boundary layer above the plant canopy. The thickness of the boundary layer depends on factors such as air currents and the structure of the plant canopy, and affects micrometeorological factors such as air temperature, vapor pressure, CO_2 concentration, and air currents within the canopy (Kitaya et al., 1998). In the small volume of the seedling canopy, these factors and the structure of the plant canopy interdependently determine gas exchange, which reciprocally affects the micrometeorology. It is important to clarify this relationship for controlling gas exchange and consequent plant growth.

As the leaf area index (LAI, total leaf area per unit ground area) of the plant canopy increases, the resultant mutual shading reduces the photosynthetic rate per unit leaf area. Reduced gas exchange between leaves and the atmosphere due to the decrease in CO_2 concentration and air currents around the leaves also limits photosynthesis (Kitaya et al., 2003). Thus, controlling air currents is essential to improve gas exchange. Along with the air current speed, the direction of air flow is equally important for the control of gas exchange. Shibuya et al. (2006) investigated the effects of air current direction on photosynthesis of tomato transplant canopy in a forced-ventilation system (Figure 19.11). When the LAI increased from 1.2 to 2.4, conventional horizontal airflow above the canopy reduced the net photosynthetic rate per unit leaf area by 20%, whereas downward airflow reduced it by only 5% (Figure 19.12). The greater reduction in the horizontal system was probably due to limited CO_2 supply into the canopy through the boundary layer.

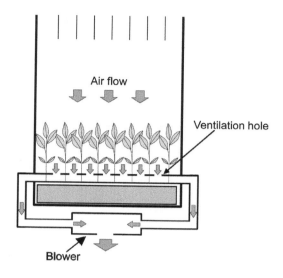

FIGURE 19.11

Plant culture system with forced ventilation within the canopy.

Modified from Shibuya et al. (2006).

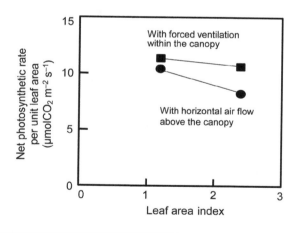

FIGURE 19.12

Effect of forced ventilation on photosynthesis of tomato transplant canopy at different leaf area indexes.

Modified from Shibuya et al. (2006).

Photosynthesis can conversely increase with increasing LAI when photosynthesis is mainly limited by a decrease in the stomatal aperture resulting from water stress. For example, Shibuya et al. (2009b) showed that photosynthesis increased with increasing LAI under high VPD (low humidity) conditions. This phenomenon can be explained by changes in transpiration rate and stomatal conductance (Figure 19.13). As plant density increases, the VPD within the canopy tends to decrease, probably

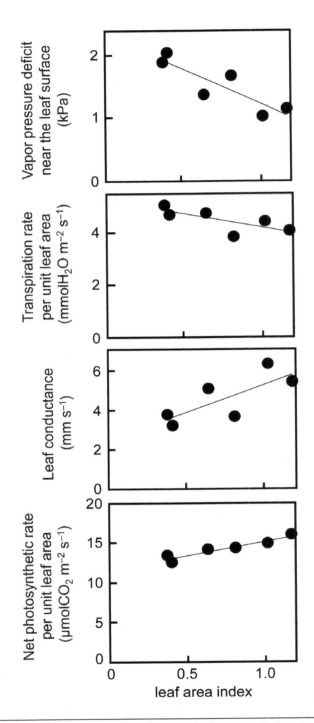

FIGURE 19.13

Relationships between leaf area index and vapor-pressure deficit, transpiration, leaf conductance, and photosynthesis under low humidity (the vapor-pressure deficit above the canopy was 3.7 kPa).

Modified from Shibuya et al. (2009b).

owing to a thickening boundary layer above the canopy, decreasing the transpiration rate per unit leaf area, which mitigates water stress. In summary, as LAI increases, the thickening boundary layer decreases the VPD within the canopy, mitigating the stomatal closure that occurs due to excessive transpiration, thus increasing photosynthesis. This is a good example of the complex interdependence among gas exchange by plants, micrometeorological factors, and the structure of the plant canopy.

EFFECTS OF LIGHT QUALITY ON LIGHT COMPETITION BETWEEN NEIGHBORING PLANTS AND CONSEQUENT EQUALITY OF PLANT GROWTH

The effects of light quality on plant growth have been thoroughly investigated, but those on the structure of plant canopy less so. Changes in light quality due to competition between neighboring plants strongly affect plant growth and morphology; for example, a reduction in the R:FR increases shoot elongation. Competition also affects the structure of a plant community: plants regulate their growth to achieve a height similar to that of their neighbors (Ballaré et al., 1994; Nagashima and Hikosaka, 2011, 2012). This process is called "height convergence" (Nagashima and Terashima, 1995). The pattern of height convergence is affected by the R:FR of background illumination, because the effects of reduced R:FR by neighboring vegetation on the shade avoidance response differ with the value of R: FR: Shibuya et al. (2013) investigated the effects of the R:FR of background illumination on the height convergence pattern in dense plant canopy by growing cucumber seedlings of the same age but with different initial heights under MHL or FL (Figure 19.14). The shoot heights of the initially taller and shorter seedlings under MHL converged within 4 days, whereas those under FL did not reach convergence within 8 days (Figure 19.15). The rapid height convergence under MHL occurred as a result of the shade avoidance response of the initially shorter seedlings stimulated by the decreased R:FR. The slower height convergence under FL is due to an insufficient R:FR level to stimulate the shade avoidance response of the shorter seedlings because of the much higher R:FR of the background illumination. These results suggest that unequal growth of plants is likely to occur under illumination with a high R:FR, because height convergence is hindered. The even plant growth needed for commercial transplant production results from light competition between neighboring plants, which simultaneously causes shoot elongation. However, excess shoot elongation can decrease the commercial value of transplants, because plants with long and narrow stems are easily damaged by environmental stresses and by the handling that occurs during transportation and transplantation of the transplants. So there may be a

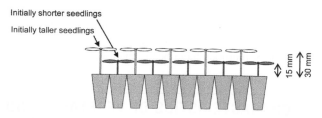

FIGURE 19.14

Cucumber seedlings of the same age but with different initial heights (30 and 15 mm) used to evaluate height convergence.

Modified from Shibuya et al. (2013).

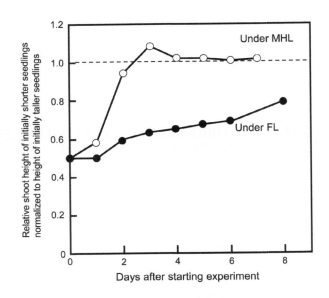

FIGURE 19.15

Time courses of shoot height of the initially shorter seedlings of cucumber in the plant canopy, normalized to the value of the initially taller seedlings, under illumination from metal halide lamps (MHL) with spectra similar to that of natural light (R:FR = 1.2) or fluorescent lamps (FL) with high R:FR (=11).

Modified from Shibuya et al. (2013).

trade-off between even plant growth and shorter stems, both of which are valuable for transplant growers, but which are both inversely controlled by the same ecophysiological mechanism.

CONCLUSIONS

In natural ecosystems, plants grow by optimizing the allocation of their limited biomass for acclimatization to environmental conditions. In plant production, however, this allocation does not always give optimal results. In addition, there is a trade-off between plant growth and plant quality. Thus, growers may have to balance conflicting benefits through environmental control in order to optimize plant growth and quality. Ecophysiological responses to plant–environment and plant–plant interactions must be considered in improving plant production systems, particularly now that many environmental control devices and plant control methods have been developed.

PHOTOSYNTHETIC CHARACTERISTICS OF VEGETABLE AND MEDICINAL TRANSPLANTS AS AFFECTED BY LIGHT ENVIRONMENT

Dongxian He

College of Water Resources and Civil Engineering, China Agricultural University, Beijing, China

INTRODUCTION

Light is one of the most important environmental factors in a PFAL, influencing the growth, morphogenesis, and other physiological processes of plants. The three most important elements in the light environment for plant growth are light intensity or photosynthetic photo flux (PPF), photoperiod, and light quality. High-quality transplants can be produced by appropriately controlling the light environment. This section describes the photosynthetic characteristics of cucumber and tomato transplants and a medicinal plant, *Dendrobium officinale*, as affected by light environment.

INFLUENCE OF LIGHT ENVIRONMENT ON VEGETABLE TRANSPLANT PRODUCTION

The lighting cost in a PFAL accounts for a significant percentage of the total production cost (Kozai, 2007, 2013a). Characterizing the photosynthetic response of transplants to different PPFs will help determine the most cost-effective PPF for vegetable transplant production. Typically, net photosynthetic rate (P_n) increases with increasing PPF within the range between the LCP and the light saturation point (LSP). The LCP is the PPF at which P_n is equal to the respiration rate, and the LSP is the PPF at which P_n no longer increases with increases in PPF.

In Figure 19.16, P_n of tomato (*Lycopersicon esculentum* Mill., cv. Zhongza No. 9) and pepper (*Capsicum annuum* L., cv. CAU No. 24) transplants increases linearly as PPF increases at PPF below 300 $\mu mol\ m^{-2}\ s^{-1}$. The slope of the increase of P_n becomes smaller when PPF is higher than 300 $\mu mol\ m^{-2}\ s^{-1}$ and lower than 600 $\mu mol\ m^{-2}\ s^{-1}$, then reaches a plateau phase. Therefore, from an economic point of view and light use efficiency (for more information, refer to Chapter 4), increasing PPF to higher than 300 $\mu mol\ m^{-2}\ s^{-1}$ would be less efficient for vegetable transplant production in a PFAL. Similar to PPF, P_n of the above tomato and pepper transplants increases linearly with CO_2 concentration. However, for tomato transplants, when CO_2 concentration is higher than 1000 $\mu mol\ mol^{-1}$, P_n does not increase significantly with further increase in CO_2 concentration. For pepper transplants, the CO_2 saturation concentration is about 800 $\mu mol\ mol^{-1}$. Therefore, the PPF and CO_2 concentration in a PFAL should be controlled according to their photosynthetic characteristics and resource use efficiency (Figure 19.16).

Effects of PPF and photoperiod on the growth of vegetable transplants

In order to identify the most cost-effective light environment for vegetable transplant production in a PFAL, cucumber (*Cucumis sativus* L., cv. CAU No. 26) transplants were grown under different combinations of PPF and photoperiod: PPF of 100, 200, 250, and 300 $\mu mol\ m^{-2}\ s^{-1}$ and photoperiod of 12 and 16 h day^{-1} (Table 19.1). Air temperature in the photoperiod and dark period, relative humidity, and CO_2 concentration were maintained at 26.0 ± 1.0°C/22.0 ± 1.0°C, 65 ± 5%, and 800 ± 100 $\mu mol\ mol^{-1}$, respectively. Biomass at the PPF of 100 $\mu mol\ m^{-2}\ s^{-1}$ and photoperiod of 12 h day^{-1} was less than those under all the other conditions. Shoot mass was the greatest at the PPF of 200 $\mu mol\ m^{-2}\ s^{-1}$ and photoperiod of 16 h day^{-1}, whereas root mass was the greatest under the PPF of 200 and 250 $\mu mol\ m^{-2}\ s^{-1}$ and photoperiod of 16 h day^{-1}. In conclusion, the PPF of 200–250 $\mu mol\ m^{-2}\ s^{-1}$ is effective for transplant production, and extending the photoperiod from 12 to 16 h day^{-1} can increase the biomass and growth of cucumber transplants in a PFAL. For commercial transplant production in a PFAL, the PPF could be designed and controlled up to 250 $\mu mol\ m^{-2}\ s^{-1}$.

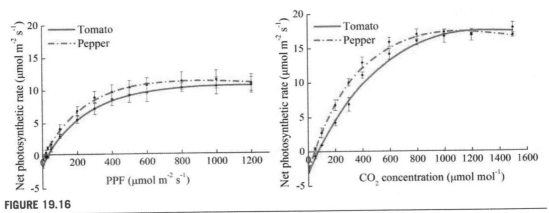

FIGURE 19.16

The response curve of net photosynthetic rates of tomato and pepper transplants to photosynthetic photon flux (PPF) and CO_2 concentration. Environmental conditions of air temperature, relative humidity, PPF, and photoperiod were maintained at $24.0 \pm 1.5°C$, $60 \pm 5\%$, 250 μmol m^{-2} s^{-1}, and 12 h day^{-1}, respectively.

Table 19.1 Influence of Photosynthetic Photon Flux (PPF) and Photoperiod on Fresh Mass and Dry Mass of Cucumber Transplants

Treatments		Fresh Mass (g plant⁻¹)		Dry Mass (mg plant⁻¹)	
Photoperiod (h day⁻¹)	PPF (µmol m⁻² s⁻¹)	Shoot	Root	Shoot	Root
12	100	$2.60 \pm 0.22d^a$	$0.60 \pm 0.15d$	$190.0 \pm 30.3e$	$21.7 \pm 4.1d$
	200	3.77 ± 0.27 cd	$1.18 \pm 0.44c$	$318.3 \pm 23.2d$	$40.0 \pm 16.7c$
	250	3.47 ± 0.46 cd	$1.75 \pm 0.29b$	$311.7 \pm 63.1d$	$60.0 \pm 15.5b$
	300	$4.39 \pm 0.69c$	$1.55 \pm 0.43bc$	$441.7 \pm 104.8c$	$66.7 \pm 25.8b$
16	100	$4.59 \pm 0.80c$	$1.00 \pm 0.35c$	341.7 ± 78.9 cd	$38.3 \pm 16.0c$
	200	$8.60 \pm 1.46a$	$2.42 \pm 0.70a$	$760.0 \pm 175.5a$	$88.3 \pm 21.4ab$
	250	$7.80 \pm 0.61ab$	$2.72 \pm 0.21a$	$635.0 \pm 86.7b$	$106.7 \pm 10.3a$
	300	$7.17 \pm 0.81b$	$2.46 \pm 0.63a$	$618.3 \pm 69.1b$	$100.0 \pm 30.1a$
ANOVA					
PPF		*	*	*	*
Photoperiod		*	*	*	*
PPF*Photoperiod		*	NS	*	NS

*Means with different letters in the same column are significantly different tested by SPSS multiple comparison at P=0.05.

Effects of light quality on growth of vegetable transplants

The spectral characteristics of artificial lighting should meet the physiological requirements of plants for photosynthesis and morphological development (Fraszczak, 2014; Vu et al., 2014; Wesley and Lopez, 2014). The influence of different spectral distributions of light at the PPF of 200 μmol m^{-2} s^{-1} at the canopy level on the growth of vegetable transplants were studied using: white LED lamps (WL),

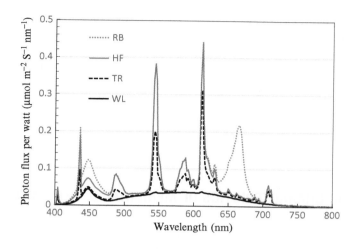

FIGURE 19.17

Spectral distributions of photon flux of different light sources, including white LED lamps (WL, R:B = 1), triphosphate fluorescent lamps (TR, R:B = 2), high-frequency fluorescent lamps (HF, R:B = 3), and red and blue LED lamps (RB, R:B = 4) at the photosynthetic photon flux of 200 μmol m^{-2} s^{-1} at the canopy surface in a PFAL. R:B is red to blue light ratio.

triphosphate fluorescent lamps (TR), high-frequency fluorescent lamps (HF), and red and blue LED lamps (RB) to determine the suitable spectral characteristics (Figure 19.17). Air temperature, relative humidity, CO_2 concentration, PPF, and photoperiod were maintained at 25.0 ± 1.5°C, 60 ± 5%, 600 ± 30 μmol mol^{-1}, 200 μmol m^{-2} s^{-1}, and 12 h day^{-1}, respectively.

The net photosynthetic rate of tomato transplants (*L. esculentum* Mill., cv. Zhongza No. 9) under the RB LED lamps was 12.5–21.2% higher than those under other broad-band white light sources (Table 19.2). However, the stomatal conductance, intercellular CO_2 concentration, and transpiration

Table 19.2 Influence of Different Light Sources (LS): White LED Lamps (WL), Triphosphate Fluorescent Lamps (TR), High-Frequency Fluorescent Lamps (HF), and Red and Blue LED Lamps (RB) on Photosynthetic Parameters of Tomato Transplants Grown at Air Temperature of 25.0 ± 1.5°C, Relative Humidity of 60 ± 5%, CO_2 Concentration of 600 ± 30 μmol mol^{-1}, Photosynthetic Photon Flux (PPF) of 200 μmol m^{-2} s^{-1}, and Photoperiod of 12 h day^{-1}

LS	R:B[a]	Net Photosynthetic Rate (μmol m^{-2} s^{-1})	Stomatal Conductance (mol m^{-2} s^{-1})	Intercellular CO_2 Concentration (μmol mol^{-1})	Transpiration Rate (mmol m^{-2} s^{-1})
WL	1	6.9±0.7b[b]	0.209±0.072a	329±20a	4.17±1.16a
TR	2	6.3±0.8b	0.121±0.052b	287±54b	2.58±0.93b
HF	3	7.0±0.9b	0.204±0.093a	327±30a	3.89±1.47a
RB	4	8.0±0.4a	0.117±0.051b	279±49b	2.13±1.08b

[a]R:B is red to blue ratio.
[b]Means with different letters in the same column are significantly different tested by SPSS multiple comparison at P=0.05.

rate were higher under the WL (R:B of 1) and HF (R:B of 3) lamps than under the TR (R:B of 2) and RB (R:B of 4) lamps. Broad-band white light can increase plant height and promote stem elongation of tomato transplants more effectively than RB LED light. Different light quality had no influence on seedling index [height × (shoot dry mass+root dry mass)/stem diameter] of tomato transplants. The results revealed that RB LED can promote photosynthesis and increase instantaneous water use efficiency of tomato transplants, whereas broad-band white light can increase plant height and promote stem elongation. However, there are big differences in the growth response to light quality among horticultural crops (van Ieperen, 2012). Research is needed to investigate the appropriate light environment for the production of different vegetable transplants in PFALs.

PHOTOSYNTHETIC CHARACTERISTICS OF MEDICINAL *D. OFFICINALE*

D. officinale, a medicinal plant endemic to China, is a rare and endangered orchid called "Plant Gold." Although the cultivated area of *D. officinale* is gradually expanding, its yield cannot meet the market demand because of the slow growth rate. Transplants of *D. officinale* are typically propagated through tissue culture, while mature plants are produced in protected structures such as high tunnels and greenhouses. It takes 4-5 years from tissue culture to mature plant. Due to its high economic value, slow growth, and diminishing wild population, cultivation in PFALs has great potential to promote the growth and development of *D. officinale* transplants. According to the literature (Ritchie and Bunthawin, 2010) and our research results (Zhang et al., 2014), *D. officinale* is a facultative crassulacean acid metabolism (CAM) plant and its photosynthetic pathway shifts between C3 and CAM, depending on environmental conditions.

Using a continuous photosynthesis measurement system, CO_2 exchange rates of the whole shoots of *D. officinale* plants were measured for two consecutive days. In the PFAL with a relatively constant environment (PPF of 164 $\mu mol\ m^{-2}\ s^{-1}$ and daily average air temperature of $26.2 \pm 1.2°C$), the maximum net CO_2 exchange rate of *D. officinale* reached 3.5 $\mu mol\ m^{-2}\ s^{-1}$, then decreased slightly and maintained a positive net CO_2 exchange rate throughout the photoperiod (Figure 19.13). Positive, low net CO_2 exchange rates were detected during the dark period. Net CO_2 exchange rates of *D. officinale* plants grown in a commercial greenhouse in Zhejiang province were measured on a sunny day (the first day, 23 April 2012, maximum PPF inside the leaf chamber reached 168 $\mu mol\ m^{-2}\ s^{-1}$) and a rainy day (the second day, 24 April 2012, maximum PPF inside the leaf chamber reached 85 $\mu mol\ m^{-2}\ s^{-1}$). The net CO_2 exchange rate reached a maximum of 4.0 and 1.6 $\mu mol\ m^{-2}\ s^{-1}$ at 2 h after sunrise on both days and the percentages from the dark period were 31 and 29%, respectively (Figure 19.18). *D. officinale* showed a typical CAM photosynthetic pattern on the sunny day, whereas the duration of positive CO_2 exchange rates during the daytime was longer on the rainy day.

Water deficit is one of the main factors inducing photosynthetic pathway switching, especially for facultative CAM plants. Figure 19.19 shows the time course of net CO_2 exchange rates and the percentage of net CO_2 exchange during the dark period to the daily amount of *D. officinale* in a PFAL for 18 days. Irrigation was withheld for the first 12 days. At the end of the photoperiod on day 12, plants were rewatered. With decreasing substrate water content, the net CO_2 exchange rate of *D. officinale* decreased, and it increased again instantly after rewatering. The percentage of net CO_2 exchange rate during the dark period to the daily amount on day 12 was 51%, 189% higher than on day 1. By day 18, this percentage decreased by 76% compared to that on day 12. Under drought stress, a dominating CO_2

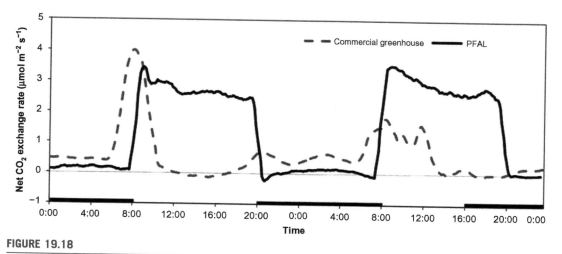

FIGURE 19.18

Time course of net CO_2 exchange rates of *D. officinale* in a PFAL and a commercial greenhouse. Dark bars on the *x*-axis indicate the dark periods in the PFAL.

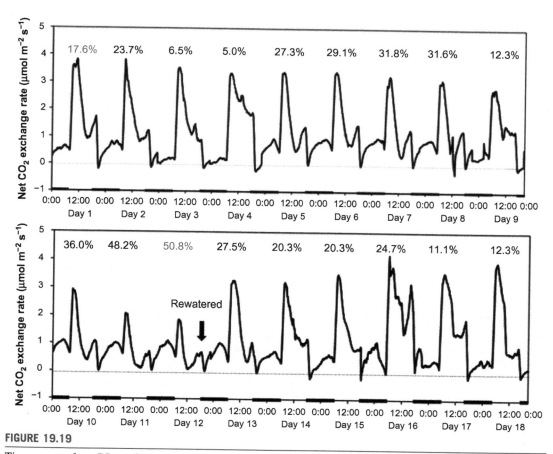

FIGURE 19.19

Time course of net CO_2 exchange rates and the percentages of net CO_2 exchange during the dark period to the daily amount of *D. officinale* in a PFAL for 18 days. Dark bars on the *x*-axis indicate the dark periods.

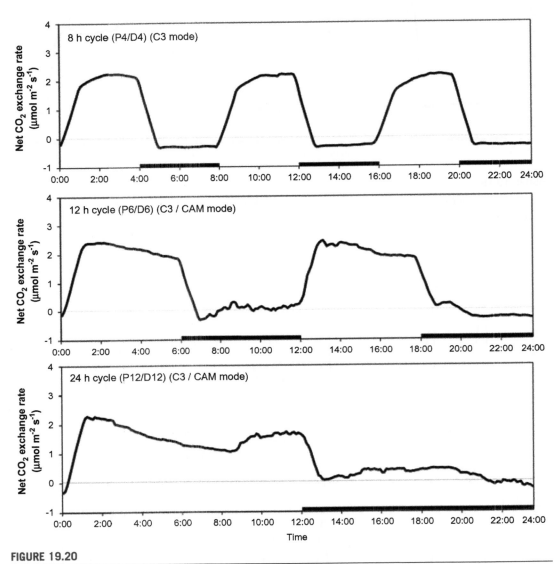

FIGURE 19.20

Time course of net CO_2 exchange rates of *D. officinale* in 8 h (4 h photoperiod, 4 h dark period), 12 h (6 h photoperiod, 6 h dark period), and 24 h (12 h photoperiod and 12 h dark period) light-dark cycles. Dark bars on the *x*-axis indicate the dark periods.

exchange pattern for *D. officinale* was CAM, and concomitance of C3 and CAM patterns were found again when the plants were rewatered.

The net CO_2 exchange rate of *D. officinale* increased rapidly after the onset of the photoperiod and decreased again in 4-6 h in the PFAL (Figure 19.20). Therefore, shortening the light-dark cycles from

24 to 8 h or 12 h (8 and 12 h cycles with equal length of photoperiod and dark period, that is, 4 h light and 4 h dark and 6 h light and 6 h dark, respectively) may increase the daily net CO_2 exchange rate and promote biomass accumulation. When photoperiod and dark periods were shortened, net CO_2 exchange rates in the dark period changed significantly. At the 8 h cycle, a typical C3 photosynthetic pattern was observed and there was little positive net CO_2 exchange rate in the dark period. The positive net CO_2 exchange rate in the dark period indicated a higher degree of CAM photosynthetic pattern under the 24 h cycle compared to the 12 h cycle. In conclusion, adjusting light–dark cycles can induce CO_2 assimilation pattern switching between C3 and CAM in D. *officinale*.

BLUEBERRY

Sma Zobayed

SHIRFA Biotech, Pitt Meadows, British Columbia, Canada
JRT Research and Development, Aldergrove, British Columbia, Canada

Blueberries are perennial flowering plants from the genus *Vaccinium*; the fruits are proven to be rich in disease-fighting nutrients. Canada is the world's number-one producer and exporter of fresh and frozen wild blueberries, with exports valued at $196 million in 2013. Blueberry cultivation is expanding rapidly throughout the world (Eck, 1988), which has led to an increased demand for planting stocks. Conventional propagation has limitations in meeting this increasing demand as propagation by stem cuttings is slow and rooting is difficult (Nickerson, 1978). Lyrene (1978) observed only 7% rooting of stem cuttings, but 95% rooting of tissue culture-derived microcuttings. Consequently, micropropagation has become increasingly popular for rapid multiplication of virus indexed cultivars (Wolfe et al., 1983). However, tissue culture is generally an expensive process (Dunstan and Turner, 1984; Kozai and Iwanami, 1987); therefore, innovative technology needs to be developed that can reduce the costs and make tissue culture more attractive to nursery operators. Much research has been conducted in recent years on the development of protocols for *in vitro* propagation of different varieties of blueberry plants (Brissette et al., 1990; Meiners et al., 2007).

A closed controlled environment system has recently been described as an alternate method of plant propagation (Zobayed et al., 2005). Although the system has many benefits over the conventional method of greenhouse propagation (Kozai, 2005), it is still limited mostly to propagation through seeds. On the other hand, tissue culture derived transplants have many advantages such as rapid, true-to-type, and disease-free transplant propagation with bushy appearance of the final plant products. In this article, a new method of transplant production using tissue culture derived microcuttings of blueberry transplants (*Vaccinium corymbosum* "Duke") under a closed controlled environment system is described. Microcuttings obtained *in vitro* were transplanted in 432-multicell trays filled with perlite (20%) and peat moss (80%) previously sterilized at 200°C for 3 h using a soil sterilizer (Gothic Arch Greenhouses, USA). Trays were placed under controlled environment conditions with a relative humidity of 90%; 16 h photoperiod with a light intensity (PPF) of 100 μmol m^{-2} s^{-1} for the first 7 days followed by 200 μmol m^{-2} s^{-1}. The room temperature was maintained at 23°C throughout the dark and light periods and CO_2 concentration was elevated at 1500 ppm. From day 15, nutrient solution containing N:P:K (ratio of 20:20:20) added with some growth regulators (combination of auxin and cytokinin) was circulated around the root system for 10 min every 3 days by using an even-flow hydroponic system.

A well-established root system developed within 4 weeks of transplanting (Figure 19.21); on week 5 the tips of each transplant were extracted and replanted following the same procedure described above. These transplants usually take about 5 weeks to establish and the tips were cut and replanted again from both old and new batches. This procedure can be continued and within 6 months approximately 1 million transplants can be obtained from 2000 tissue culture derived shoot cuttings in a floor area of 1800 sq ft (eight layers of shelves) (Figure 19.22). The multiplication ratio (number of

FIGURE 19.21

Blueberry transplants growing under a closed controlled environment system.

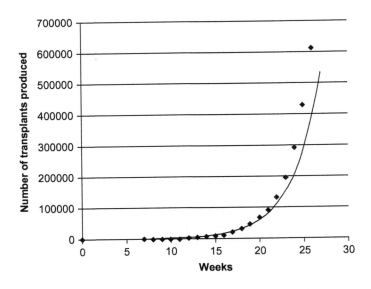

FIGURE 19.22

Number of transplants in a closed controlled environment system (an exponential growth curve).

transplants produced from each mother plant in a month) of the plants through this procedure is higher than that by tissue culture (Figure 19.23). Transplants were eventually moved to a greenhouse for further establishment. Biochemical analysis showed that these plants have similar auxin (indole-3-acetic acid) concentration in their shoot tissues compared with that of the tissue culture derived plants directly established in a greenhouse (Figure 19.24). Their morphological appearance (bushy) and growth after transplanting under greenhouse conditions were similar to tissue culture derived plants.

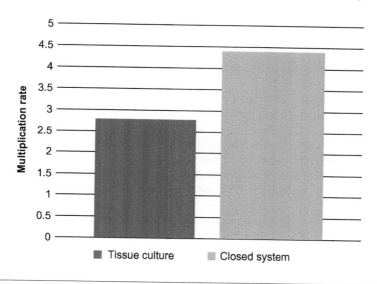

FIGURE 19.23

Multiplication ratio of blueberry plants grown under tissue culture and closed control environment system.

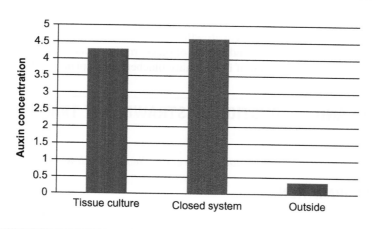

FIGURE 19.24

Auxin concentration (µg/g dry mass) of blueberry plant shoot tissues obtained from tissue culture derived plants, transplants grown under closed controlled environment system and from field derived plants (collected on May 2013; Lynden, WA 98264, USA).

Advantages of this system of propagation are:

1. production of high-quality transplants,
2. uniform growth of transplants under uniformly controlled environments,
3. grown in a protected area free from pests, insects, and pathogens and thus transplants are disease-free,
4. virus-free transplant propagation as the initial microcuttings are originated through tissue culture,
5. no disturbance by outside weather, enabling year-round production,
6. low production cost.

The low production cost can be explained as follows:

(1) the very high productivity per unit floor area per year, mainly due to the use of multilayered shelves (8 shelves),
(2) a high planting density per unit tray area,
(3) a high percentage of salable transplants (>95%), and shorter production period (30–70% compared with greenhouse),
(4) higher utilization efficiencies of water, CO_2, and fertilizers than a greenhouse mainly due to the minimized ventilation and recycling of water,
(5) virtually no heating cost even in the winter because of its thermally insulated structure,
(6) lower labor cost due to the smaller floor area, worker-friendly shelves, and comfortable working environment,
(7) easier control of plant development such as stem elongation, flower bud initiation, bolting, and root formation (Kozai et al., 1998; Kozai, 1998; Kozai et al., 1999; Kozai et al., 2000a,b,c; Kozai et al., 2004),
(8) unlike the tissue culture system, there is no need for any special medium preparation, sterilization of tools and medium, and laminar flow (HEPA filtered air flow),
(9) rooting percentage in this system is almost 100%.

Commercial application of the closed controlled environment system for blueberry transplant propagation requires the development of additional protocols for optimum production. Potential approaches include manipulation of many physical and chemical growth conditions such as light, temperature, and nutrition, which influence the multiplication ratio and growth of blueberry transplants. This technology is likely to play a significant role in the production of high-quality blueberry transplants in the near future.

PROPAGATION AND PRODUCTION OF STRAWBERRY TRANSPLANTS

Changhoo Chun

Department of Plant Science, Seoul National University, Seoul, South Korea

VEGETATIVE PROPAGATION

Fragaria x *ananassa* Duch. ($2n = 8x = 56$), a cultivated strawberry, is an octaploid and interspecific hybrid between the wild species *F. chiloensis* L. ($2n = 8x = 56$) and *F. virginiana* Duch. ($2n = 8x = 56$) (Darrow, 1966). In addition to its polyploidy, allogamous behavior in *F.* x *ananassa* contributes further complexity to the genome structure. Intraspecific crosses have been extensively used to obtain new cultivars with improved agronomic traits. As a result, seed propagation has become unreliable in strawberry production, while limited numbers of seed-propagated cultivars, mainly everbearing, are commercially available using their near isogenic lines.

Most commercial strawberry cultivars are reproduced asexually, mainly using vegetative cuttings made by burying the vegetative stolons, called runners. Runners are the plants' means of vegetative propagation, as runner plants arise from the crown during long days with warm temperatures, beginning in late spring and continuing until autumn (Kirsten, 2013). The crown is a compressed modified stem where leaves, runners, branch crowns, and flower clusters arise. After the development of numerous lateral roots, the runner plants become independent and are cut free from the mother plants.

Strawberry nurseries produce runner plants rooted in substrate in a cell tray (called plug transplants) or a small plastic pot to be supplied as transplants to growers. Some of the advantages of strawberry plug transplants are reduction in pesticide requirement and soil-borne diseases, ease of transplanting, reduced water requirement, and improved plant survival (Durner et al., 2002). The continual growing and cutting of plants from previous ones is the job of the commercial strawberry nursery. Propagules are transplanted in greenhouses at the end of March for early forcing culture and at the end of April for semi-forcing culture to grow into mother plants for producing transplants needed around the middle of September and middle of October, respectively, in the cases of Korea and Japan.

LICENSING AND CERTIFICATION

To meet increasing customer demands for fruit quality and year-round supply and to adapt to changes in local climates and progress of production technologies, new strawberry cultivars have been continuously bred. These new cultivars are protected by plant patents in the United States and Canada, while the most common form of intellectual property protection in many parts of the world is UPOV (Union for the Protection of New Varieties of Plants), which specifies the rights of compliant plant breeders. Forms of licensing varieties and systems for collecting royalty differ from country to country. In the United States and Canada, varieties are licensed on a nonexclusive basis directly to plant nurseries. In the EU, Japan, and Korea on the other hand, the licensing program relies on business partners including private enterprises, nonprofit organizations, and local government bodies with exclusive rights within a defined territory.

The licensed organizations have their own systems for the production, propagation, and distribution of disease-tested stocks. In terms of certified strawberry materials, there are several different grading systems, e.g., nuclear stock, foundation 1 stock, foundation 2 stock, super elite stock, elite stock, and A grade stock under the UK system; nuclear stock (SEE), propagation stock I (SE), propagation stock II (EE), and certified stock (E) under the Dutch system; and nuclear transplant, elite transplant, pre-basic transplant, and basic transplant under the Korean system, and so on (FERA (Food and Environment Research Agency), 2006). The highest-grade transplants are grown under very strict conditions, are propagated under careful control, and receive field inspections for major pathogenic viruses and nematodes.

To eliminate viruses and other pathogens and to reinvigorate strawberry plants for increased runner production, tissue culture is used (Rancillac and Nourrisseau, 1989). One meristem plant can bear 300–800 daughter plants in one season in a screened greenhouse (FPS (Foundation Plant Services), 2008). However, micropropagation of strawberry is not widely used in many countries because of problems with variant types, especially a hyperflowering trait, as documented by Jemmali et al. (1995).

PLUG TRANSPLANTS

Under the UK system, nurseries purchase super elite or elite graded stocks from licensed organizations, to propagate them to produce elite graded stocks or A grade transplants, respectively. Plug transplants are small containerized plants produced from runner tips, which is an alternative to conventional field-grown strawberry transplants (Durner et al., 2002). These are grown in greenhouses and high tunnels in less time than field-produced bare-root transplants.

Plug transplants afford flexible transplanting dates, opportunity of mechanical transplanting, and allow greater water utilization efficiency for transplant establishment compared to fresh bare-root plants. Uses for plug transplants have dramatically increased in recent years, especially in Korea and Japan where most strawberries are cultivated in greenhouses. Technology for producing plug transplants has progressed, resulting in improved cost efficiency, uniformity, and elimination of viruses. Further well-coordinated efforts among horticulturists, engineers, and industry could lead to the development of more efficient and useful production systems for strawberry plug transplants.

TRANSPLANT PRODUCTION IN A PFAL

The novel propagation method for producing strawberry plug transplants in a PFAL developed by Chun et al. (2012) differs from conventional methods in: (1) small size of propagules, (2) high plant density, (3) early fixation of unrooted runner tips, (4) use of generated runner plants as propagules for subsequent propagation cycles, (5) fast propagation cycle, (6) year-round propagation, (7) uniformity of propagules, runner plants, and harvested plug transplants, and (8) simultaneous growth of propagules to attain a sufficient size of plug transplants for shipment or cold storage. Figure 19.25 shows a schematic diagram of the newly developed propagation method.

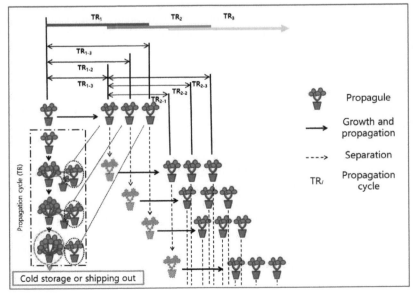

FIGURE 19.25

Schematic diagram of a novel propagation method for producing rooted plug transplants of strawberry in a PFAL developed by Chun et al. (2012). TC*i* is the propagation cycle of the *i*-th generation.

Configuration of S-PFAL

S-PFAL (a PFAL for strawberry transplant production, Figure 19.26a and b basically has the same structural elements as the PFAL reported by Kozai (2013b): (1) thermally insulated warehouse-like structure, (2) multiple racks (30–40 cm between racks vertically) equipped with lighting devices, (3) air-conditioning facilities, (4) CO_2 delivery system, (5) fans for circulating room air, and (6) nutrient solution delivery system. The configuration of an S-PFAL is also similar to a practical commercial system initially developed for seedling production.

FIGURE 19.26

S-PFAL for production of strawberry plug transplants (a) and propagules and runner plants growing on each rack (b).

Environmental control

The use of a CPPS reduces the transplant production time and improves the quality and yield of the final product by optimizing environmental conditions (Chun and Kozai, 2001). Kubota and Kozai (2001) also reported that precise control of plant growth and multiplication was an advantage of vegetative propagation under artificial lighting.

Chun et al. (2003) found that not only the growth of mother plants but also the formation of runners and growth of runner plants of 'Hoko-wase' were promoted under elevated CO_2 concentration (800–1200 mg L^{-1}). Kim et al. (2010) reported that the propagation rate of 'Maehyang' cultivar increased as PPF increased (220–280 μmol m^{-2} s^{-1}) in an S-PFAL. Long-day condition (e.g., photoperiod of 16 h day^{-1}) and relatively warm air temperature (e.g., photo-/dark periods of 28/24°C) promote runner formation, runner elongation, and initiation and growth of runner tips and runner plants, and consequently increase the rate of propagation. Both propagules and runner plants are subirrigated with Yamazaki nutrient solution for strawberry (2 N, 1.5 P, 3 K, 2 Ca, 1 Mg, and 1 S me L^{-1}; 0.7 dS cm^{-1}) for 10 min, twice a day.

Small propagules with high planting density

Mother plants in conventional propagation methods in greenhouses or nursery fields are usually too big to be contained in a rack (30–40 cm high) of an S-PFAL. In the present propagation method, the initial propagules were selected from runner plants of virus-free nuclear stocks. Small propagules (Figure 19.27a) that had a crown diameter of 5 mm and two leaves are used. Each of six propagules is transplanted on a cell of one side row of a 32-cell plastic tray. The distance between two

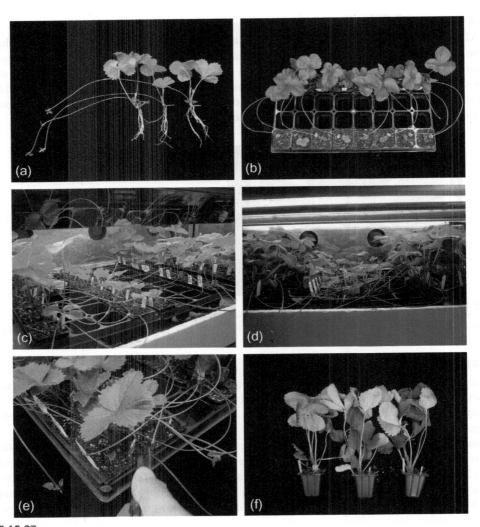

FIGURE 19.27

Smaller propagules (a) are transplanted on plastic plug trays (b) to be cultivated for producing rooted plug transplants. Using U-shaped staples (c) three runner tips from a propagule are fixed in sequence (d) and the rooted runner plants are separated when they reach the same size as initial propagules (e) to be used as propagules in the next propagation cycle. At the end of each propagation cycle, the initial propagules are harvested for shipment or cold storage, of which crown diameter and leaf number are similar to those of stocks used in conventional propagation methods (f).

adjacent propagules is 5.9 cm and each cell is filled with 125 mL of growing medium (Figure 19.27b), resulting in a planting density of 44.4 propagules/m^2.

Fixing of runner tips

Runners from propagules are placed on the medium surface of other cells of the tray and runner tips are fixed 10–12 days after runner initiation using U-shaped staples (Figure 19.27c). Within 2 days after fixation, new roots of 2–3 cm in length develop and an adventitious root zone is consequently established from the bottom of runner tips. In one propagation cycle (32 days), three runner tips from a propagule are fixed in sequence (Figure 19.27d).

Separation of runner plants

Three runner plants are separated from propagules by snipping the runners 22, 26, and 32 days after transplanting of propagules (Figure 19.27e). When they are separated, the crown diameter is about 5 mm and the leaf number is about 2, which is the same as those of initial propagules. The plants are used as propagules in the subsequent propagation cycle.

Simultaneous growth of propagules

When the third runner plant is separated from propagules, the crown diameter and leaf number of initial propagules become about 12 mm and 7, respectively (Figure 19.27f), which is similar to the size of stocks used for conventional propagation in greenhouses and nursery fields. It is also similar to the size of transplants used for soilless culture. Hence, they can be harvested as rooted plug transplants and shipped out. After short and simple acclimatization they can be transplanted for fruit harvest, yielding advantages in dormancy break and promotion of flowering. They can also be stored at $-2°C$ since they are continuously produced year-round in the S-PFAL, while stocks for propagation and transplants for cultivation are needed only at certain times of the year.

Productivity of S-PFAL

The productivity of this new propagation method in an S-PFAL can be evaluated from the cumulative number of runner plants produced. Under optimized culture conditions determined from a series of experiments as described previously, Chun et al. (2012) produced 3497 runner plants in 365 days, starting the propagation process with nine propagules and a cultivation area of 3.6 m^2 in an S-PFAL, which is about 110–140 times greater than that by conventional propagation methods. In this trial, the maximum number of propagules cultivated at the same time was 160, and the cumulative number of runner plants produced increased exponentially until it reached this number, and then increased linearly after exceeding the maximum capacity. Simulation results showed that the annual cumulative number of runner plants produced could be increased to 15,444 and 28,804 if the cultivation area were 18 and 36 m^2, respectively.

APPLICATION OF S-PFAL IN KOREA

The cultivation area of strawberry in Korea is 6800 ha and its yield is 205,000 tons. The Korean strawberry industry was highly dependent on foreign cultivars and the domestic cultivars accounted for less than 4% of the total cultivation area in 2003. However, as new domestic cultivars with high fruit quality have been sequentially bred and introduced, the percentage dramatically increased to 83% in 2014.

Transplant production of strawberry is slow, laborious, and expensive with many limitations. About 25–30 runner plants per mother plant can be propagated in a year with conventional propagation methods. About one billion strawberry rooted plug transplants are needed every year and it takes 5–6 years to diffuse the transplants of newly developed cultivars to Korean growers.

Several S-PFALs using the novel propagation method have been built on a trial basis at national and provincial research agencies where major domestic cultivars are bred and nuclear stocks are maintained. The trials were satisfactory, and so scaling up is currently being reviewed for propagating mainly elite transplants in the Korean national propagation program.

REFERENCES

Ballaré, C., 2009. Illuminated behaviour: phytochrome as a key regulator of light foraging and plant anti-herbivore defence. Plant Cell Environ. 32, 713–725.

Ballaré, C.L., Scopel, A.L., Jordan, E.T., Vierstra, R.D., 1994. Signaling among neighboring plants and the development of size inequalities in plant populations. Proc. Natl. Acad. Sci. USA 91, 10094–10098.

Bazzaz, F.A., Chiariello, N.R., Coley, P.D., Pitelka, L.F., 1987. Allocating resources to reproduction and defense. Bioscience 37, 58–67.

Berlinger, M.J., 1986. Host plant resistance to *Bemisia tabaci*. Agric. Ecosyst. Environ. 17, 69–82.

Boccalandro, H.E., Rugnone, M.L., Moreno, J.E., Ploschuk, E.L., Serna, L., Yanovsky, M.J., Casal, J.J., 2009. Phytochrome B enhances photosynthesis at the expense of water-use-efficiency in *Arabidopsis*. Plant Physiol. 150, 1083–1092.

Brissette, L., Tremblay, L., Lord, D., 1990. Micropropagation of lowbush blueberry from mature field-grown plants. Hortscience 25, 349–351.

Cerrudo, I., Keller, M.M., Cargnel, M.D., Demkura, P.V., de Wit, M., Patitucci, M.S., Pierik, R., Pieterse, C.M.J., Ballaré, C.L., 2012. Low red/far-red ratios reduce *Arabidopsis* resistance to *Botrytis cinerea* and jasmonate responses via a COI1-JAZ10-dependent, salicylic acid-independent mechanism. Plant Physiol. 158, 2042–2052.

Chu, C.C., Cohen, A.C., Natwick, E.T., Simmons, G.S., Henneberry, T.J., 1999. *Bemisia tabaci* (Hemiptera: Aleyrodidae) biotype B colonisation and leaf morphology relationships in upland cotton cultivars. Aust. J. Entomol. 38, 127–131.

Chun, C., Kozai, T., 2000. Closed transplant production system at Chiba University. In: Kubota, C., Chun, C. (Eds.), Transplant Production in the 21st Century. Kluwer Academic Publishers, Dordrecht, The Netherlands, pp. 20–27.

Chun, C., Kozai, T., 2001. A closed type transplant production system, a hybrid of scaled-up micropropagation system and plant factory. J. Plant Biotechnol. 3, 59–66.

Chun, C., Takagi, M., Kozai, T., Kato, M., 2003. Effect of CO2 concentration and PPF on number of strawberry propagules produced and their growth in a closed system (written in Japanese: Heisa-gata ichigo ikubyo shisutemu ni okeru nisanka tanso oyobi kogosei yuko koryoshisoku ga zoshokutai seisansu oyobi seiiku ni oyobosu eikyo). In: Proceedings of 2003 Annual Meeting of Japanese Society of Agricultural, Biological and Environmental Engineers and Scientists. 135.

Chun, C., Park, S.W., Jeong, Y.W., Ko, K.D., 2012. Strawberry propagation method using closed transplant production systems. Korean patent 10-1210680.

Darrow, G.M., 1966. The Strawberry. Holt, Rinehart & Winston, New York. pp. 49-114.

Dunstan, D.I., Turner, K.E., 1984. The acclimatization of micropropagated plants. In: Vasil, I.K. (Ed.), Cell Culture and Somatic Cell Genetics of Plants, vol. 1. Academic, New York, pp. 123–129.

Durner, E.F., Poling, E.B., Maas, J.L., 2002. Recent advances in strawberry plug transplant technology. HortTechnology 12, 545–550.

Eck, P., 1988. Blueberry Science. Rutgers University Press, New Brunswick, N.J..

FERA (Food and Environment Research Agency), 2006. Plant Health Propagation Scheme. https://www.gov.uk/government/uploads/system/uploads/attachment_data/file/386077/phps-soft-fruit.pdf.

FPS (Foundation Plant Services), 2008. Guide to the Strawberry Clean Plant Program. http://fpms.ucdavis.edu/websitepdfs/articles/fpsstrawberrybrochure08.pdf.

Franklin, K.A., 2008. Shade avoidance. New Phytol. 179, 930–944.

Fraszczak, B., 2014. The effect of fluorescent lamps and light-emitting diodes on growth of dill plants. Z. Arznei.-Gewurzpfla. 19 (1), 34–39.

Itagaki, K., Shibuya, T., Tojo, M., Endo, R., Kitaya, Y., 2014. Atmospheric moisture influences on conidia development in *Podosphaera xanthii* through host-plant morphological responses. Eur. J. Plant Pathol. 138, 113–121.

Jemmali, A., Boxus, P., Kevers, C., Gaspar, T., 1995. Carry-over of morphological and biochemical characteristics associated with hyperflowering of micropropagated strawberries. Plant Physiol. 147, 435–440.

Kim, S.K., Jeong, M.S., Park, S.W., Kim, M.J., Na, H.Y., Chun, C., 2010. Improvement of runner plant production by increasing photosynthetic photon flux during strawberry transplant propagation in a closed transplant production system. Korean J. Hortic. Sci. Technol. 28, 535–539.

Kirsten, A., 2013. The Mid-Atlantic Berry Guide. The Pennsylvania State University, University Park, PA, USA.

Kitaya, Y., Shibuya, T., Kozai, T., Kubota, C., 1998. Effects of light intensity and air velocity on air temperature, water vapor pressure and CO_2 concentration inside a plant canopy under an artificial lighting condition. Life Support Biosph. Sci. 5, 199–203.

Kitaya, Y., Tsuruyama, J., Shibuya, T., Yoshida, M., Kiyota, M., 2003. Effects of air current speed on gas exchange in plant leaves and plant canopies. Adv. Space Res. 31, 177–182.

Koch, K., Hartmann, K.D., Schreiber, L., Barthlott, W., Nienhuis, C., 2006. Influences of air humidity during the cultivation of plants on wax chemical composition, morphology and leaf surface wettability. Environ. Exp. Bot. 56, 1–9.

Kozai, T., 1998. Transplant production under artificial light in closed systems. In: Lu, H.Y., Sung, J.M., Kao, C.H. (Eds.), Asian Crop Science 1998, Proc. of the 3rd Asian Crop Science Conference. Taichung, Taiwan, pp. 296–308.

Kozai, T., 2005. Closed system with lamps for high quality transplant production at low costs using minimum resources. In: Kozai, T., Afreen, F., Zobayed, S. (Eds.), Photoautotrophic (Sugar-Free Medium) Micropropagation as a New Micropropagation and Transplant Production System. Springer, The Netherlands, pp. 275–311.

Kozai, T., 2006. Closed systems for high quality transplants using minimum resources. In: Gupta, S.D., Ibaraki, Y. (Eds.), Plant Tissue Culture Engineering. Springer, Berlin, pp. 275–312.

Kozai, T., 2007. Propagation, grafting and transplant production in closed systems with artificial lighting for commercialization in Japan. Propag. Ornam. Plants 7 (3), 145–149.

Kozai, T., 2013a. Resource use efficiency of closed plant production system with artificial light: concept, estimation and application to plant factory. Proc. Jpn. Acad. Ser. Phys. Biol. Sci. 89 (10), 447–461.

Kozai, T., 2013b. Sustainable plant factory: closed plant production systems with artificial light for high resource use efficiencies and quality produce. Acta Hortic. 1004, 27–40.

Kozai, T., Iwanami, Y., 1987. Effects of CO_2 enrichment and sucrose concentration under high photon fluxes on plant growth of carnation (*Dianthus caryophylus* L.) in the tissue culture during the preparation stage. J. Jpn. Soc. Hortic. Sci. 57, 279–288.

Kozai, T., Kubota, C., Heo, J., Chun, C., Ohyama, K., Niu, G., Mikami, H., 1998. Towards efficient vegetative propagation and transplant production of Sweetpotato (*Ipomoea batatas* (L.) Lam.) under artificial light in closed systems. In: Proc. of International Workshop on Sweetpotato Production System toward the 21st Century. Miyazaki, Japan, pp. 201–214.

Kozai, T., Ohyama, K., Afreen, F., Zobayed, S., Kubota, C., Hoshi, T., Chun, C., 1999. Transplant production in closed systems with artificial lighting for solving global issues on environmental conservation, food, resource and energy. In: Proc. of ACESYS III Conference. Rutgers University, CCEA (Center for Controlled Environment Agriculture), pp. 31–45.

Kozai, T., Kubota, C., Chun, C., Ohyama, K., 2000a. Closed transplant production systems with artificial lighting for quality control, resource saving and environment conservation. In: Proceedings of The XIV Memorial CIGR World Congress 2000, November 28-December 1, Tsukuba, Japan, pp. 103–110.

Kozai, T., Chun, C., Ohyama, K., Kubota, C., 2000b. Closed transplant production systems with artificial lighting for production of high quality transplants with environment conservation and minimum use of resource. In: Proceedings of The 15th Workshop on Agricultural Structures and ACESYS (Automation, Culture, Environment and System). Conference December 4–5, Tsukuba, Japan, pp. 110–126.

Kozai, T., Kubota, C., Chun, C., Afreen, F., Ohyama, K., 2000c. Necessity and concept of the closed transplant production system. In: Kubota, C., Chun, C. (Eds.), Transplant Production in the 21st Century. Kluwer Academic Publishers, Dordrecht, The Netherlands, pp. 3–19.

Kozai, T., Chun, C., Ohyama, K., 2004. Closed systems with lamps for commercial production of transplants using minimal resources. Acta Hortic. 630, 239–254.

Kozai, T., Ohyama, K., Chun, C., 2006. Commercialized closed systems with artificial lighting for plant production. Acta Hortic. 711, 61–70.

Kubota, C., Kozai, T., 2001. Mathematical models for planning vegetative propagation under controlled environments. HortScience 36, 15–19.

Lichtenthaler, H.K., Buschmann, C., Döll, M., Fietz, H.J., Bach, T., Kozel, U., Meier, D., Rahmsdorf, U., 1981. Photosynthetic activity, chloroplast ultrastructure, and leaf characteristics of high-light and low-light plants and of sun and shade leaves. Photosynth. Res. 2, 115–141.

Lyrene, P., 1978. Blueberry callus and shoot-tip culture. Proc. Fla. State Hortic. Soc. 91, 171–172.

McAuslane, H.J., 1996. Influence of leaf pubescence on ovipositional preference of Bemisia argentifolii (Homoptera: Aleyrodidae) on soybean. Environ. Entomol. 25, 834–841.

McGuire, R., Agrawal, A.A., 2005. Trade-offs between the shade-avoidance response and plant resistance to herbivores? Tests with mutant Cucumis sativus. Funct. Ecol. 19, 1025–1031.

Meiners, J., Schwab, M., Szankowski, I., 2007. Efficient in vitro regeneration systems for Vaccinium species. Plant Cell Tissue Organ Cult. 89, 169–176.

Moreno, J.E., Tao, Y., Chory, J., Ballaré, C.L., 2009. Ecological modulation of plant defense via phytochrome control of jasmonate sensitivity. Proc. Natl. Acad. Sci. USA 106, 4935–4940.

Nagashima, H., Hikosaka, K., 2011. Plants in a crowded stand regulate their height growth so as to maintain similar heights to neighbours even when they have potential advantages in height growth. Ann. Bot. 108, 207–214.

Nagashima, H., Hikosaka, K., 2012. Not only light quality but also mechanical stimuli are involved in height convergence in crowded Chenopodium album stands. New Phytol. 195, 803–811.

Nagashima, H., Terashima, I., 1995. Relationships between height, diameter and weight distributions of Chenopodium album plants in stands: effects of dimension and allometry. Ann. Bot. 75, 181–188.

Nickerson, N.L., 1978. In vitro shoot formation in lowbush blueberry seedling explants. HortScience 13, 698.

Rancillac, M., Nourrisseau, J.G., 1989. Micropropagation and strawberry plant quality. Acta Hortic. 265, 343–348.

Ritchie, R.J., Bunthawin, S., 2010. The use of pulse amplitude modulation (PAM) fluorometry to measure photosynthesis in a CAM orchid, Dendrobium spp. (D. cv. Viravuth Pink). Int. J. Plant Sci. 171 (6), 575–585.

Shibuya, T., Tsuruyama, J., Kitaya, Y., Kiyota, M., 2006. Enhancement of photosynthesis and growth of tomato seedlings by forced ventilation within the canopy. Sci. Hortic. 109, 218–222.

Shibuya, T., Hirai, N., Sakamoto, Y., Komuro, J., 2009a. Effects of morphological characteristics of Cucumis sativus seedlings grown at different vapor pressure deficits on initial colonization of Bemisia tabaci (Hemiptera: Aleyrodidae). J. Econ. Entomol. 102, 2265–2267.

Shibuya, T., Sugimoto, A., Kitaya, Y., Kiyota, M., 2009b. High plant density of cucumber (*Cucumis sativus* L.) seedlings mitigates inhibition of photosynthesis resulting from high vapor-pressure-deficit. HortScience 44, 1796–1799.

Shibuya, T., Endo, R., Hayashi, N., Kitamura, Y., Kitaya, Y., 2010a. Potential photosynthetic advantages of cucumber (*Cucumis sativus* L.) seedlings grown under fluorescent lamps with high red:far-red light. HortScience 45, 553–558.

Shibuya, T., Komuro, J., Hirai, N., Sakamoto, Y., Endo, R., Kitaya, Y., 2010b. Preference of sweetpotato whitefly adults to cucumber seedlings grown under two different light sources. HortTechnology 20, 873–876.

Shibuya, T., Itagaki, K., Tojo, M., Endo, R., Kitaya, Y., 2011. Fluorescent illumination with high red-to-far-red ratio improves resistance of cucumber seedlings to powdery mildew. HortScience 46, 429–431.

Shibuya, T., Endo, R., Hayashi, N., Kitaya, Y., 2012. High-light-like photosynthetic responses of *Cucumis sativus* leaves acclimated to fluorescent illumination with a high red:far-red ratio: interaction between light quality and quantity. Photosynthetica 50, 623–629.

Shibuya, T., Takahashi, S., Endo, R., Kitaya, Y., 2013. Height-convergence pattern in dense plant stands is affected by red-to-far-red ratio of background illumination. Sci. Hortic. 160, 65–69.

Shibuya, T., Endo, R., Yuba, T., Kitaya, Y., 2015. The photosynthetic parameters of cucumber as affected by irradiances with different red:far-red ratios. Biol. Plant. 59, 198–200. In press.

Smith, H., Whitelam, G.C., 1997. The shade avoidance syndrome: multiple responses mediated by multiple phytochromes. Plant Cell Environ. 20, 840–844.

Szwacka, M., Tykarska, T., Wisniewska, A., Kuras, M., Bilski, H., Malepszy, S., 2009. Leaf morphology and anatomy of transgenic cucumber lines tolerant to downy mildew. Biol. Plant. 53, 697–701.

van Ieperen, W., 2012. Plant morphological and developmental responses to light quality in a horticultural context. Acta Hortic. 956, 131–140.

Vu, N.T., Kim, Y.S., Kang, H.M., 2014. Influence of short-term irradiation during pre- and post-grafting period on the graft-take ratio and quality of tomato seedlings. Hortic. Environ. Biotechnol. 55 (1), 27–35.

Wang, H., Yu, J.Q., Jiang, Y.P., Yu, H.J., Xia, X.J., Shi, K., Zhou, Y.H., 2010. Light quality affects incidence of powdery mildew, expression of defence-related genes and associated metabolism in cucumber plants. Eur. J. Plant Pathol. 127, 125–135.

Waring, G.L., Cobb, N.S., 1992. The impact of plant stress on herbivore population dynamics. In: Bernays, E.A. (Ed.), In: Insect–Plant Interactions, vol. 4. CRC Press, Boca Raton, FL, USA, pp. 167–226.

Wesley, C.R., Lopez, R.G., 2014. Comparison of supplemental lighting from high-pressure sodium lamps and light-emitting diodes during bedding plant seedling production. HortScience 49 (5), 589–595.

Wolfe, D.E., Eck, P., Chin, C., 1983. Evaluation of seven media for micropropagation of highbush blueberry. HortScience 18, 703–705.

Zhang, Z., He, D., Niu, G., Gao, R., 2014. Concomitant CAM and C3 photosynthetic pathways in *Dendrobium officinale* plants. J. Am. Soc. Hortic. Sci. 139 (3), 209–298.

Zobayed, S., Afreen, F., Kozai, T., 2005. Necessity and production of medicinal plants under controlled environments. Environ. Control Biol. 43, 243–252.

PHOTOAUTOTROPHIC MICROPROPAGATION

20

Quynh Thi Nguyen[1], Yulan Xiao[2], Toyoki Kozai[3]

Institute of Tropical Biology, Vietnam Academy of Science and Technology, Hochiminh City, Vietnam[1] Yangtze Delta Region Institute of Tsinghua University, Jiaxing, China[2] Japan Plant Factory Association, c/o Center for Environment, Health and Field Sciences, Chiba University, Kashiwa, Chiba, Japan[3]

INTRODUCTION

Photoautotrophic micropropagation (PAM) narrowly refers to the propagation and growth of explants or plants under disease-free conditions on medium containing no supplemental organic components as nutrients. In PAM or sugar-free medium micropropagation, chlorophyllous explants having photosynthetic ability are used to enhance their photoautotrophic growth. Environmental factors such as light intensity and carbon dioxide concentration need to be properly controlled for promoting photosynthesis of explants and/or plants. PAM, which improves both the *in vitro* aerial and root zone environments, significantly promotes the growth of *in vitro* plants, increases the multiplication rate, and thus shortens the multiplication period of the *in vitro* stage. Moreover, *in vitro* plants grown under photoautotrophic conditions show better acclimatization with higher survival rate during the *ex vitro* stage. This chapter reviews the special features of PAM and its practical application over the last decade, and the potential for scaling up large culture vessels to aseptic culture rooms for closed transplant production systems.

DEVELOPMENT OF PAM

Intensive studies on the PAM method started in the late 1980s at Chiba University, Japan, with several experiments on environmental factors surrounding *in vitro* plants in small vessels, such as CO_2 concentration, photosynthetic photon flux (PPF), photo- and dark periods, etc. All findings revealed a high net photosynthetic rate for *in vitro* plants cultured on media containing only inorganic components when the *in vitro* environment is properly controlled.

There have been more than 100 articles reporting successful PAM for more than 50 plant species (Kozai et al., 2005), and there have been about eight review papers on this new method (Kozai, 1991; Jeong et al., 1995; Kubota et al., 1997; Kozai and Nguyen, 2003; Zobayed et al., 2004; Kozai and Xiao, 2005; Kozai, 2010; Xiao et al., 2011). There have also been four books containing chapters on PAM (Kurata and Kozai, 1992; Aitken-Christie et al., 1995; Kubota and Chun, 2000; Kozai et al., 2005). PAM was the focus of one symposium (Kozai et al., 1995).

These publications stated that most chlorophyllous explants/plants *in vitro* including somatic embryos at the cotyledonary stage are able to grow photoautotrophically, and that the low or negative

Plant Factory. http://dx.doi.org/10.1016/B978-0-12-801775-3.00020-2

net photosynthetic rate of plants cultured photomixotrophically is due not only to poor photosynthetic ability, but also to the low CO_2 concentration in the air-tight culture vessel during the photoperiod. Furthermore, it has been proved that the photoautotrophic growth of many plant species can be significantly promoted by increasing the CO_2 concentration and light intensity in the vessel, by decreasing the relative humidity in the vessel, and by using a fibrous or porous supporting material with high air porosity instead of gelling agents such as agar.

These studies have shown that for successful PAM, an understanding of the *in vitro* environment and basic theories of environmental control is essential. From our own experience, another important aspect for successful application is a physio-ecological knowledge of plant species. To enhance photosynthesis of *in vitro* plants, it is necessary to understand the status of the *in vitro* environment inside the culture vessel and know how to create optimum environmental conditions for maximizing *in vitro* plant photosynthetic growth. Without a good knowledge of the *in vitro* environment and the interaction between plants and the *in vitro* (and *ex vitro*) environments, the PAM method is difficult to apply.

ADVANTAGES AND DISADVANTAGES OF PAM FOR GROWTH ENHANCEMENT OF *IN VITRO* PLANTS

PAM has many advantages over conventional (photomixotrophic) micropropagation using sugar-containing medium with respect to both the biological aspects of *in vitro* plants and engineering aspects. Biological advantages include: (1) promotion of growth and photosynthesis of *in vitro* plants; (2) prevention of morphological and physiological disorders; (3) decrease in microbial contamination inside the culture vessel; (4) shortening of the *in vitro* plant multiplication cycle; and (5) high survival rate of *in vitro* plants when transferred to an *ex vitro* environment. Engineering advantages include: (1) simplification of the micropropagation system; (2) flexibility in the design of the culture vessel for large-scale *in vitro* plant production; (3) year-round increase in production yield per unit floor area; (4) reduction of labor cost; and (5) easier automation of the culture system.

However, PAM also has disadvantages such as: (1) knowledge required of the physical *in vitro* environment; (2) higher costs for CO_2 enrichment or use of gas-permeable filter discs to increase CO_2 concentration in the vessels, lighting, and cooling; and (3) limited application to multiplication system using multi-shoots or plants having C4 or CAM photosynthesis pathway.

In PAM, PPF, and CO_2 concentration during the photoperiod play the most important roles. Since increasing PPF alone cannot raise the net photosynthetic rate for *in vitro* plants at their CO_2 compensation point, the CO_2 concentration in the vessel needs to be increased to at least 400 μmol mol^{-1} (standard atmospheric CO_2 concentration in the culture vessel, not in the culture room) either by natural or forced ventilation.

NATURAL VENTILATION SYSTEM USING DIFFERENT TYPES OF SMALL CULTURE VESSELS

The natural ventilation method is mainly described as an air exchange (CO_2 and water vapor), caused by the difference in partial pressure in water vapor and CO_2 gas and the difference in air pressure between inside and outside the vessel caused by the air current surrounding the vessel. One simple way to enhance the natural ventilation of a vessel is to use lids having gas-permeable filters, or to use vessels having improved ventilation properties. Therefore, the vessel ventilation rate or number of air exchanges (N) will be increased in accordance with the number of microporous gas filter discs attached to the vessel. In the natural ventilation method, by attaching one, two, and three Millipore filter discs (10 mm in diameter each) on the vessel lid or sidewalls, the CO_2 concentration inside a Magenta GA-7 box-type vessel containing *in vitro* plants reaches about 150, 200, and 250 μmol mol^{-1}, respectively, during the photoperiod when CO_2 concentration in the culture room is maintained at 350–400 μmol mol^{-1}. By increasing the ventilation of the culture vessel, on the other hand, the concentration of ethylene accumulating in airtight vessels can also be reduced, which in turn avoids the adverse effects of ethylene on plant development, such as shoot regeneration, leaf expansion, and shoot growth (Jackson et al., 1991; Biddington, 1992).

The benefits of using gas-permeable film or discs for increasing the vessel ventilation rate, plant photosynthesis, and growth have been shown by Kozai et al. (1988) in potatoes, Kozai and Iwanami (1988) in carnation, Nguyen et al. (1999) in coffee, Cui et al. (2000) and Kubota and Kozai (2001) in tomato, Nguyen and Kozai (2001) in banana (*Musa* spp.), Lucchesini et al. (2001) in myrtle (*Myrtus communis* L.), Xiao et al. (2003) in sugarcane, Couceiro et al. (2006a) in St. John's wort (*Hypericum perforatum*), Xiao and Kozai (2006) in statice, Zhang et al. (2009) in *Momordica grosvenori*, and by many other authors. Several woody plants, such as *Thyrsostachys siamensis*, *Gmelina arborea*, *Coffea arabusta*, *Azadirachta indica*, *Paulownia fortunei*, *Acacia*, *Eucalyptus*, etc., were also proved to have greater growth on sugar-free medium in ventilated vessels than on sugar-containing medium in airtight vessels (Nguyen and Kozai, 2005). Simulations for estimating the CO_2 concentration inside the culture vessel (C_i), net photosynthetic rate (P_n) of *in vitro* plants and increase in plant dry weight as functions of N, PPF, and photoperiod were also conducted intensively by Niu et al. (1996, 1997).

When cultured in Magenta vessels on two supporting materials, gelrite (G) or Florialite (F), under ambient CO_2 concentration (400 μmol mol^{-1}) and PPF of 100 μmol m^{-2} s^{-1}, *in vitro* strawberry (*Fragaria × ananassa* Duch.) plants showed an increase in P_n over time under FG and FF treatments on sugar-free medium, as compared with SG and SF treatments on sugar-containing medium, at which P_n became very low, nearly zero (Nguyen and Nguyen, 2008). After 15 days in the *ex vitro* stage, the percent of survival was the highest, 100%, in the FF treatment, and was the lowest, 85%, in the SG treatment (Figure 20.1).

Nguyen et al. (2012) also proved the photoautotrophic growth of thyme plants (*Thymus vulgaris* L.) in ventilated Magenta vessels ($N = 3.9$ h^{-1}) by using nodal cuttings. On day 35, thyme plants cultured on sugar-free medium under PPF of 95 μmol m^{-2} s^{-1} showed the greatest increase in fresh and dry weights, as well as the longest shoots and roots. P_n of *in vitro* thyme plants cultured on sugar-free medium was ten times higher than those on sugar-containing medium under the same PPF (2.6 vs. 0.23 μmol h^{-1} per plant).

There are numerous commercially available gas-permeable membrane and vessel systems with high ventilation characteristics, such as polycarbonate Magenta box-type vessels having Millipore gas-permeable filter discs as shown in Figure 20.2a. However, gas-permeable filter discs are still costly, hindering the greater commercialization of PAM. Flasks capped with thin, white paper or plastic (polypropylene) bag-type vessels having paper filter discs (1.3 cm in diameter) attached on the hole of the vessel sidewalls have been used for PAM in Vietnam recently because of their low cost (Figure 20.2b and c). Nguyen et al. (2010a) proved that, on day 45, nodal cuttings of *Plectranthus amboinicus*, cultured on a sugar-free half-strength MS (Murashige and Skoog, 1962) medium inside

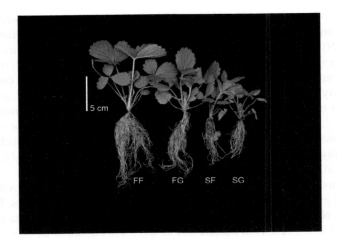

FIGURE 20.1

Strawberry *in vitro* plants as affected by different sucrose concentrations and supporting materials after 15 days in the *ex vitro* stage. For treatment codes, F or S on the left denotes sugar-free or sugar-containing medium, F or G on the right denotes Florialite™ or GELRITE® used as supporting material.

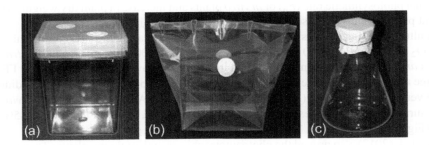

FIGURE 20.2

Vessels often used in photoautotrophic micropropagation: (a) Magenta box-type vessel having Millipore® filter discs fixed on each of the holes of the vessel cap; (b) Plastic (polypropylene) bags having paper filter discs with the opening of the bag folded two times and fixed with five paper clips; and (c) Flasks capped with sterilized paper.

plastic bag-type vessels (12 cm wide, 19 cm high, 8 cm deep at the bottom; air volume: 1.2 L when inflated with air) with two or four ventilated paper filter discs ($N = 4.2$ or 5.5 h^{-1}, respectively) grew much better than those cultured on sugar-containing medium inside bags without filter discs ($N = 0.7 \text{ h}^{-1}$). A high ventilation rate ($N = 5.5 \text{ h}^{-1}$) in accordance with high light intensity of 170 µmol m^{-2} s^{-1}, moreover, significantly increased fresh and dry weights of *P. amboinicus in vitro* plants (Nguyen et al., 2010a).

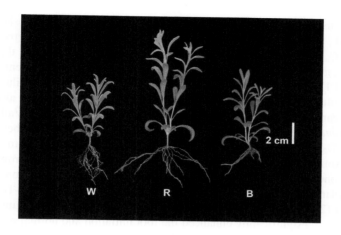

FIGURE 20.3

Lavender plants cultured photoautotrophically in plastic bags under different light qualities on day 42. For treatment codes, W, R, or B denotes *in vitro* plants grown under white, red, or blue fluorescent lamps, respectively.

Plastic bag-type vessels ($V = 1$ L when inflated with air) having two ventilated paper filter discs were also used to prove the effect of light quality, produced by white, red, or blue fluorescent lamps, on the biomass accumulation of lavender (*Lavandula angustifolia*) plants cultured photoautotrophically (on MS medium with only mineral source) under ambient CO_2 concentration, PPF of 140 μmol m^{-2} s^{-1} and photoperiod of 12 h day^{-1}. The number of primary roots and shoot length were the highest when lavender plants were cultured under red fluorescent lamps on day 42 (Figure 20.3). However, under white fluorescent lamps, fresh and dry weights together with the number of secondary roots were significantly greater than those under blue or red fluorescent lamps (unpublished data).

The use of ventilated vessels obviously increases plant transpiration and medium evaporation, which is beneficial for plant growth because nutrient uptake of *in vitro* plants is also enhanced (Kozai, 2010). Explants, previously living in airtight vessels with poor cuticular wax layer formation and stomatal malfunction, may wilt for the first few days after being transplanted to the ventilated vessel. This phenomenon can be solved by covering all ventilated membranes with clear tape and reducing the PPF at the culture shelf to a level equal to that of the previous culture (30–50 μmol m^{-2} s^{-1}) for the first few days or a week. The tape needs to be removed and the PPF should be raised gradually when small new shoots and/or new roots appear. There is no wilting problem if explants are derived from the photoautotrophic condition. The use of a porous or fibrous substrate with dry substrate surface is obviously advantageous for promoting the root growth and is described in many publications. The transplanting of explants onto the medium in culture vessels or plastic bags will be easier to perform if the substrate is shaped into a cube block rather than a loose substrate (Kozai, 2010).

CO_2 in the ventilated vessel (C_{in}) in the natural ventilation method can be increased by enhancing the air diffusion (CO_2 and water vapor) through increasing the number of gas-permeable filters attached to the lid or sidewalls of the culture vessel. However, this cannot meet the plant demand for adequate CO_2 and will increase the cost of *in vitro* plants produced by this method (Kozai et al., 2005). Several publications showed beneficial effects of increasing carbon dioxide concentration higher than the ambient level on the growth of *in vitro* plants (e.g., Desjardins et al., 1988; Mosaleeyanon et al., 2004). By keeping the CO_2 concentration in the culture room (C_{out}) at around 1000 $\mu mol\ mol^{-1}$, C_{in} can be increased to about 500 $\mu mol\ mol^{-1}$; thus, P_n in the ventilated vessels is roughly 3.3 times greater in a CO_2 enriched culture room than in a CO_2 nonenriched culture room; whereas P_n is about 17 times greater in a CO_2 enriched culture room with ventilated vessels than in a CO_2 nonenriched culture room with nonventilated vessels (Kozai, 2010). This also explains why sucrose is needed for carbon assimilation in conventional micropropagation. Sha Valli Khan et al. (2003) showed that *P. fortunei* plants *in vitro* also developed a significantly large number of shoots under the photoautotrophic CO_2-enriched condition. Stomata of the leaf abaxial surface remained widely opened under the photomixotrophic condition when compared with those having a narrow opening under the photoautotrophic condition. High CO_2 levels can also improve the acclimation and growth rate of micropropagated plants when transferred to *ex vitro* conditions (Kapchina-Toteva et al., 2014). Kozai and Nguyen (2003) demonstrated that high CO_2 concentration (1600 $\mu mol\ mol^{-1}$) and high PPF (250 $\mu mol\ m^{-2}\ s^{-1}$) significantly enhanced the root initiation and normal root vascular development of *Paulownia* plants cultured photoautotrophically on a vermiculite-based medium for 28 days, and their growth continued increasing significantly during 15 days in the *ex vitro* stage. In another research, strawberry explants, under CO_2 of 1000 $\mu mol\ mol^{-1}$ and PPF of 200 $\mu mol\ m^{-2}\ s^{-1}$, developed shoots and roots that were two times larger than those under ambient CO_2 and PPF of 100 $\mu mol\ m^{-2}\ s^{-1}$ after 4 weeks of culture. The high level of CO_2 and PPF in the *in vitro* stage continued improving the plant growth in the *ex vitro* stage (28 days) by producing the longest and greatest number of runners (Nguyen et al., 2008).

Wu and Lin (2013) also proved that *Protea cynaroides* plants showed a significant increase in shoot growth when cultured on sugar-free medium but in the presence of some vitamins and plant growth substances under the CO_2 enriched condition. Particularly, the shoot dry weight of *P. cynaroides* plants cultured in 5000 and 10,000 $\mu mol\ mol^{-1}$ CO_2 were 2.1 and 4.2 times, respectively, higher than those cultured on sugar-containing medium in ambient CO_2 concentration.

For commercialization, C_{out} in the culture room can be controlled by using an infrared-type CO_2 controller connected with an injection unit of pure CO_2 gas in a container (Kozai et al., 1995). This controller is commonly used by growers for enriching CO_2 in greenhouses. Nearly 10,000 transplants can be produced using 1.5 kg of CO_2, provided the culture room is fairly airtight with little CO_2 escaping to the outside (Kozai, 2010). Methods and mass balance analysis of CO_2 enrichment in the culture room were described by Jeong et al. (1993).

In the natural ventilation system, C_{in} can be maintained at a higher level by increasing the carbon dioxide of the culture room. However, in large-scale production using numerous small culture vessels, C_{in} and other gas concentrations in the vessels having gas-permeable filter discs tend to be different from one another and are difficult to regulate during the culture period. Besides, it is not easy to obtain a high natural ventilation rate for a large vessel. This problem can be overcome by using forced ventilation.

FORCED VENTILATION SYSTEM FOR LARGE CULTURE VESSELS

In forced ventilation, C_{in} and air movement inside the vessel can be easily increased during the production process by flushing a particular gas mixture directly into the vessel, and the ventilation rate can be easily controlled using an airflow controller. C_{in} in a large vessel with an air volume of up to 100 L or

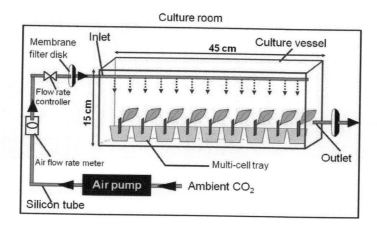

FIGURE 20.4

Schematic diagram of the forced ventilation system with polycarbonate vessel ($45 \times 25 \times 15$ cm^3) used for photoautotrophic micropropagation at the Institute of Tropical Biology, Hochiminh City, Vietnam.

greater can be maintained at around the ambient CO_2 of the culture room or gradually increased to a high level of over 1000 µmol mol^{-1} depending on the photosynthetic ability of *in vitro* plants. In this case, pure CO_2 or CO_2 diluted with air is directly injected into the vessel. When *in vitro* plants have high photosynthetic ability, large vessels with forced ventilation should be more effective with P_n increased when increasing the air current speed up to 50 and 100 mm s^{-1} compared with small vessels with natural ventilation (Kozai, 2010). Large culture vessels, which can be used for mass micropropagation, in different sizes and designs, have been used in photoautotrophic, forced ventilation of several plant species, such as *Eucalyptus* (Zobayed et al., 2000), coffee, yam (Nguyen et al., 2001, 2002), calla lily (Xiao and Kozai, 2004), *Dendrobium* (Nguyen et al., 2010b), and *P. amboinicus* (Nguyen et al., 2011a). A simple configuration of a forced ventilation system is shown schematically in Figure 20.4.

In vitro grapevines (*Vitis vinifera*) var. Thompson seedless grew much better, in terms of shoot length, number of unfolded leaves, and increased fresh and dry weights, in a forced ventilation system using polycarbonate vessels (Nalge Nunc Bio-Safe Carrier) compared with those in a natural ventilation system using small Magenta vessels (Nguyen et al., 2011b). Moreover, the adventitious roots of grapevines developed a lot of lateral roots when the plants were grown on vermiculite, a porous supporting material (Figure 20.5).

Several articles have proved that forced ventilation is more convenient and beneficial than natural ventilation when using large vessels for commercialization (Xiao and Kozai, 2004; Kozai et al., 2006). However, in a forced ventilation system, the increase in PPF has a beneficial effect on enhancing growth and photosynthesis, provided that CO_2 concentration inside the vessel is also raised to meet the demand for carbon fixation of *in vitro* plants. Besides, it has been observed that an uneven spatial distribution of air current speed in a large vessel creates spatial variation in the growth of *in vitro* plants with forced ventilation (Kozai, 2010). Therefore, uniform air current speeds are essential for ensuring equal growth of plants in large vessels. This drawback can be overcome in a closed plant production system (CPPS) (Kozai, 2010).

FIGURE 20.5

Growth of *in vitro* grapevines (*V. vinifera* L.) var. Thompson seedless under forced ventilation system in a large vessel (Nalge Nunc Bio-Safe Carrier) containing 50 plants on day 35.

POTENTIAL FOR SECONDARY METABOLITE PRODUCTION OF *IN VITRO* MEDICINAL PLANTS BY USING PAM

As approximately 80% of the world's population uses herbal medicines, plant materials, and extracted natural products are in high demand, besides the demand for synthetic drug production; in particular, the major components of synthetic drugs are derived from natural sources (Julsing et al., 2007). Plant tissue culture is widely used for the production of many medicinal plants, which features a constant production of identical germplasm from a limited stock material. Some defined plant-derived compounds, such as artemisinin, are widely produced in industrial processes in bioreactors. However, the medicinal properties of many plant species cannot be attributed to individually defined compounds, but rather are a consequence of the specific mixture of compounds found in the respective plant species (Julsing et al., 2007). Moreover, the lack of consistency of the desired active metabolites in medicinal plants cultured *in vitro* is still a major problem (Kapchina-Toteva et al., 2014).

Sirvent et al. (2002) proved that medicinal plants grown in different environments show variations in the contents of their major secondary metabolites. Some experiments conducted under controlled conditions indicated that temperature and light might be major factors contributing to such effects (Canter et al., 2005). Zobayed and Saxena (2004) suggested that temperature, light intensity, and CO_2 concentration are not easily controlled in greenhouses and fields, causing variations in the yield of bioactive compounds and biomass of medicinal plants. However, these factors can be well adjusted when using micropropagation under a controlled environment. Therefore, feasible strategies for the commercial production of medicinal plants can be developed aiming to increase the yield of specific bioactive compounds by optimizing the environmental conditions. The major bioactive compounds of St. John's wort plants, namely hyperforin, pseudohypericin, and hypericin, were increased in plants grown photoautotrophically at high temperatures (Zobayed et al., 2005; Couceiro et al., 2006b). *In vitro* hardening of *Artemisia annua* plantlets cultured in liquid sugar-free medium with CO_2 enrichment was an effective procedure to enhance the production of artemisinin (Supaibulwattana et al., 2011). Saldanha et al. (2013) proved that sugar-free medium might increase the level of some secondary metabolites in *Pfaffia glomerata* when cultured *in vitro* under photoautotrophic conditions with

FIGURE 20.6

In vitro plants of *P. vietnamensis* cultured photoautotrophically from nodal cuttings, (a) on day 60 and (b) on day 180 with rhizome and tuber formation.

nonenriched or enriched CO_2 (360 or 720 µmol mol^{-1}, respectively). A lack of sucrose in the culture medium increased 20-hydroxyecdysone levels, regardless of CO_2 levels. In another study, Pham et al. (2012) showed the positive effect of high CO_2 concentration (1200 µmol mol^{-1}) on the accumulation of two major lignans, phyllanthin, and niranthin, in *Phyllanthus amarus* plants cultured on sugar-free medium in a natural ventilation system after 45 days of culture. Light intensity and photoperiod also had effects on the accumulation of phyllanthin, hypophyllanthin, and niranthin when *P. amarus* plants were cultured photoautotrophically (Pham and Nguyen, 2014).

In vitro shoots bearing two open leaves of *Panax vietnamensis* Ha et Grushv., when cultured in two-hole Magenta vessels containing vermiculite-based sugar-free MS medium, could easily produce rhizomes and tuber roots in the *in vitro* stage (Figure 20.6). In these organs, major saponins, including majonoside-R$_2$ (MR$_2$), existed at higher levels in the enriched CO_2 condition (1100 µmol mol^{-1}) compared with those in the nonenriched CO_2 condition on day 90 (unpublished data).

SCALING UP A PAM SYSTEM TO AN ASEPTIC CULTURE ROOM—A CLOSED PLANT PRODUCTION SYSTEM

The scaled-up forced ventilation system can be considered as an aseptic culture room, in which a large culture vessel contains many small sterilized trays. This kind of micropropagation system can also be considered as a transplant production system for growing small disease-free cuttings/seedlings or as a closed vegetative propagation system using artificial light. In this system, workers are not allowed to enter the culture room for handling the trays with plants or for regulating environmental devices in normal modes. Thus, automation must be used for both tray transportation and environmental control inside the culture room. Energy and mass exchanges between the inside and outside of the system can be minimized for maximizing water, CO_2, light, and other resource utilization efficiencies and for reducing electricity consumption (Kozai et al., 2005). The first CPPS was commercialized in Japan in 2004, and they have since been used in more than 100 locations throughout Japan as of 2010 (Kozai, 2010). Thanks to the improved control of environment and pathogen invasion, transplants grown in the CPPS show more uniform, faster, pathogen-free growth than in a greenhouse.

CONCLUSION

Since the concept of PAM was developed more than two decades ago, many studies have been conducted on improving the *in vitro* environment and enhancing the growth and development of *in vitro* plants. Although a large-scale PAM for commercialization has not been achieved worldwide, research ranging from the natural ventilation method using small culture vessels with gas-permeable filter discs to the forced ventilation method using large culture vessels has clearly proved the feasibility of scaling up the PAM system. This has led to the development of a new technology, the CPPS with artificial lighting, for high-quality transplant production. This system is not completely aseptic but is clean and pathogen-free. By combining the concepts of CPPS and PAM, a large number of high-quality transplants such as seedlings, small cuttings, and scions for grafting can be produced in a limited time with high energy and water use efficiency.

REFERENCES

Aitken-Christie, J., Kozai, T., Smith, M.A.L., 1995. Automation and Environmental Control in Plant Tissue Culture. Kluwer Academic Publishers, Dordrecht.

Biddington, N.L., 1992. The influence of ethylene in plant tissue culture. Plant Growth Regul. 11, 173–187.

Canter, P.H., Thomas, H., Ernst, E., 2005. Bringing medicinal plants into cultivation: opportunities and challenges for biotechnology. Trends Biotechnol. 23, 180–185.

Couceiro, M.A., Afreen, F., Zobayed, S.M.A., Kozai, T., 2006a. Enhanced growth and quality of St. John's wort (*Hypericum perforatum* L.) under photoautotrophic in vitro conditions. In Vitro Cell. Dev. Biol. Plant 42, 278–282.

Couceiro, M.A., Afreen, F., Zobayed, S.M.A., Kozai, T., 2006b. Variation in concentration of major bioactive compounds of St. John's wort: effects of harvesting time, temperature and germplasm. Plant Sci. 170, 128–134.

Cui, Y.Y., Hahn, E.J., Kozai, T., Paek, K.Y., 2000. Number of air exchanges, sucrose concentration, photosynthetic photon flux, and differences in photoperiod and dark period temperatures affect growth of *Rehmannia glutinosa* plantlets in vitro. Plant Cell Tissue Org. Cult. 81, 301–306.

Desjardins, Y., Laforge, F., Lussier, C., Gosselin, A., 1988. Effect of CO_2 enrichment and high photosynthetic photon flux on the development of autotrophy and growth of tissue-cultured strawberry, raspberry and asparagus plants. Acta Hortic. 230, 45–53.

Jackson, M.B., Abbott, A.J., Belcher, A.R., Hall, K.C., Butler, R., Camerson, J., 1991. Ventilation in plant tissue cultures and effects of poor aeration on ethylene and carbon dioxide accumulation, oxygen depletion and explant development. Ann. Bot. 67, 229–237.

Jeong, B.R., Fujiwara, K., Kozai, T., 1993. Carbon dioxide enrichment in autotrophic micropropagation: methods and advantages. HortTechnology 3, 332–334.

Jeong, B.R., Fujiwara, K., Kozai, T., 1995. Environmental control and photoautotrophic micropropagation. Hortic. Rev. 17, 125–172.

Julsing, M.K., Quax, W.J., Kayser, O., 2007. The engineering of medicinal plants: prospects and limitations of medicinal plant biochemistry. In: Kayser, O., Quax, W.J. (Eds.), Medicinal Plant Biotechnology: From Basic Research to Industrial Application. WILEY-VCH Verlag GmbH & Co. KGaA, Weinheim, pp. 3–8.

Kapchina-Toteva, V., Dimitrova, M.A., Stefanova, M., Koleva, D., Kostov, K., Yordanova, Z.P., Stefanov, D., Zhiponova, M.K., 2014. Adaptive changes in photosynthetic performance and secondary metabolites during white dead nettle micropropagation. J. Plant Physiol. 171, 1344–1353.

Kozai, T., 1991. Autotrophic micropropagation. In: Bajaj, Y.P.S. (Ed.), High-Tech and Micropropagation I, Biotechnology in Agriculture and Forestry, vol. 17. Springer-Verlag, New York, pp. 313–343.

Kozai, T., 2010. Photoautotrophic micropropagation: environmental control for promoting photosynthesis. Propag. Ornam. Plants 10 (4), 188–204.

Kozai, T., Iwanami, Y., 1988. Effects of CO_2 enrichment and sucrose concentration under high photon fluxes on plantlet growth of carnation (*Dianthus caryophyllus* L.) in tissue culture during the preparation stage. J. Jpn. Soc. Hortic. Sci. 57, 279–288.

Kozai, T., Nguyen, Q.T., 2003. Photoautotrophic micropropagation of woody and tropical plants. In: Jain, S.M., Ishii, K. (Eds.), Micropropagation of Woody Trees and Fruits. Kluwer Academic Publishers, Dordrecht, pp. 757–781.

Kozai, T., Xiao, Y., 2005. A commercialized photoautotrophic micropropagation system. In: Gupta, S., Ibaraki, Y. (Eds.), Plant Tissue Culture Engineering. Springer, Berlin, pp. 355–371.

Kozai, T., Koyama, Y., Watanabe, I., 1988. Multiplication and rooting of potato plantlets in vitro with sugar medium under high photosynthetic photon flux. Acta Hortic. 230, 121–127.

Kozai, T., Zimmerman, R., Kiyata, Y., Fujiwara, K., 1995. Environmental effects and their control in plant tissue culture. International Society for Horticultural Science, Kyoto. Acta Hortic. 393.

Kozai, T., Afreen, F., Zobayed, S.M.A., 2005. Photoautotrophic (Sugar-Free Medium) Micropropagation as a New Propagation and Transplant Production System. Springer, Dordrecht.

Kozai, T., Nguyen, Q.T., Xiao, Y., 2006. A commercialized photoautotrophic micropropagation system using large vessels with forced ventilation: plant growth and economic benefits. Acta Hortic. 725, 279–292.

Kubota, C., Chun, C., 2000. Transplant Production in the 21st Century. Kluwer Academic publishers, Dordrecht.

Kubota, C., Kozai, T., 2001. Growth and net photosynthetic rate of tomato plantlets during photoautotrophic and photomixotrophic micropropagation. HortScience 36, 49–52.

Kubota, C., Fujiwara, K., Kitaya, Y., Kozai, T., 1997. Recent advances in environmental control in micropropagation. In: Goto, E., Kurata, K., Hayashi, M., Sase, S. (Eds.), Plant Production in Closed Ecosystems. Springer Science + Business Media, Dordrecht, pp. 153–169.

Kurata, K., Kozai, T., 1992. Transplant Production Systems. Kluwer Academic Publishers, Dordrecht.

Lucchesini, M., Mensuali-Sodi, A., Massai, R., Gucci, R., 2001. Development of autotrophy and tolerance to acclimatization of *Myrtus communis* transplants cultured in vitro under different aeration. Biol. Plant. 44, 167–174.

Mosaleeyanon, K., Cha-um, S., Kirdmanee, C., 2004. Enhanced growth and photosynthesis of rain tree (*Samanea saman*) plantlets in vitro under CO_2 enrichment with decreased sucrose concentration in the medium. Sci. Hortic. 103, 51–63.

Murashige, T., Skoog, E., 1962. A revised medium for rapid growth and bioassays with tobacco tissues. Physiol. Plant. 15, 473–497.

Nguyen, Q.T., Kozai, T., 2001. Growth of in vitro banana (*Musa* spp.) shoots under photomixotrophic and photoautotrophic conditions. In Vitro Cell. Dev. Biol. Plant 37, 824–829.

Nguyen, Q.T., Kozai, T., 2005. Photoautotrophic micro-propagation of woody species. In: Kozai, T., Afreen, F., Zobayed, S.M.A. (Eds.), Photoautotrophic (Sugar-Free Medium) Micropropagation as a New Propagation and Transplant Production System. Springer, Dordrecht, pp. 119–142.

Nguyen, M.T., Nguyen, Q.T., 2008. Effects of sucrose, ventilation and supporting materials on the growth of strawberry (*Fragaria ananassa* Duch.) plants cultured in vitro and survival rate of plantlets ex vitro. J. Biol. 30 (2), 45–49 (in Vietnamese with English abstract).

Nguyen, Q.T., Kozai, T., Nguyen, U.V., 1999. Effects of sucrose concentration, supporting material and number of air exchanges of the vessel on the growth of in vitro coffee plantlets. Plant Cell Tissue Org. Cult. 58, 51–57.

Nguyen, Q.T., Kozai, T., Heo, J., Thai, D.X., 2001. Photoautotrophic growth response of in vitro coffee plantlets to ventilation methods and photosynthetic photon fluxes under carbon dioxide enriched condition. Plant Cell Tissue Org. Cult. 66, 217–225.

Nguyen, Q.T., Le, H.T., Thai, D.X., Kozai, T., 2002. Growth enhancement of in vitro yam (*Dioscorea alata*) plantlets under photoautotrophic condition using a forced ventilation system. In: Nakatani, M., Komaki, K. (Eds.), Potential of Root Crops for Food and Industrial Resources: Twelfth Symposium of the International Society for Tropical Root Crops (ISTRC), September 10-16, 2000. International Society for Tropical Root Crops, Tsukuba, Tsukuba, Japan, pp. 366–368.

Nguyen, M.T., Nguyen, Q.T., Nguyen, U.V., 2008. Effects of light intensity and CO_2 concentration on the in vitro and ex vitro growth of strawberry (*Fragaria ananassa* Duch.). J. Biotechnol. 6 (1), 233–239 (in Vietnamese with English abstract).

Nguyen, Q.T., Hoang, T.M., Nguyen, H.N., 2010a. Effects of sucrose concentration, ventilation rate and light intensity on the growth of country borage (*Plectranthus amboinicus* (Lour.) Spreng.) cultured photoautotrophically in nylon bags having ventilated membranes. In: Proceedings of National Conference on Plant Biotechnology for Southern Area, held in Hochiminh City in October 24-25, 2009. Science and Technology Publishing House, Hochiminh City, (in Vietnamese with English abstract), pp. 297–301.

Nguyen, Q.T., Hoang, T.V., Nguyen, H.N., Nguyen, S.D., Huynh, D.H., 2010b. Photoautotrophic growth of *Dendrobium* 'Burana White' under different light and ventilation conditions. Propag. Ornam. Plants 10 (4), 227–236.

Nguyen, H.N., Pham, D.M., Nguyen, A.H.T., Hoang, N.N., Nguyen, Q.T., 2011a. Study on plant growth, carbohydrate synthesis and essential oil accumulation of *Plectranthus amboinicus* (Lour.) Spreng cultured photoautotrophically under forced ventilation condition. J. Biotechnol. 9 (4A), 605–610 (in Vietnamese with English abstract).

Nguyen, Q.T., Nguyen, H.N., Hoang, N.N., Pham, D.M., Nguyen, M.T., Huynh, D.H., 2011b. Photoautotrophic micropropagation for sustainable production of plant species. J. Sci. Technol. 49 (1A), 25–32 (in English with Vietnamese abstract).

Nguyen, D.P.T., Hoang, N.N., Nguyen, Q.T., 2012. A study on growth ability of *Thymus vulgaris* L. under impact of chemical and physical factors of culture medium. J. Biol. 34 (3se), 234–241 (in Vietnamese with English abstract).

Niu, G., Kozai, T., Kitaya, Y., 1996. Simulation of the time courses of CO_2 concentration in the culture vessel and net photosynthetic rate of *Cymbidium* plantlets. Trans. ASAE 39 (4), 1567–1573.

Niu, G., Kozai, T., Hayashi, M., Tateno, M., 1997. Simulation of the time courses of CO_2 concentration in the culture vessel and net photosynthetic rate of potato plantlets cultured photoautotrophically and photomixotrophically in vitro under different lighting cycles. Trans. ASAE 40 (6), 1711–1718.

Pham, D.M., Nguyen, Q.T., 2014. Growth and lignin accumulation of *Phyllanthus amarus* (Schum. & Thonn.) cultured in vitro photoautotrophically as affected by light intensity and photoperiod. J. Biol. 36 (2), 203–209 (in Vietnamese with English abstract).

Pham, D.M., Nguyen, H.N., Hoang, N.N., Nguyen, S.D., Nguyen, Q.T., 2012. Growth promotion and secondary metabolite accumulation of *Phyllanthus amarus* cultured photoautotrophically under carbon dioxide enriched condition. J. Biol. 34 (3se), 249–256 (in Vietnamese with English abstract).

Saldanha, C.W., Otoni, C.G., Notini, M.M., Kuki, K.N., da Cruz, A.C.F., Neto, A.R., Dias, L.L.C., Otoni, W.C., 2013. A CO_2-enriched atmosphere improves in vitro growth of Brazilian-ginseng (*Pfaffia glomerata* (Spreng.) Pedersen). In Vitro Cell. Dev. Biol. Plant 49, 433–444.

Sha Valli Khan, P.S., Kozai, T., Nguyen, Q.T., Kubota, C., Vibha, D., 2003. Growth and water relations of *Paulownia fortunei* under photomixotrophic and photoautotrophic conditions. Biol. Plant. 46 (2), 161–166.

Sirvent, M.T., Walker, L., Vance, N., Gibson, D.M., 2002. Variation in hypericins from wild populations of *Hypericum perforatum* L. in the Pacific Northwest of the USA. Econ. Bot. 56, 41–48.

Supaibulwattana, K., Kuntawunginn, W., Cha-um, S., Kirdmanee, C., 2011. Artemisinin accumulation and enhanced net photosynthetic rate in Qinghao (*Artemisia annua* L.) hardened *in vitro* in enriched-CO_2 photoautotrophic conditions. Plant Omics J. 4 (2), 75–81.

Wu, H.C., Lin, C.C., 2013. Carbon dioxide enrichment during photoautotrophic micropropagation of *Protea cynaroides* L. plantlets improves in vitro growth, net photosynthetic rate, and acclimatization. HortScience 48, 1293–1297.

Xiao, Y., Kozai, T., 2004. Commercial application of a photoautotrophic micropropagation system using large vessels with forced ventilation: plantlet growth and production cost. HortScience 39 (6), 1387–1391.

Xiao, Y., Kozai, T., 2006. In vitro multiplication of statice plantlets using sugar-free media. Sci. Hortic. 109, 71–77.

Xiao, Y., Lok, Y., Kozai, T., 2003. Photoautotrophic growth of sugarcane in vitro as affected by photosynthetic photon flux and vessel air exchanges. In Vitro Cell. Dev. Biol. Plant 39, 186–192.

Xiao, Y., Niu, G., Kozai, T., 2011. Development and application of photoautotrophic micropropagation systems. Plant Cell Tissue Org. Cult. 105, 149–158.

Zhang, M., Zhao, D., Ma, Z., Li, X., Xiao, Y., 2009. Growth and photosynthethetic capability of *Momordica grosvenori* plantlets grown photoautotrophically in response to light intensity. HortScience 44 (3), 757–763.

Zobayed, S.M.A., Saxena, P.K., 2004. Production of St. John's wort plants under controlled environment for maximizing biomass and secondary metabolites. In Vitro Cell. Dev. Biol. Plant 40, 108–114.

Zobayed, S.M.A., Afreen, F., Kubota, C., Kozai, T., 2000. Mass propagation of *Eucalyptus camaldulensis* in a scaled-up vessel under in vitro photoautotrophic condition. Ann. Bot. 85, 587–592.

Zobayed, S.M.A., Afreen, F., Xiao, Y., Kozai, T., 2004. Recent advancement in research on photoautotrophic micropropagation using large culture vessels with forced ventilation. In Vitro Cell. Dev. Biol. Plant 40, 450–458.

Zobayed, S.M.A., Afreen, F., Kozai, T., 2005. Temperature stress can alter the photosynthetic efficiency and secondary metabolite concentrations in St. John's wort. Plant Physiol. Biochem. 43, 977–984.

BIOLOGICAL FACTOR MANAGEMENT

INTRODUCTION

Biological factor management is crucial in plant factory with artificial lightings (PFALs) and other closed plant production systems (CPPSs) producing high-quality plants without using pesticides. However, there are few reports on biological factors such as algae, microorganisms, and pest insects. These biological factors may cause the degradation of plant quality and/or suppression of rapid plant growth as well as effects on human health. Much more research and development are required in this field.

The first section focuses on controlling algae growth, which is a critical maintenance task in PFAL crop production. The second section concerns food microbiological tests consisting of the environmental test and the quality test. Methods of measuring the total numbers of bacteria and fungi are described.

CONTROLLING ALGAE

Chieri Kubota

School of Plant Sciences, The University of Arizona, Tucson, Arizona, USA

Controlling algae is a critical maintenance task in PFAL crop production. Wet surfaces in contact with nutrients and light can develop algae quickly. Examples include the upper surface of plant growing substrate plugs/cubes used in hydroponics, gaps between floating panels over nutrient solution, the inside of plumbing made of light transmitting material (tubing) and the nutrient reservoir exposed to light. Algae can house pests (such as fungus gnats) and therefore diseases and must be eliminated. Best practice is covering such surfaces with an opaque material to exclude light, but there are a few options to control algae chemically.

HYDROGEN PEROXIDE

The availability of a commercial product and its use may depend on government regulations, but if such a product containing stabilized hydrogen peroxide is available, it is a useful product for controlling algae with minimal or no phytotoxicity. Some commercial products available in the United States are also registered for organic crop production use. The effective concentration may need to be carefully selected as potential phytotoxicity seems to be dependent on the different nutrient delivery systems. Substrate-based culture systems seem to have higher buffering capacity, whereas use of hydrogen peroxide product in deep flow technique (DFT) systems requires carefully selected concentration and frequency of application to avoid possible damage to the roots.

OZONATED WATER

A small system to generate ozonated water may be useful in PFALs. The ozone concentration should be carefully selected to be effective for algae control yet not cause phytotoxicity. Graham et al. (2012) reported that 3 mg/L was an upper limitation for tomato seedlings grown in rockwool cubes. It is of interest that in some cases the application of ozonated water could improve overall plant productivity (Graham et al., 2011). Freshly made ozonated water could be sprayed around the problematic area developing algae as a control.

CHLORINE

Use of a low dose of chlorine has been commercially practiced in hydroponic greenhouses. However, use of chlorine (or hypochlorous acid) for algae control in hydroponic systems of PFALs may need more research before further use is recommended. Caution may be needed as residual chlorine could form a highly phytotoxic compound, chloramine, when reacted with a low concentration of ammonium ions (NH_4^+) in the nutrient solution (Date et al., 2005). A typical symptom of chloramine phytotoxicity is root browning. Mixing sodium thiosulfate in the source water (tap water) at 2.5 mg/L is recommended to remove residual chlorine (S. Date, personal communication). However, when the system has many organic compounds (such as in the nutrient solution of recirculating systems or organic substrate-based culture systems), addition of chlorine may not be an issue, as organic matter would convert chlorine to chloride. Date et al. (2005) showed that lettuce growth was not inhibited in the range of 0–1 mg/L chlorine concentration when the nutrient solution did not contain NH_4-N. A small amount of NH_4^+ (5.2×10^{-3} mM) could inhibit lettuce growth when 0.5 mg/L chlorine was in the solution (and therefore forming chloramine) (Date et al., 2005). Given that tap water containing 1-2 mg/L of chlorine is typically used in PFALs in urban areas and that commercial hydroponic grade fertilizers contain trace amounts of NH_4-N, a cost-effective technology to remove NH_4^+ may need to be developed to allow effective usage of chlorine for algae control in PFALs.

SUBSTRATES

Leaf crop production in hydroponics starts with small substrate plugs to grow seedlings. Substrates with relatively high water-holding capacity (such as rockwool plugs) tend to have wet surfaces all the time and create favorable growing conditions for algae. In Asia, use of polyurethane foam as a substrate is common in hydroponics and polyurethane seems to have a relatively dry surface due to its low water-holding capacity. However, algae can still develop near the nutrient solution line inside the foam. Use of darker colored polyurethane foams that exclude light may better control algae.

MICROORGANISM MANAGEMENT

Miho Takashima

Japan Research Promotion Society for Cardiovascular Diseases, Sakakibara Heart Institute, Tokyo, Japan

Methodologies for microbiological testing of food should be selected according to purpose (Japan Food Research Laboratories (JFRL), 2013). The cultivation method takes time to obtain results, and requires experience and technology, but it is an international microbial testing method and currently the microbial testing standard. Meanwhile, the rapid method is a way to obtain results by abbreviating the culture method or replacing it in part or in whole with other methods (Igimi, 2013). Compared with conventional methods, the rapid method is one that enables quantitative determination regarding the presence of microorganisms and bacteria numbers in a short time. However, there are many methods with unconfirmed validity, and it is important to select and implement a methodology after confirming its rapidity, efficacy, sensitivity, reproducibility, and reliability.

MICROBIOLOGICAL TESTING

This section focuses on microorganism control in a PFAL. Microorganism management in a plant factory is split into environment management and quality management. Figure 21.1 shows the classifications for microbiological testing.

Environmental testing is largely divided into two methods. One is the testing of microorganisms in the air or attached to a surface, and the other is the testing of raw water and a culture solution (Figure 21.1). In a PFAL, surface-attached microorganisms are considered to be the same as suspended ones, and in this instance, microbiological testing of raw water and culture solutions will also be considered as being included in the quality testing.

In quality testing, it is important to examine and understand the number of bacteria and fungi. These are the standards and indicators for microorganisms in food and the environment. It is also important to examine for contaminant indicator bacteria such as *Coliform* groups and *Staphylococcus aureus*, if necessary. The conventional plate count method is described in the following section.

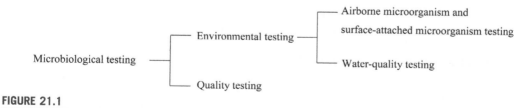

FIGURE 21.1

Classifications for microbiological testing.

ENVIRONMENTAL TESTING—AIRBORNE MICROORGANISMS

Numerous microorganisms exist in the air, and are mainly bacteria, fungi, and viruses. A measurement methodology for viruses is yet to be established, as they are small and easily become inactive, which makes them difficult to culture and measure. Accordingly, measurement of airborne bacteria and fungi is what is conducted. Testing for bacteria means a viable bacteria count test, usually by testing using a standard agar medium. The bacteria count is used as a typical indicator of microorganism contamination (Japan Food Research Laboratories (JFRL), 2013). A viable bacteria count is determined after the most suitable incubation period, and displayed in terms of colony forming units (CFUs) (Buttner et al., 1997). The concentration of airborne microorganisms (CFU per volume) is found by dividing the number of CFU per sample by the volume of air sampled (Buttner et al., 1997). Testing for fungi means testing for molds and yeasts. Fungi are ubiquitous in nature and occasionally cause sickness in people (Yang and Johanning, 1997). Some reports also indicate that they have become a major problem in buildings where moisture control is poor or water intrusion is common (Yang and Johanning, 1997). Fungi such as molds and yeasts are factors that should be understood in PFALs, which always have highly humid environments.

Airborne microorganisms usually mean the bacteria (microorganisms) floating in the air, which attach to the surfaces of products or fall down. There are two methods for measuring airborne bacteria: testing for fallen bacteria and testing for airborne bacteria suspended in a fixed volume of air.

The effect of airborne microorganisms on humans and environments has been reported by Stetzenbach (1997), in a report in which he describes the adverse effects brought about in connection with airborne microorganisms (e.g., algae, bacteria, endotoxins, fungi, mycotoxins, protozoa, and viruses). According to the report, the environmental effect most airborne microorganisms have, except for algae and endotoxins, is to cause reductions in agricultural productivity. The report also describes occasional effects on human health (e.g., infections, allergic reactions).

Measurement of the atmospheric environment evaluates quality, concentration, and quantity. In PFALs, it is important to generally know the overall total for bacteria and fungi counts. For this reason, two kinds of testing methodology are described in detail here.

Measurement of fallen bacteria using plate method

Sampling is conducted by opening the lid of a plate for a fixed period of time and allowing the culture medium to be freely exposed, after which the lid is closed and the culture medium inverted and incubated. The measurement methodology finds the number of developed colonies per plate after incubation, as illustrated in Figure 21.2. Standard agar medium is used to test for bacteria, while a potato dextrose agar medium is used to test for fungi. This is the most widespread measurement methodology for airborne bacteria in Japan since it is extremely simple and can be carried out at low cost.

Measurement of airborne microorganisms

This is the typical method in Western countries, and one that has also received recognition in recent years in Japan (Yamazaki et al., 2001). The methodology is split into a number of methods based upon air sampling. As shown in Figure 21.3, an impactor-type air sampler is used. This is recommended by the International Organization for Standardization (ISO) and is commonly used to collect airborne microorganisms. Moreover, a variety of impactor-type samplers are commercially available (Buttner et al., 1997). Sampling methods need to consider air flow, and assess the average in the

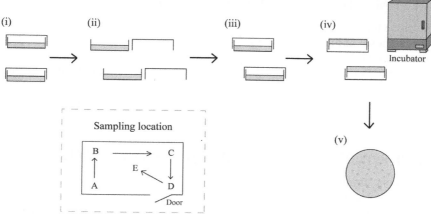

FIGURE 21.2

Flowchart: Measuring microorganisms using the plate method. (i) Placement of two culture plates; (ii) The plates are placed at the measurement location with their lids left off—5 min for general bacteria, and 20 min for fungi; (iii) The plates have their lids put on; (iv) The plates are incubated at $35 \pm 1°C$ for 48 ± 3 h for general bacteria, and $23 \pm 2°C$ for 7 days for fungi; (v) The number of colonies is counted. The data are expressed as the number of colony forming units (CFU) per unit time—CFU per 5 min for fallen bacteria, and CFU per 20 min for fungi. *Sample locations should be in accordance with the monitoring strategy. For example, as illustrated in steps A to D in the diagram, sampling locations are determined according to the flow of humans and objects.

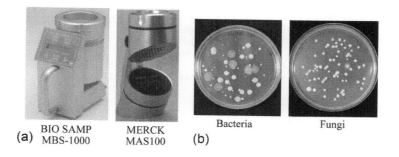

FIGURE 21.3

Monitoring devices and examples of airborne microorganisms. (a) Air samplers from the hygiene environmental microorganism measuring instruments on the Marusan Pharma Biotech Corporation website; (b) Measurement of PFAL airborne microorganisms at Sakakibara Heart Institute. An air sampler (BIO SAMP MBS-1000) was used. Pictures provided by SHARP.

sampling area. Figure 21.3 shows an example of measurement of bacteria and fungi. Not only does this method provide a quantitative evaluation, it also has the advantage of enabling the comparison of measurements under varying environmental conditions. The media used are the same as in the settle plate method.

QUALITY TESTING—TESTING FOR BACTERIA AND FUNGI

The general procedure from sample preparation (Figure 21.4a) to colony number determination (Figure 21.4c) is shown in the diagram. Two kinds of tests are described in Figure 21.4b, and are basically conducted in the same way but with one differing from the other in how the solution is added. The testing for bacteria is conducted using either of two methods, the coating method and the mixing method, with the coating method generally being used for fungi. Incubation is carried out using optimized conditions, times, and temperatures. Figure 21.4c shows the method for calculating the number of microorganisms per sample.

EXAMPLES OF REPORTS OF MICROBIOLOGICAL TESTING IN PFALs

Much literature concerning the production of food exists, yet initiatives and research reports focusing on hygiene management in PFALs remain few in number (Uehara et al., 2011), as do similar studies in PFALs themselves (Hayakumo et al., 2013).

A study does exist concerning microbiological monitoring of airborne microorganisms and surface-attached microorganisms (Hayakumo et al., 2013). This study compares the number of bacteria and fungi in greenhouses and in PFALs. It reports a lower number of airborne microorganisms (bacteria at 0–4 CFU/m^3, fungi at 16–24 CFU/m^3), as well as surface-attached microorganisms (bacteria at 100 $CFU/25\ m^2$, fungi at 1–51 $CFU/25\ m^2$) in PFALs.

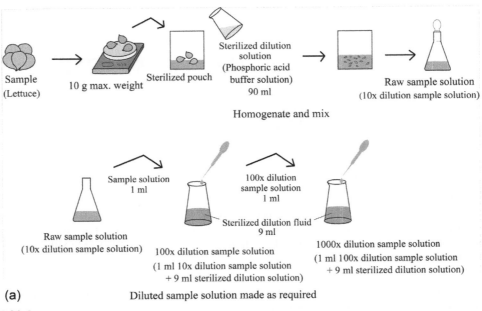

(a)

FIGURE 21.4

Outline of the testing procedure for bacteria and fungi. Sampling and testing was carried out as shown in (a) to (c): (a) Sample preparation;

(Continued)

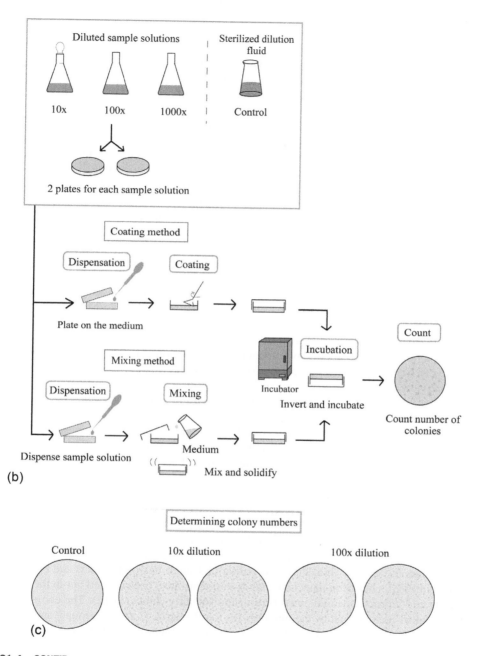

(b)

(c)

FIGURE 21.4—CONT'D

(b) Two testing methods; (c) Counting of colony numbers. Plates with 30–300 colonies are selected for measurement. The average of the two plates is taken as the viable bacteria number. This is calculated from the total number of colonies on the two plates × 1/2 × dilution factor. For example, at 100× dilution and with colony numbers of 168 and 183, the number of colony forming units can be calculated using the following formula: (168 + 183) × 1/2 × 100 = 17,550 = 1.755 × 10^4 = 1.8 × 10^4 CFU/g.

There is a report on microbiological testing of vegetables, including products from PFALs, covering the past 12 years (Uehara et al., 2011). It reports the number of bacteria for each kind of vegetable and production procedure. The distribution of bacteria per unit (1 g) is 10^2–10^6 CFU in PFALs. The numbers for PFAL vegetables are values two to three orders lower when compared to those for other growing environments (outdoor cultivation: 10^2–10^{10} CFU, hydroponic cultivation: 10^3–10^9 CFU) (Uehara et al., 2011). In such PFALs, the environmental control and hygiene management make the presence of fewer microorganisms possible. Other reports show bacteria at less than 300 (=3.0×10^2) CFU (Ooshima et al., 2013), 10^2–10^4 CFU (Sasaki, 1997), and 1.2×10^6 CFU (Nagayama, 2012). Vegetables grown in PFALs are mostly leaf vegetables and tend to be eaten raw, which is the reason why environment control and hygiene management are important.

Some reports also exist concerning trials on culture solutions (Sasaki, 1997) and storage trials (Tokyo Metropolitan Institute of Public Health, 2011). One report analyzes changes in culture solutions over time and describes the importance of culture solution hygiene management (Sasaki, 1997). The results of vegetable storage trials (Tokyo Metropolitan Institute of Public Health, 2011) demonstrate temperature-based influence. More so than outdoor-cultivated vegetables, PFAL vegetables show sudden increases in their bacteria numbers as temperature rises. Moreover, PFAL vegetables obtained directly from the factory have bacteria at 10^1–10^5 CFU (Tokyo Metropolitan Institute of Public Health, 2011). With regards to lettuce, there is a report that acknowledges significant differences in predistribution and postdistribution lettuce bacteria numbers (Sasaki, 1997). According to the results of a questionnaire about environment and production controls, no standardized temperature for delivery in the distribution process exists, with the information indicating a nonuniform state of affairs that varies with each PFAL, including cold storage only in summer, delivery at less than 5°C, delivery at less than 10°C (Tokyo Metropolitan Institute of Public Health, 2011). In addition, a correlation between bacteria numbers and distribution temperatures was found. A report found that vegetable bacteria numbers only decreased by a factor of 1–2 log after washing (International Commission on Microbiological Specifications for Foods (ICMSF), 2011). Therefore, it is known that improved individual hygiene awareness and control of distribution processes and storage temperatures are important.

It is said that, when compared with other cultivation methods, bacteria numbers are lower in vegetables produced in PFALs. However, if appropriate temperature control does not occur in the distribution process, then the advantage vegetables produced in PFALs have of being produced in a biologically clean airspace can be easily lost. For this reason, management of hygiene at all stages of the food chain, from production to distribution, is of greater importance in PFALs.

In recent years in Japan, a Third Party Accreditation Committee on Products and Product Systems (TPAC-PPS) in PFALs has been launched, and assessments have been carried out in the four fields of environment, safety/peace-of-mind, universality, and sociality (Inoue, 2012). The committee focuses on microbiological factors such as the numbers for live bacteria, *Coliform* groups, and *E. coli* O157 in PFALs. Microbiological standards have also been established, with three certification levels (Gold, Silver, Bronze) created.

In recent years, the safety of PFAL products and product systems is being emphasized. Consequently, there will most likely be increased need and progress with respect to research in the field in the future.

CONCLUDING REMARKS

This section has focused on PFAL microbiological testing and described examples of testing methods and reports for the same. As also mentioned previously, no rules concerning microbiological testing exist. Kozai (2012) states that bacteria in PFALs usually number less than 300, which is less than 1/100 the number for greenhouse plants, and less than 1/500 the number for outdoor cultivation. Other than this report, it can be said that no other literature exists and that no particular attention has been paid with respect to microorganism numbers. Therefore, the current state is one in which each producer has their own management standardized protocols (Pak et al., 2014).

In recent years, interest has grown in microbiological testing. In Japan, discussions aimed at bringing Japan into accord internationally with testing methodologies such as the broadly internationally recognized ISO and Bacteriological Analytical Manual (BAM) methods are proceeding. With the recent emphasis on food safety, there is an increasing level of priority concerning microbiological testing. Rapid methods with confirmed validity will become indispensable, and further progress will surely be made in future research.

REFERENCES

Buttner, M.P., Willeke, K., Grinshpun, S.A., 1997. Sampling and analysis of airborne microorganisms. In: Hurst, C.J., Knudsen, G.R., McInerney, M.J., Stetzenbach, L.D., Walter, M.V. (Eds.), Manual of Environmental Microbiology. ASM Press, Washington, DC, pp. 629–639.

Date, S., Terabayashi, S., Kobayashi, Y., Fujime, Y., 2005. Effects of chloramines concentration in nutrient solution and exposure time on plant growth in hydroponically cultured lettuce. Sci. Hortic. 103, 257–265.

Graham, T., Zhang, P., Woyzbun, E., Dixon, M., 2011. Response of hydroponic tomato to daily applications of aqueous ozone via drip irrigation. Sci. Hortic. 129, 464–471.

Graham, T., Zhang, P., Dixon, M.A., 2012. Closing in on upper limits for root zone aqueous ozone application in mineral wool hydroponic tomato culture. Sci. Hortic. 143, 151–156.

Hayakumo, M., Sawada, Y., Takasago, Y., Tabayashi, N., Aoki, T., Muramatsu, K., 2013. Hygiene management technology for the production of medical materials in PFAL (written in Japanese). Japanese society of Agricultural, Biological and Environmental Engineers and Scientists 2013

Igimi, S., 2013. Basics of microbial test (Chapter 1). In: Igimi, S., Ezaki, T., Takatori, K., Tsuchido, T. (Eds.), Guidebook of Easy and Rapid Microbial Test Methods (Written in Japanese: Biseibutu no kan'i jinsoku kensaho). Technosystem Co. Ltd, Tokyo, pp. 3–7.

Inoue, T., 2012. Third party accreditation committee on products and product system (written in Japanese). Energy Resour. 33 (4), 27–31.

International Commission on Microbiological Specifications for Foods (ICMSF), 2011. Microorganism in foods 6 (written in Japanese: Shokuhin biseibutsu no seitai biseibutsu seigyo no zenbo). In: Yamamoto, S., Maruyama, T., Kasuga, F. (Eds.), Microbial Ecology of Food Commodities. Chuo Hoki, Tokyo, pp. 328–386.

Japan Food Research Laboratories (JFRL), 2013. Visual Version of Food Sanitary Test Method (Written in Japanese: Bijuaru-ban shokuhin eisei shikenho tejun to pointo). Chuo Hoki, Tokyo.

Kozai, T., 2012. Plant factory with artificial light: Japanese agricultural revolution spread around the world (written in Japanese: Jinkoko-gata shokubutsu kojo sekai ni hirogaru nihon no nogyo kakumei) Ohmsha, Tokyo.

Nagayama, M., 2012. The integrative medical project of agromedicine in Sakakibara Heart Institute (written in Japanese). In: The 54th Business Workshops in Japan Plant Factory Association (JPFA) Handout, JPFA (Chiba University Plant Factory Project).

Ooshima, T., Hagiya, K., Yamaguchi, T., Endo, T., 2013. Commercialization of LED used plant factory (written in Japanese). Nishimatsu Construction's Technical Research Institute Report 36, 1-6.

Pak, J., Nakamura, K., Harada, T., Wada, K., 2014. Hygiene control in plant factory (written in Japanese). Japanese Society of Agricultural Biological and Environmental Engineers and Scientists 2014.

Sasaki, H., 1997. The number of bacteria in vegetable produced in plant factory (written in Japanese). SHITA Report 13, 53-61.

Stetzenbach, L.D., 1997. Introduction to aerobiology. In: Hurst, C.J., Knudsen, G.R., McInerney, M.J., Stetzenbach, L.D., Walter, M.V. (Eds.), Manual of Environmental Microbiology. ASM Press, Washington, DC, pp. 619–628.

Tokyo Metropolitan Institute of Public Health in Department of Regional Food and Pharmaceutical Safety Control, 2011. Food hygiene association in the production of vegetables in plant factory (written in Japanese) 61, 8.

Uehara, S., Kishimoto, Y., Ikeuchi, Y., Katoh, R., Arai, T., Hirai, A., Nakama, A., Kai, A., 2011. Surveys of bacterial contaminations in vegetables in Tokyo from 1999 to 2010 (written in Japanese). Annual Report of Tokyo Metropolitan Institute of Public Health 62, 151-156.

Yamazaki, S., Kano, F., Takatori, K., Sugita, N., Aoki, M., Hattori, K., Hosobuchi, K., Mikami, S., 2001. Measurement and assessment of airborne microorganism (written in Japanese: Kankyo biseibutsu no sokutei to hyoka). In: Yamazaki, S. (Ed.), Measurement and Assessment of Environmental Microorganism. Ohmsha, Tokyo, pp. 95–116.

Yang, C.S., Johanning, E., 1997. Airborne fungi and mycotoxins. In: Hurst, C.J., Knudsen, G.R., McInerney, M.J., Stetzenbach, L.D., Walter, M.V. (Eds.), Manual of Environmental Microbiology. ASM Press, Washington, DC, pp. 651–660.

DESIGN AND MANAGEMENT OF PFAL

22

Toyoki Kozai[1], Shunsuke Sakaguchi[2], Takuji Akiyama[2], Kosuke Yamada[2], Kazutaka Ohshima[2]

Japan Plant Factory Association, c/o Center for Environment, Health and Field Sciences, Chiba University, Kashiwa, Chiba, Japan[1] PlantX Corp., Kashiwa, Japan[2]

INTRODUCTION

In order to design and manage a plant factory with artificial lighting (PFAL) properly, it is necessary to understand various scientific, technical, and business fields related to PFALs. These fields include light sources and lighting systems (LSs), environmental measurement and control, management of energy, materials and workers, sanitation control, plant ecophysiology, hydroponics, plant nutrition, automation, cost/benefit analysis, finance, fresh food marketing, and sales promotion.

A variety of computer-assisted tools for design and management (D&M) of manufacturing factories are available in many industries. However, there are few such tools available for PFALs, which are an emerging technology and business. Accordingly, a computer-assisted D&M tool for PFALs needs to be developed based on methods and techniques common to computer-assisted tools used in other industries. It should be noted, however, that a PFAL is used to produce living organisms, i.e. plants, and so a core part of the PFAL D&M is the plant–environment interaction model.

In Part 1 of this chapter, the features, structure, functions, and usefulness of our PFAL-D&M system are introduced. In Part 2, some typical outputs for the hourly and monthly electricity consumptions, coefficient of performance (COP) of air conditioners, light and temperature distributions, plant growth measurements, and analysis of the PFAL-D&M are presented.

STRUCTURE AND FUNCTION OF THE PFAL-D&M SYSTEM

The PFAL-D&M system consists of three parts: design, management, and database (Figure 22.1). The PFAL-D&M software is stored in the "cloud service" area for use via the Internet, except for the "data acquisition and upload (Data)" part of "production management (PM)" in the PFAL-M subsystem (Figure 22.2).

There are two types of data acquisition and upload part (Figure 22.3). Type A is designed to be installed at an existing PFAL, in which the data acquisition and upload part (sensors and data uploader) of PFAL-M-PM-Data is basically separated from the measurement and control system of the existing PFAL. In Type A, there may be a small difference in measured temperatures between the two sensors, because the two temperature sensors may have different accuracies and may be installed at different

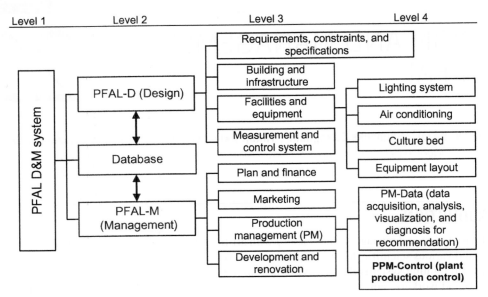

FIGURE 22.1

Structure of PFAL-D&M (plant factory with artificial lighting - design and management) system.

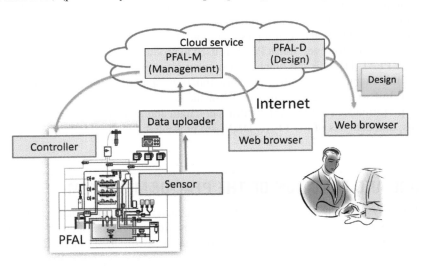

FIGURE 22.2

Structure of PFAL-D&M system, in the cloud and PFAL.

places (e.g., one by the air conditioner and the other at the center of the culture room). In this case, PFAL-M-PM-Data cannot collect the temperature data used for its control. The same applies for other environmental factors. Type B is designed to be installed at a newly built PFAL, in which the data acquisition and upload part is embedded in the measurement and control system of the PFAL. Then,

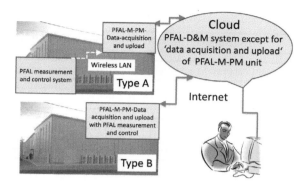

FIGURE 22.3

PFAL-D&M is used via Internet. Type A: Installed at an existing PFAL; Type B: Installed at a newly built PFAL.

the air temperature used for its control is transmitted to the database of the PFAL-M-PM Data part and the air temperature used for its control can be stored.

PFAL-D (DESIGN) SUBSYSTEM

The PFAL-D subsystem consists of the parts "requirements, constraints, and specifications," "building, and infrastructure," "facilities and equipment," and "measurement and control" (Figure 22.4). The whole PFAL-D subsystem is stored in the "cloud service" area.

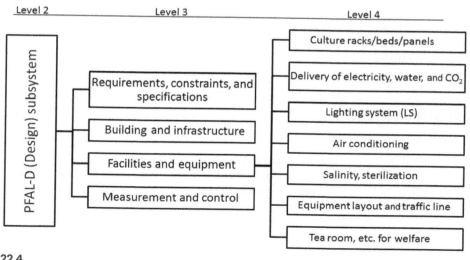

FIGURE 22.4

Structure of PFAL-D (design) system.

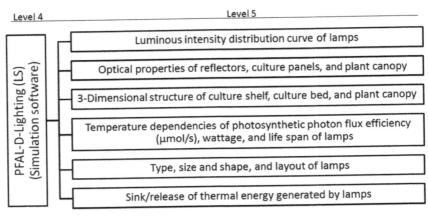

FIGURE 22.5

Parameters used to simulate the light environment (time courses of spectral and spatial distributions) using PFAL-D-LS (design of lighting system).

LIGHTING SYSTEM (LS)

The cost for electricity accounts for 25–30% of the total production cost while the cost for lighting accounts for 70–80% of the electricity cost. Since it is difficult to improve the hardware of the LS after constructing the PFAL, a good hardware design of the LS is essential.

The software of the "LS" part is customized for the PFAL to simulate the light environment as affected by the parameters shown in Figure 22.5. The simulated light environment using mathematical equations based on optical physics is, in general, fairly accurate and useful.

When designing the LS, it should be noted that the spatial and spectral distributions of photosynthetic photon flux density (PPFD) above the culture shelf are affected by the three-dimensional structure of the culture shelf and the optical properties of the plant community. Also, the design of the heat energy source/sink of lamps such as LEDs affects the temperatures of the LEDs, and thus the photosynthetic photon efficiency of the lamps.

PFAL-M SUBSYSTEM
STRUCTURE OF SOFTWARE

The PFAL-M subsystem consists of the "plan and finance," "marketing," "PM," and "development and renovation" parts. "Plan and finance" are executed in line with the vision, mission, and goals of the PFAL. Marketing includes customer development, product development, sales promotion, and product and service delivery.

The PFAL-M-PM part is divided into PM-Data and PM-Control. PM-Data is further divided into three parts: (1) data acquisition and upload, (2) data processing, analysis, and visualization, and (3) diagnosis and recommendation for improvements. Data collected by the PFAL-M-PM-Data part are processed, analyzed, and visualized for each factor and/or combinations of the factors as shown, for example, in Figure 22.6. The data collected by PFAL-M-Data are used to calculate various indices such

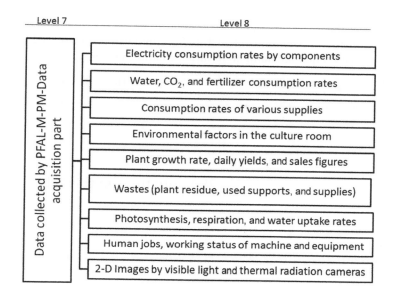

FIGURE 22.6

Data collected by PFAL-M-PM Data acquisition part.

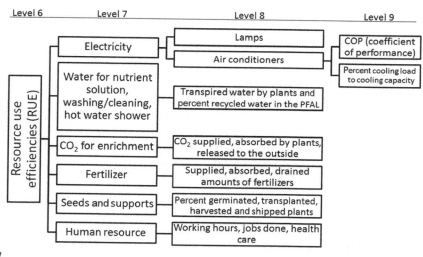

FIGURE 22.7

Resource use efficiencies (RUE) to be displayed on the screen. RUE is the ratio of the amount of each resource fixed in the salable part of produce to the resource input.

as resource use efficiencies (RUE) to express the cost performance of the plant production system in the PFAL (Figure 22.7). The PFAL-M-PM-C (control) part consists of: (1) production costs, sales, and revenues; (2) seeding, seedling production, and cultivation; (3) harvesting, packing, cooling, and shipping; and (4) order receipt, order, and stock management.

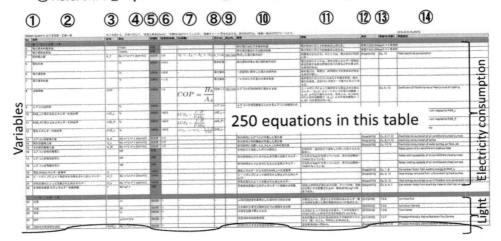

①Serial number ②Variable name ③Symbol ④Unit ⑤Category ⑥Time interval
⑦ Equations ⑧Equations in TEX ⑨ URL ⑩Definition ⑪Explanations
⑫Reference⑬Equation number ⑭ Variable name in English

FIGURE 22.8

Tabulated list of variables used for electricity consumption and light environment together with their attributes (1) to (14) stored in PFAL-M. The variables used for CO₂, water, heat energy, plant growth, etc. are also stored in the same way.

LOGICAL STRUCTURE OF EQUATIONS

All the variables, equations, and constants/coefficients used in PFAL-D&M are systematically tabulated as shown in Figure 22.8. Figure 22.9 shows a part of the structured map in PFAL-D&M, which includes 72 measured data, 73 indices, 5 set points, 22 constants/coefficients, and many equations. The most important equations with explanations are given in Chapter 4 of this book and Kozai (2013). This software can be used to optimize the environmental conditions and thus maximize the cost performance of the PFAL, for example (Figure 22.10).

DESIGN OF LIGHTING SYSTEM
PPFD DISTRIBUTION

The PPFD distribution on the culture panel can be well simulated by PFAL-D&M (LS). Figure 22.11 shows a simulated result of the two-dimensional PPFD distribution on the culture panel with and without a light reflector set above the light sources. The results show that the average PPFD is increased by 38% using a white reflector, compared with the one without a reflector.

Figure 22.12 shows a simulated result of the three-dimensional PPFD distribution over model crisp lettuce heads planted in a grid-like fashion on the culture panel. Using this software, the ratio of light

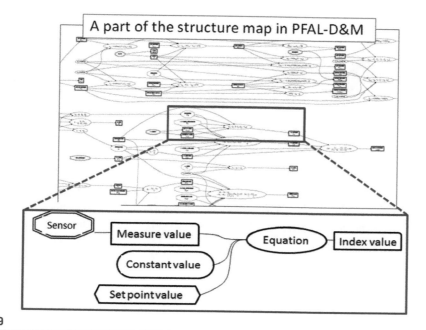

FIGURE 22.9

A part of the structure map in PFAL-D&M showing how to obtain the index value from the measured value, constant value, set point value, and equation.

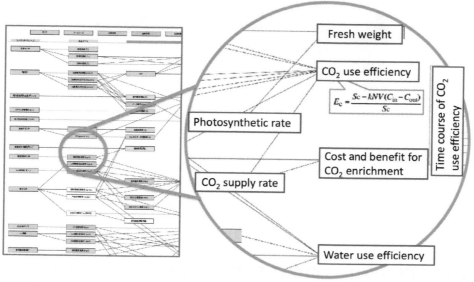

FIGURE 22.10

A logical structure of the equations stored in PFAL D&M. The equations listed in Figure 22.8 are logically connected as shown here (left). The variables in the circle (right) show the indices such as CO_2 and water use efficiencies and rates of photosynthesis and CO_2 supply. (See also Figure 22.7.)

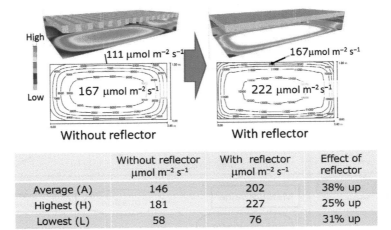

	Without reflector μmol m⁻² s⁻¹	With reflector μmol m⁻² s⁻¹	Effect of reflector
Average (A)	146	202	38% up
Highest (H)	181	227	25% up
Lowest (L)	58	76	31% up

FIGURE 22.11

An output example of PFAL-D for light environment improvement by use of a light reflector.

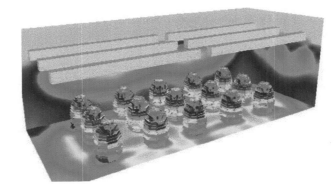

FIGURE 22.12

A simulated result using PFAL-D of the three-dimensional PPFD distribution on crisp lettuce heads planted on the culture panels.

energy received by plants to that emitted by the light source can also be estimated. Thus, the software tool is useful for designing the planting density, vertical distribution of PPFD, etc.

SCHEDULING THE LIGHTING CYCLES TO MINIMIZE ELECTRICITY CHARGE

The lighting time schedule considerably affects the "ad valorem" electricity charge, which is basically proportional to the integral of electricity consumption (kWh). However, the hourly ad valorem charge depends on the time zone and season in countries such as Japan.

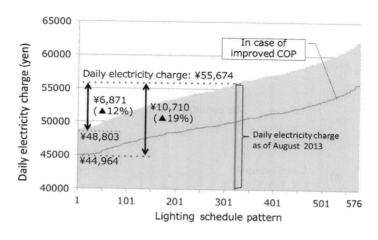

FIGURE 22.13

Daily "ad valorem" electricity charge as affected by lighting time schedule in Tokyo (1 US$ = 120 yen).

If the lamps are divided into two groups and each lamp group is turned on for 16 continuous hours starting on the hour, every hour (0:00, 1:00, 2:00, and so on), there will be 576 (24 × 24) combinations of lighting time schedule. Figure 22.13 shows the daily ad valorem electricity charges of 576 combinations assuming the charge system for the summer of Tokyo Electric Power Co., Inc.

The daily electricity charge of an existing PFAL was 55,674 yen in August 2013. However, based on simulation results, it can be reduced to 48,803 yen by simply changing the lighting time schedule (the daily integral of electricity consumption remains unchanged).

The daily electricity charge shown in Figure 22.13 is affected also by the COP of air conditioners as discussed in the next section. If the number of air conditioners in operation is changed to keep the COP high, the daily electricity charge is reduced to 44,964 yen.

In the case of three lamp groups, there are 13,824 (=24 × 24 × 24) combinations of lighting time schedule. By using the PFAL-D (lighting), the lighting time schedule providing the lowest electricity charge for lighting and air conditioning can be found among a large number of candidates.

ELECTRICITY CONSUMPTION AND ITS REDUCTION

The PFAL-M-PM-Data part handles the data acquisition, processing, and visualization regarding the rate and state variables in a PFAL, including the electricity consumption.

DAILY CHANGES IN ELECTRICITY CONSUMPTION

The hourly ad valorem electricity charge depends on the time zone and season in Japan. It is highest in summer during 13:00–16:00 and is lowest during 22:00–8:00 the next morning (nearly 50% discount at nighttime). Thus, electricity cost can be significantly reduced by setting the light period to 22:00–8:00 (10 h) and to the other 6 h before 22:00 and/or after 8:00, thus avoiding the period of 13:00–16:00.

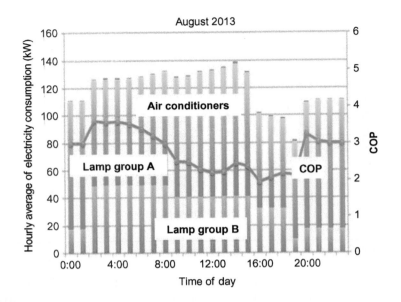

FIGURE 22.14

An example of daily changes in electricity consumption by lamp groups A and B, and COP (coefficient of performance or electrical energy use efficiency) of air conditioners in a PFAL. Lamps of each tier are turned on together for 16 continuous hours a day, but the light period is shifted. COP of air conditioners is higher at lower outside air temperature.

The monthly sum of daily electricity consumption is mostly determined by the lighting time schedule, COP (for the definition, see the next section) of air conditioners, and the time zone-dependent electricity charge system.

Figure 22.14 shows an example of hourly changes in electricity consumption of the culture room with nine tiers in August for lamp groups A (five tiers) and B (four tiers) and air conditioners of a PFAL in Japan. All the lamps are turned on for 16 continuous hours a day but the starting time of the light period for each tier out of the nine tiers is different. The hourly change in averaged COP of the seven air conditioners is given by the line in Figure 22.14, which is higher when the outside air temperature is lower.

COP AS AFFECTED BY THE TEMPERATURE DIFFERENCE BETWEEN INSIDE AND OUTSIDE

The COP (the ratio of the cooling load of the culture room to the electricity consumption of the air conditioners), or electrical energy use efficiency of air conditioners, is influenced by several factors.

Figure 22.15 shows the COPs in summer as affected by the difference in air temperature between inside and outside the culture room. Since the room temperature is always 22°C, the COP decreases with increasing outside temperature. The COP increases to around 10 in winter, because air temperature is 5–25°C higher inside than outside (not shown in Figure 22.15).

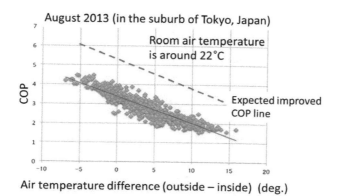

FIGURE 22.15

COPs of air conditioners in August as affected by air temperature difference between inside and outside. Electricity cost for air conditioning is halved when COP is doubled. The dashed line indicates the maximum possible COP which is achieved when the cooling load is around 70% of the cooling capacity. (See also Figure 22.16.)

The lower the COP, the higher the electricity consumption of air conditioners, and thus the electricity cost, especially during 13:00–16:00. It is important to set the light period at nighttime (when air temperature outside is lower than daytime) to achieve a higher COP, even if a nighttime discount of electricity charge is not available.

COP AS AFFECTED BY THE ACTUAL COOLING LOAD

The COP is also affected by the ratio of actual cooling load to the cooling capacity of the air conditioners (Figure 22.16). The COP increases with increasing ratio up to 0.7–0.8, but decreases when it is over 0.8 (Sekiyama and Kozai, 2015). The actual cooling load was estimated as the total electricity consumed by the lamps, air circulation fans, and nutrient solution pumps. The variations of COP

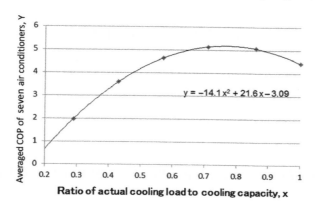

FIGURE 22.16

Averaged COP of seven air conditioners as affected by the ratio of actual cooling load to full cooling capacity. Total electricity consumption of seven air conditioners at full cooling capacity is 58.6 kW (Sekiyama and Kozai, 2015).

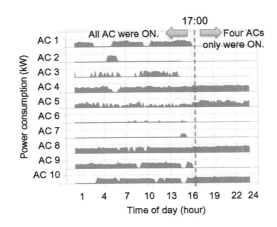

FIGURE 22.17

Diurnal courses of power consumptions of air conditioners (AC). From 1:00 to 17:00, all (10) ACs were turned on, but four ACs only were turned on after 17:00. In either case, air temperature was controlled at the set point of 22°C. Average COP of ACs in operation was 20–25% higher after 17:00 than before 17:00, showing the importance of the control of ACs in operation.

for the same temperature difference in Figure 22.15 are caused mainly by the ratio of actual cooling load to the cooling capacity, indicating that the number of air conditioners in operation needs to be adjusted to keep the ratio in the range of 0.7–0.8.

Figure 22.17 shows the diurnal time course of power consumption of each of 10 or 4 air conditioners. From 1:00 to 17:00, 10 air conditioners were turned on, but 4 air conditioners only were turned on after 17:00. In either case, air temperature was controlled at the set point of 22°C. Average COP of air conditioners in operation was 20–25% higher after 17:00 than before 17:00. Thus, it is important to control the number of air conditioners in operation depending on the actual cooling load in order to achieve a high COP and thus reduce the electricity consumption.

MONTHLY CHANGES IN ELECTRICITY CONSUMPTION

Figure 22.18 shows the monthly electricity charge for a PFAL producing about 100,000 leaf lettuce heads per month. The percent basic charge is lowest (12%) in summer and highest in winter (16%) with an average of 16%. In Japan, the monthly basic charge is determined by the maximum 30-min power consumption (kW) during the past 1 year. In Figure 22.18, the basic charge can be reduced by reducing the maximum 30-min power consumption during 3 months in summer. Reducing the annual maximum consumption by smoothing it over time is essential to reduce the yearly total of the monthly basic charge.

VISUALIZATION OF POWER CONSUMPTION BY COMPONENTS ON THE DISPLAY SCREEN

Figure 22.19 shows an example of visualized power consumption by components on the computer display screen as a daily report automatically generated by the PFAL-M&D-M. This kind of report can be generated weekly, monthly and yearly, or upon request by the PFAL manager. With this report, the

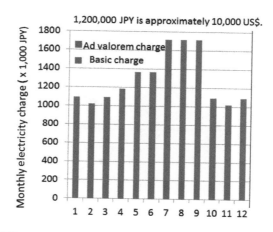

FIGURE 22.18

An example of monthly electricity charge for a PFAL in Japan, producing about 100,000 leaf lettuce heads per month. Percent basic charge is lowest (12%) in summer and highest in winter (16%) with an average of 16%. In Japan, the monthly basic charge is determined by the maximum 30-min power consumption (kW) during the past 1 year.

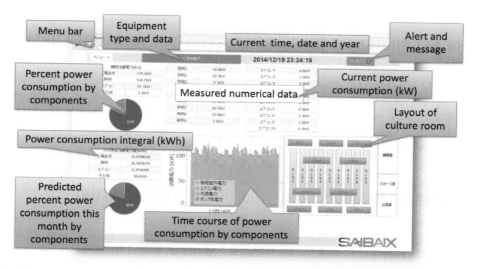

FIGURE 22.19

Visualized daily report of power consumption by components on the computer display screen for the PFAL manager as a daily report.

PFAL manager can grasp the status and progress of the PFAL, and easily identify problems that need to be solved or improved. A radar chart is also useful to understand the overall performance of the PFAL at a glance. Figure 22.20 is a typical radar chart showing the overall performance of electricity cost management.

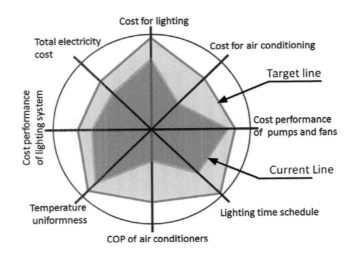

FIGURE 22.20

Radar chart showing the overall performance of electricity charge management to clarify the points to be improved. Each axis is automatically scaled.

RATES OF NET PHOTOSYNTHESIS, DARK RESPIRATION, AND WATER UPTAKE BY PLANTS

Based on the CO_2, water, and energy balance equations for the culture room, the hourly and daily changes in net photosynthetic rate, dark respiration rate, and water uptake rate of plants in the culture room can be estimated and visualized. These data can be related to the total projected leaf area and can be used to determine the parameter values for mathematical models of plant photosynthesis, dark respiration, and transpiration (for details of equations, see Chapter 4).

THREE-DIMENSIONAL DISTRIBUTION OF AIR TEMPERATURE

Plant growth is directly affected by temperature. Thus, achieving uniform spatial air and plant temperature distributions in the entire culture room is essential to ensure uniform plant growth. PFAL-M-PM-Data collects temperature data in the culture room at 100–200 different measuring points, and presents the three-dimensional distribution of temperature in the culture room.

Figures 22.21 and 22.22 show the layout of culture tiers together with the three-dimensional air temperature distributions (can be hourly, daily, and weekly averages). These uneven distributions of air temperature are mainly caused by the uneven air flows from the air circulation fans of air conditioners, natural convection due to heat energy generated from lamps, and air flow resistance due to culture shelves with reflectors. Once the three-dimensional uneven distributions of air temperature are visualized, it is relatively easy to improve the distributions by changing the direction and number of air circulation fans in operation. A thermal imaging camera for measuring the surface temperatures of walls, floor, culture shelves, and plants is also useful to find the points that may cause uneven temperature distribution and water condensation.

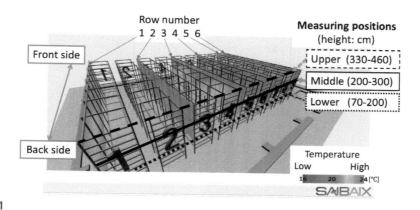

FIGURE 22.21

An example of three-dimensional air temperature distribution in the culture room with 9 rows and 10 tiers in the PFAL.

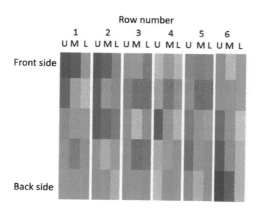

Shelves U : Upper, M: Middle, L : Lower (three shelves each)

FIGURE 22.22

Three-dimensional air temperature distributions in the culture room of PFAL. U, M, and L denote the upper, middle, and lower shelves. For row numbers, floor layout, and heights of shelves, see Figure 22.21.

PLANT GROWTH MEASUREMENT, ANALYSIS, AND CONTROL
DETERMINATION OF PARAMETER VALUES FOR PLANT GROWTH CURVE

Plant growth (fresh/dry weight, leaf area, volume, etc.) in a plant community is well expressed by a logistic growth curve as shown in Figure 22.23. Based on the measurement of plant growth, PFAL-M-PM finds the parameter values of the plant growth curve. In the equation in Figure 22.23, the parameter values t, k, r, and S_{max} denote, respectively, time, initial value at t, relative growth rate, and maximum (saturated) value ("e" denotes the exponential function). The photographs on the right of the graph were taken from the top to estimate the projected leaf area of plants on the culture panel. The

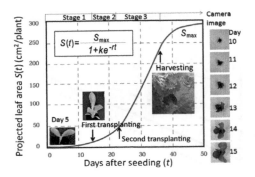

FIGURE 22.23

Plant growth curve expressed as logistic growth equation. Photos on the right were taken from the top. Stage 1: from seeding to first transplanting, Stage 2: from first transplanting to second transplanting, and Stage 3: from second transplanting to harvesting.

value r in the equation can be estimated from the photographs. The parameter values r and S_{max} are determined by planting density, environmental factors, genetic characteristics of plants, etc.

DETERMINATION OF DATES FOR TRANSPLANTING

Figure 22.24 shows the effects of the first and second transplanting dates on the growth of projected leaf area. The growth is delayed and saturated after the percent projected area reaches 100%. If the first and/or second transplanting is conducted after these dates, the productivity decreases. In addition, the work hours for transplanting increase because of overlapping with surrounding plants.

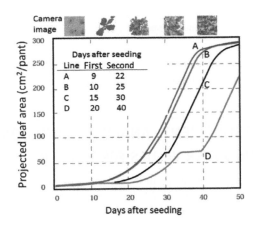

FIGURE 22.24

Projected leaf areas as affected by the days after seeding of first and second transplanting.

On the other hand, if the transplanting is conducted when the percent projected area is significantly lower than 100%, the efficiency of using the culture space and thus productivity decrease. PFAL-M-PM helps the production manager to decide the transplanting dates to optimize the productivity.

DETERMINATION OF THE NUMBER OF CULTURE PANELS FOR DIFFERENT GROWTH STAGES

Culture panels with different numbers of cells (holes) are used for growing plants with growth stages of 1, 2, and 3 (Chapter 18 and Figure 22.24). Table 22.1 shows an example of the relative area occupied in the culture panels for growth stages 1–3 as affected by days required for the growth stages 1, 2, and 3 (D1, D2, and D3), and the number of cells per tray (C1, C2, and C3). D1–D3 and C1–C3 basically determine the plants produced per day. Once D1, D2, and D3 are fixed based on the growth analysis, the number of panels required for each day can be calculated using the equations given in Table 22.1. Success rates of a1, a2, and a3 ($0 <$ a1, a2, a3 < 10) for each stage can be implemented in the equations. The production process in the PFAL can be reasonably managed based on this kind of theoretical consideration and computer software.

CONCLUSIONS

PFAL-D&M is expected to be useful for designing the culture room and managing the plant production process of PFALs. In the near future, PFAL-D&M will be able to cover about 70% of the software used to manage the plant production process. However, much remains to be done.

Table 22.1 Ratio of Area Occupied by the Culture Panels (30×60 cm^2 Each) for Plant Growth Stages 1–3 as Affected by Days Required and the Number of Cells (Holes) Per Tray for the Growth Stages 1, 2, and 3

Growth Stage	1	2	3	Total
Days required	D1 = 14	D2 = 10	D3 = 10	D1 + D2 + D3 = 34
No. of cells/tray	C1 = 300	C2 = 25	C3 = 6	-
Relative total number of panels required	P1 × D1 = 1 × 14 = 14	D2 × P1 × C1/C2 = 10 × 12 = 120	D3 × P1 × C1/C3 = 10 × 50 = 500	PT = P1 × (D1 + D2 × C1/C2 + D3 × C1/C3) = 634
Relative area ratio of panels occupied	P1 × D1/PT = 0.0221	D2 × P1 × (C1/C2)/PT1 = 0.1893	D3 × P1 × (C1/C3)/PT = 0.7886	PT/PT = 1.0000
No. of panels stored on the culture beds	A/(P1 × D1/PT) = 0.0221 × 63,400 = 1401	A/(P1 × (C1/C2))/PT = 0.1893 × 63,400 = 12,001	A/(P1 × (C1/C3))/PT = 0.7886 × 63,400 = 49,984	A = 63,386
No. of daily harvestable plants	-	-	49,984/10 = 4998	

The total number of panels stored on the culture beds in culture room A is assumed to be 63,400 and the success rate at each plant growth stage to be 1.0. The daily number of harvestable plants changes by changing D1–D3 and C1–C3.

It should be noted that some parts of the plant production system cannot be covered by PFAL-D&M, such as the microbial ecosystem and algae growth and propagation in the culture beds. It is therefore necessary to design and develop an entirely new PFAL cultivation system that makes plant PM much easier than existing systems.

REFERENCES

Kozai, T., 2013. Sustainable plant factory: closed plant production systems with artificial light for high resource use efficiencies and quality produce. Acta Hortic. 1004, 27–35.

Sekiyama, T., Kozai, T., 2015. Issues on reduction in cost for electricity consumption for air conditioning using heat pumps in greenhouse horticulture(written in Japanese: Engei shisetsu no kucho ni hito ponpu wo shiyosuru baai no kosuto sakugen ni kakawaru kadai). Electr. Util. Agric. (Nougyo Denka) 68 (2), 12–16.

AUTOMATED TECHNOLOGY IN PLANT FACTORIES WITH ARTIFICIAL LIGHTING

Hiroshi Shimizu[1], Kazuhiro Fukuda[2], Yoshikazu Nishida[3], Toichi Ogura[2]

Graduate School of Agriculture, Kyoto University, Kyoto, Japan[1] Osaka Prefecture University, Sakai, Osaka, Japan[2] Itoh Denki Co., Ltd., Kasai, Hyogo, Japan[3]

INTRODUCTION

Pest and weed control, which are essential in open-field agriculture, are not necessary for vegetable production in plant factories with artificial lighting (PFAL), which are completely isolated from the outside environment. Depending on the PFAL's facilities, there is typically no work until harvesting after the plants are relocated. Therefore, the work to be automated includes seeding, transplanting, moving cultivation panels, harvesting, weight checking, packaging, metal inspection, and panel cleaning. Although it is debatable whether PFALs should be completely automated, Ogura (2011) has stated that humans should be involved at strategic points in the process.

In addition, the consumables and facilities of PFALs have not yet been standardized, and the size of cultivation panels and plant density per panel differ between facilities. General equipment for automation has also become made-to-order, and the cost of introduction is high. Hence, automated equipment has not been introduced in many PFALs.

Only a single cultivation layer was available for this study because PFALs at an early stage employ high-intensity discharge lamps. Under this condition, a spacing device to reduce the plant density as the plants grow was used to improve production efficiency. However, heat from the tube surface is reduced by the light source, so this light source should be replaced by a light-emitting diode (LED) or fluorescent light (FL) from the discharge lamp. A multistage process is then possible, dramatically increasing production efficiency. Therefore, improving production efficiency by using a spacing device is no longer very meaningful.

This paper examines the automated devices used in the case study "Green Clocks Corporation (Osaka, Japan)," which was completed in September 2014.

The facility discussed here is a leading large-scale PFAL in Japan with a cultivation capacity of 5200 lettuce heads per day. It also received financial support from the Ministry of Economy, Trade, and Industry. The following three systems were used to reduce the operating cost by 40%: (1) a seedling sorting robot system (yield improvement); (2) an automated cultivation system (labor saving); and (3) LED light sources (energy saving). In particular, (1) and (2) are unique systems that were newly developed for this facility.

In this PFAL, the basic cultivation schedule is: seeding and greening, 4 days; nursery, 14 days; and cultivation, 18 days. Hence, it takes 36 days from sowing to harvesting. A transplant operation is performed

Plant Factory. http://dx.doi.org/10.1016/B978-0-12-801775-3.00023-8

twice during the production process; the first transplant is performed by a robot from the greening process to the nursery panel, and the second is manually carried out from the nursery panel to the cultivation panel. Harvesting is done manually, and the heads of lettuce are shipped after packing and metal inspection. The automated devices and equipment used in this PFAL will be discussed in the following section.

SEEDING DEVICE

The entire seeding device is shown in Figure 23.1a. The suction holes, which have the same pitch as the sowing urethane mat (300 holes with dimensions of $300 \times 600 \text{ mm}^2$), are open to the suction plate at the top (Figure 23.1b), and the air is sucked through here. The coated seeds are thrown into the suction plate

FIGURE 23.1

Seeding device. (a) Appearance, (b) suction holes, (c) suction plate, and (d) sowing urethane mat.

(Figure 23.1c), and the plate is manually moved back and forth to the left and right. Finally, the coated seeds fall into all the suction holes. At the same time, the sowing urethane mat with hemispherical holes of the same pitch as the suction plate is set on the lower table (Figure 23.1d).

Next, the suction plate is manually turned upside down and brought into close contact with the sowing urethane mat. When the air suction is stopped, the coated seeds are released from the suction plate and fall into the sowing urethane mat, thus completing the seeding. If more than one seed is sucked into one suction hole, the device is able to remove the excess seeds by vibrating the suction plate.

An operator determines whether each suction hole contains a coated seed, then inverts the suction plate and shuts off the suction.

SEEDLING SELECTION ROBOT SYSTEM

Five thousand excellent seedlings are selected from 6000 seedlings per day. Seedling selection is performed in the final stage of the seeding and greening process over 4 days in a greening room. Mechanization is essential for processing 6000 seedlings daily in terms of reliability and continuity. In addition, this facility uses the circadian rhythm in seedling diagnosis—the first such attempt worldwide to the best of the authors' knowledge. This technology aims to improve the accuracy of the diagnosis.

A versatile, high-speed image-processing system is generally adopted for large-scale seedling diagnosis. Usually, an image is only captured one time to analyze the shape and size of the seedling. However, by measuring the seedlings multiple times and using time series data, the accuracy of the diagnosis can be improved.

This process is effective because the circadian rhythm is contained in the biological information. For example, it is well known that the growth rate and photosynthetic rate fluctuate, even under constant light and temperature conditions. Therefore, it is preferable to extract the characteristic quantities, such as the amplitude and period, from the circadian rhythm; using this as an index of seedling diagnosis effectively improves diagnostic accuracy. Figure 23.2a shows an example of measurement using the (AtCCA1:LUC) transgenic lettuce circadian rhythm germination period. Even in continuous light conditions, the circadian rhythm is formed with rooting.

The following is a summary of the diagnostic method. In the seeding and greening process, sprouting is carried out in a dark room for 2 days, and then greening is performed for 4 days with an LED light. On the last day of the greening process, the diagnosis of seedlings is performed. Inspection of 6000 seedlings is carried out on the final day; from these, 5000 are selected as superior specimens. Translocation to the seedling diagnosis system from the greening stage is performed by a shuttle robot that was newly developed for this project (Figure 23.2b).

The seedling diagnosis system consists of a blue LED for chlorophyll excitation and a cooled charge-coupled device (CCD) camera with high sensitivity and high resolution. An automated conveying device carries a germination panel into the dark box, and the chlorophyll pigment of the plants is excited by the blue LED light. The fluorescence emitted from the chlorophyll is captured every 4 h, six times per day (Figure 23.2c).

The size and shape of the individual plants are also measured. From the time series data of the six points per day obtained using this method, the size, shape, chlorophyll fluorescence intensity, circadian rhythm, and cycle of the circadian rhythm are calculated. The suitability of the seedlings is quantified by a predetermined evaluation function using these data.

The aforementioned diagnostic tasks are carried out continuously for 24 h. The suitability characteristics are assigned to all individual specimens immediately prior to sorting. The identification (ID) numbers and evaluation values assigned to all individual seedlings are automatically transmitted to the

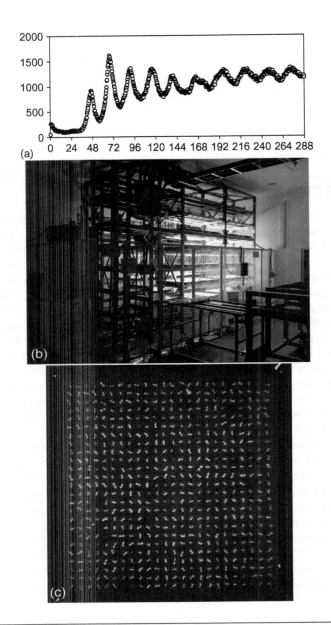

FIGURE 23.2

Seedling selection robot system. (a) Example of circadian rhythm of transgenic lettuce, (b) shuttle robot system, (c) chlorophyll fluorescent image, and (d) transplant robot.

(Continued)

FIGURE 23.2, CONT'D

transplant robot (Figure 23.2d), and the 5000 seedlings that are determined to be superior specimens are relocated to a nursery panel.

SHUTTLE-TYPE TRANSFER ROBOT

In the PFAL of Green Clocks Co., one lane (with a width of 1.5 m and a length of 27.9 m) consists of 36 cultivation units (one unit consists of a cultivation panel and tray) arranged in a row; these have 104 lanes (16 multistage: two rows; 18 multistage: four rows) (Figure 23.3a). Therefore, the total number of cultivation units is 3774. Two cultivation panels filled with seedlings after the nursery stage are carried to the upstream side of each lane. All units are moved to the downstream side two by two every day and are harvested at the other end of the lane after 18 days. This operation is performed every day. Then, plants of two units are harvested daily, and so the total production is 5200 plants per day.

A shuttle-type transfer robot is used to move the cultivation panel (Figure 23.3b; Table 23.1) for the following reasons: for safety, as the cultivation shelf is 7.8 m high; to achieve unmanned operation to minimize the amount of bacteria in the production area; and to reduce labor costs.

FIGURE 23.3

(a) Multi-stage cultivation shelves and (b) shuttle-type transfer robot.

Table 23.1 Specifications of Shuttle-Type Transfer Robot

Length	700 mm
Width	692–696 mm
Height	765 mm (fully lifted: 96.5 mm)
Weight	22 kg
Mobile velocity	2.9–14 m/min
Lifting height	20 mm
Conveying weight	10 kg
Functions	Self-propelled, battery powered Four-wheel timing belt drive Wireless communication Automatic charging function

The shuttle-type transfer robot is installed to move under the cultivation unit of each lane. For specific operation of this robot, there is a space of two cultivation units at the downstream end of the lane after the daily harvest operation.

The robot moves to the bottom of the cultivation unit (third from the end), and a cam mechanism and built-in motor elevate the tabletop of the robot by 20 mm. As a result, the cultivation unit is lifted up.

The robot moves to the lane end in this state and lowers the tabletop; then, the cultivation unit is set into position. All cultivation units are repeatedly moved by this operation, and free space for two cultivation units can be created in the upstream side of each lane. Two new cultivation units filled with seedlings after the nursery stage are carried to this space.

Because it is inefficient to use the shuttle-type transfer robot in each lane, a lifter that moves the robot vertically is set up at the end of the lane in order to transfer the operation to more than one lane by one shuttle-type transfer robot. In addition, because the robot is battery powered, it is programmed to return to the charging station to charge its battery after the daily transfer task.

CULTIVATION PANEL WASHER

A time-consuming operation in the vegetable production process in a PFAL is washing the trays and cultivation panels. The amount of blue-green algae, which is a type of phytoplankton, increases when light falls on the nutrient solution. As a sharp increase in phytoplankton worsens quality, PFALs are not designed for a large amount of light to be irradiated directly onto the nutrient solution.

However, sowing urethane sponges are included in the nutrient solution, and light directly hits the sponges during the early stage of cultivation. The population of blue-green algae increases in such a situation and adheres to the holes of the cultivation panel.

Washing the cultivation panels and trays after cultivation is essential because the dirt on them breeds bacteria. In large-scale cultivation, such as 218 panel harvests every day, automatic cleaning equipment has been introduced to reduce time-consuming hand-washing (Figure 23.4).

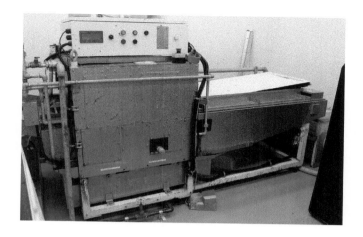

FIGURE 23.4

Cultivation panel washer.

The device is passed along the trays and panels, and a cylindrical brush rotates like a car wash and washes each panel in approximately 40 s.

REFERENCES

Ogura, T., 2011. Operation should be automated and not so in plant factory for leafy vegetables. J. SHITA 23, 37–43.

LIFE CYCLE ASSESSMENT

24

Yasunori Kikuchi

Presidential Endowed Chair for "Platinum Society", The University of Tokyo, Tokyo, Japan

STANDARD OF LCA

INTRODUCTION

Life cycle assessment (LCA) has become a useful tool to quantify environmental impacts as potential impacts based on a product's life cycle from raw material acquisition through production, use, end-of-life treatment, recycling, and final disposal, i.e., cradle-to-grave. For example, a photovoltaic (PV) power generation system can supply electricity without any input of fossil fuels or other materials. However, it requires energy and materials in its production processes, maintenance, or waste treatment. The environmental impacts attributable to such inputs must be taken into account in the life cycle of PV power generation. Biomass-derived resources can be renewable resources. They may require, however, energy for transportation and cultivation, and fossil-based fertilizers. Although the carbon contained in biomass can be regarded as the carbon fixed in the body of plants from the air, the net carbon balance includes the production of such inputs. Air conditioners consume energy for adjusting temperature and moisture of the room. The environmental impacts originating in the life cycle of air conditioners are dominant in their use phase, which means that the efficiency of air conditioners, i.e. coefficient of performance (COP), may be the most sensitive parameters to the environmental impacts. This discussion can be conducted on the basis of LCA results. Considering life cycle, the environmental aspects of product or services can be accurately analyzed and interpreted in decision making.

LCA was defined in the ISO 14040 series, which indicates the role of LCA as follows. Using LCA, the environmental aspects of a product or service life cycle are evaluated. LCA can facilitate the identification of the points in a life cycle to be improved from the viewpoint of the environmental performance of products, and thus can be a help for decision makers in industry, government, or other organizations to design a system or decide on a strategy. At that time, the relevant parameters and indicators on environmental performance can also be specified. Obtained LCA results can be utilized in marketing of products or services shown as an ecolabel, or an environmental product declaration. LCA study has four phases in its practice according to ISO14040/44, which are goal and scope definition, life cycle inventory analysis (LCI), life cycle impact assessment (LCIA), and interpretation, as shown in Figure 24.1. Based on the ISO definition, various textbooks have been published all over the world, e.g., Bauman and Tillman (2004) and Haes et al. (2002).

Plant Factory. http://dx.doi.org/10.1016/B978-0-12-801775-3.00024-X

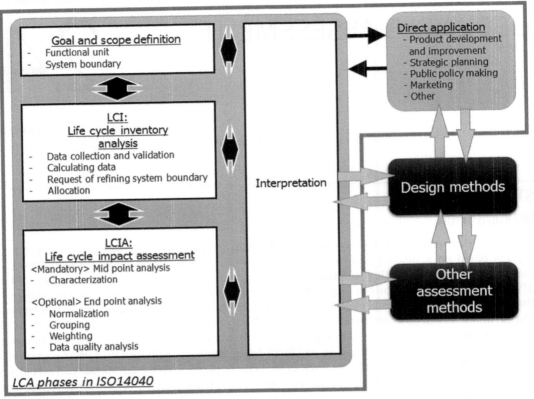

FIGURE 24.1

System design and assessment centering LCA stages in ISO 14040.

GOAL AND SCOPE DEFINITION

"Goal and scope definition" is the phase of setting conditions on LCA. The system boundary, functional unit, impact categories to be evaluated, and objective of LCA execution are specified in this phase. System boundary is the target scope of LCA where the unit processes to be evaluated in the LCA study are included. For example, the crude oil sourcing can or cannot be included in this system boundary, based on the existence of any change in that process. If the amount of crude oil used in the compared scenarios is the same, it can be excluded. The system boundary should be set considering functional units at the same time. Functional units are the basic common aspects of products or services to be compared in LCA, e.g., production of a car, 1 kg of chemicals, the transportation of a person by 1 km, 1 day living in a house in Japan, and so on. Figure 24.2 shows the overview of system analysis in LCA considering a functional unit with examples on products 1 and 2. Product 1 has the functions A, B, and C, while product 2 has only functions 1 and 2. For example, three-color pens and two-color ones are the same situation as products 1 and 2, respectively. To compare these products, the difference of functions must be addressed. A third

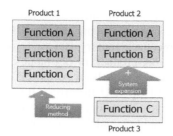

FIGURE 24.2

System analysis in LCA considering functional unit.

product with the missing function C only can be additionally taken into account in this comparison. The system should be expanded to include this product or save the same amount of environmental loads as product 3 in those of product 1. Hence, the system boundary may be modified, to adjust the functional unit.

As mentioned previously, the premises and conditions are set in goal and scope definition. In this regard, this goal and scope definition phase may be conducted with feedback information from other phases through interpretation. As shown in Figure 24.1, the boxes are connected by the arrows with multiple directions. This means that the phases can be iterated if necessary. For example, the one-way plastic bags used in purchase behavior can be used as waste bags in Japan, which means that the fresh waste bags must be considered in the system boundary of their life cycle because the waste bags must be consumed if the one-way plastic bags are not used. As well as system boundary and functional unit, the investigation settings on inventory data should also be specified, as introduced in the following discussion. The impact category in an impact assessment is also an important setting. As shown in Figure 24.1, the ISO defines complete LCA as the quantification of environmental impacts by characterization step, in other words mid-point analysis (see also the impact assessment part).

LIFE CYCLE INVENTORY ANALYSIS

"LCI" quantifies the total environmental loads generated within life cycle stages defined in the goal and scope definition. The procedure to collect inventory data starts on the basis of the data requirement for each life cycle stage, especially which type of data is needed for the stage: foreground or background data. Foreground data mean that the data are attributable to the target products accurately, while background data can be extracted from special or temporal averages. For example, the detail pathways of consumed electricity from power generation plant to the demand such as transmission or distribution must be investigated to obtain the foreground data of 1 kWh of electricity in the system boundary, while the average data in the database can be applicable for background data. Generally, the production data of target products or services may be investigated as foreground data to quantify the environmental loads. The grid power data may be extracted from a database even if it is consumed in the foreground processes. In other representation, the foreground data are the data of foreground processes which are under control of the decision-maker for which an LCA is carried out. Note that the definition of foreground and background data is more complicated in the actual LCA study field. Please refer to the textbooks on LCA.

Inventory data is available by converting on-site process data, in other words operation results. For example, energy and materials are consumed in a process, e.g., (J/month) or (kg/month) to produce the product, e.g., (kg/month). At this time, the inventory data can be obtained by dividing input energy and materials by output products, e.g., (J/kg-product) or (kg-input/kg-product).

As for the database, various databases have been developed and opened such as ecoinvent (Swiss Center for Life Cycle Inventories), ELCD (EU-JRC-IES), JLCA-LCA database (JLCA), and so on. Each database has data sets on life cycle inventory data, the properties of which are different on geographical or temporal applicability, qualities, and the boundary of the data set. In extracting data from a database, such differences in data sets should be carefully taken into account. Existing LCA software tools such as SimaPro (PRé Consultants), or MiLCA (JEMAI) can be a support for LCA practitioners.

LIFE CYCLE IMPACT ASSESSMENT

"LCIA" quantifies environmental impacts by multiplying the results of LCI (mass-environmental loads/functional unit) with impact factors (environmental impact/environmental load). As an example, the impact categories and their endpoints, areas of protection (AoP), and single index are shown in Figure 24.3, which is the LCIA method based on endpoint modeling (LIME) second version

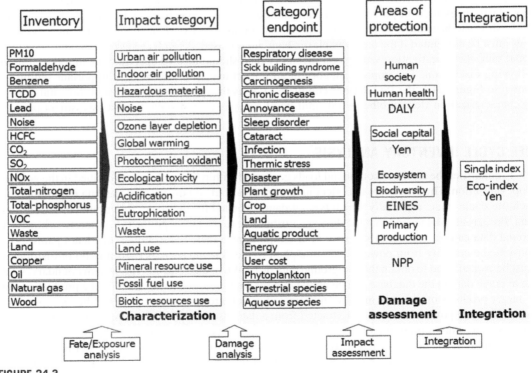

FIGURE 24.3

Overview of LIME 2 (Itsubo and Inaba, 2010).

(Itsubo and Inaba, 2010). The explanation following is based on this LIME2. As discussed before, based on Figure 24.1, the characterization of environmental impacts must be conducted in ISO standard. The fate and exposure analyses for environmental loads can be addressed using characterization factors quantified in LCIA methods or other standards. For example, the global warming potential (GWP) is one of the major indicators for climate change, which is originally quantified in IPCC technical reports. In addition to the characterization, the damages on AoP can be quantified using damage factors, which are developed by various types of analyses. For example, the damages on human health have been quantified using research and investigation on the dose-response of chemical substances, such as epidemiological studies. Disability adjusted life years (DALY) has been adopted as an indicator of human health in some LCIA methods to integrate different types of endpoints attributable to damages on human health. As well as this endpoint analysis, integration is an optional procedure of LCIA. In LIME2, the indicators on environmental impacts are aggregated into LIME indicators (JPY).

As mentioned already, the LCIA method can decrease the number of indicators using analyses on mechanisms of the adverse effects caused by environmental impacts. This may avoid some types of trade-off relationships between environmental impacts. In this regard, however, the meanings of each indicator such as the emission amount of carbon dioxide or the consumption of phosphorus resources must be carefully interpreted for considering the improvement of the target system.

INTERPRETATION

The "interpretation" phase has two main roles in LCA at the least: the intermediate or the decision making phases. As intermediate phase, a tentative result or situation of LCI may be returned to the goal and scope definition as a feedback requesting redefinition of the conditions of LCA studies. The expansion of system boundary, for example, can be considered in this phase based on such feedback. Impact assessment methods can also be reselected on the basis of the available impact factors. The practitioners of the LCA study should be able to find out the necessity of such redefinition by checking the intermediate outcomes from the LCA study. In the decision-making phase, on the other hand, the "direct application" of LCA results must be taken into account. For making a decision on such direct application, the practitioners should be able to understand all phases of LCA, existing uncertainties in the quantified LCI and LCIA results, and so on. There may be trade-off relationships between environmental impacts, which can also be found in the analysis with other aspects such as economic or social ones. Multi-objective decision making must be conducted in actual system design frameworks (Sugiyama et al., 2008; Kikuchi and Hirao, 2009).

REMARKS FOR THE ASSESSMENT OF PFAL

An example of life cycles of plant factory with artificial lighting (PFAL) and their products is shown in Figure 24.4. Plant and product life cycles intersect each other at the manufacturing stage. An LCA study on PFALs should consider these two types of life cycles. In the following sections, the points to be taken into account in the LCA of PFALs are overviewed. At this time, the goal of LCA is supposed to be the design of a PFAL.

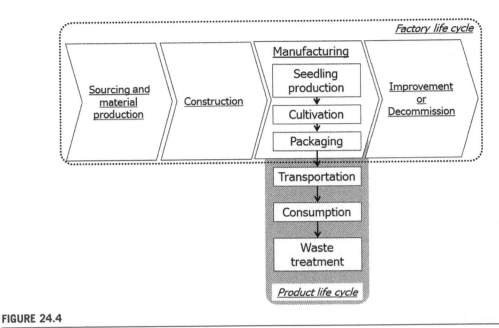

FIGURE 24.4

An example of the life cycles of indoor farming plants and products.

INVENTORY DATA COLLECTION/IMPACT ASSESSMENT

The construction and manufacturing data of PFAL should be collected as foreground data. As in the construction of a factory, base metals and materials such as steels and cement are taken into account. The exterior and interior of factories must be considered as the required input for the construction of the factory in Figure 24.4. The cumulative production data of each metal and material can be a background one, when any special ones are not used, e.g., local ecomaterials, or reuse of existing buildings or construction wastes.

As for the processes supplying energy, the sources of energy should be carefully taken into account; for example, some factories install PV power systems or other distributed energy technologies for their own processes. At that time, the environmental loads originating from the installation of such technologies must be considered as foreground data, because the initial environmental loads for installing energy technologies have a high sensitivity with the operation ratio of them. The life time or actual usage years of the technologies are also highly sensitive to the total environmental loads.

Regarding the operation data, as well as energy supply, the conditions of fertilizers are also important and should be carefully examined. In particular, the resource use impacts of phosphorus or nitrogen cycles can be one of the most important topics in agriculture. According to the material flows of phosphorus, it is consumed without an effective recycling system (Matsubae et al., 2011). Needless to say, phosphorus is one of the most important and inevitable elements for fertilizers. If it was depleted to some degree, the fertilizers cannot be produced and the capacity of food cultivation might be seriously decreased. So the efficiency of the usage of phosphorus—in other words, the consumption of phosphorus per production—must be carefully checked and assessed. In the comparison with the general farming land, the LCA study of indoor farming plants should also take into account the land use inventories, such as area, time of occupation, or the existence of transformation. The impact categories on land use can be the GHG (greenhouse gas) emission from soil or other emission. At the same time, the products from PFAL, such as lettuces, are cultivated in clean rooms. It means that there is no loss of lettuce just before eating them because even the outside leaf is clean. In many cases, the vegetables grown outdoors have loss of leaves due to dirt. Such differences should be taken into account by including the use or consumption phase of products.

FUNCTIONAL UNIT

Indoor farming plants can play various roles for agriculture. Even though the most commonly usable functional unit can be the production of agricultural products, they have multiple functions to be considered for comparison with outdoor cultivation; the stability of production even if there is unusual weather such as huge typhoons, unusual rain, low temperature, and so on, or the considerably high productivity on land, water, and fertilizers use. Some of these points can be quantified by LCA as environmental impacts, such as resource depletion impacts. However, it is not easy to consider all aspects of indoor farming plants by general LCA. For example, LCA utilizes the steady-state LCI data generally. The stability of production against unusual weather can be an indicator on the risk hedge for such events. The precautionary function must be addressed by setting the goal and scope of an LCA study.

INTERPRETATION

In the interpretation, the practitioners of the LCA study should take into account not only the results of LCA, but also other aspects. For example, indoor farming plants can become new and huge power demand users on the power grid. Because PV and WT (wind turbine) power systems may cause a non-negligible fluctuation in the power system, resulting in the limitation of usage of power during its shortage, indoor farming plants may play the role of a stabilizer of power demand/supply by changing the power load. In other aspects, indoor farming plants can effectively utilize unused waste heat from other manufacturing plants such as plastics molding, food processing, or other production processes. It may increase the efficiency of resource consumption using low temperature heat. In urban areas, indoor farming plants can be a workplace for elderly people. These points cannot be easily addressed in an LCA study. At that time, required assessment methods should be integrated at this phase and final decisions should be carefully made.

SUMMARY AND OUTLOOK

LCA can quantify the environmental impacts through the analysis of inventories of included processes in a target life cycle. Because the LCI inventory database strongly facilitates the collection of inventory data, the practitioners can conduct a LCA study by investigating important foreground data. Impact assessment methods enable the practitioners to consider direct and indirect effects by omitting environmental loads from the target life cycles. Functional unit and system boundary have many open points to be considered in the interpretation, which means that various types of assessment can be performed by setting hue conditions appropriately.

LCA has some limitations originating from the characteristics in the framework and the data availability. For example, the settings of system boundary and functional unit should be decided by the practitioners, which may result in a subjective decision, even though such settings can cause large differences in the results. LCA has a critical review phase to check such subjectively decided settings or parameters, justify assumptions the practitioners could not avoid, and approve the transparency and completeness of the LCA study. This phase can contribute to making up for the limitations of LCA. On the other hand, the system dynamics and nonlinearity of process inventory and environmental impacts are the examples of the points which are difficult to be considered by LCA study. Such limitations must be taken into account by the LCA practitioners. Note that some discussion on the limitation or unresolved problems in LCA have been conducted, e.g., Reap et al. (2008a,b).

In the evaluation of indoor farming by LCA, the various functions achievable by the technology should be taken into account. As well as the quality of products, that of processes as a source of foods, e.g., production stability, adaptivity for climate, high productivity on land, water, and fertilizer, must be addressed as differences from cultivation in open farming land. The technology development, at the same time, should also be considered as the parameters sensitive to the final results of LCA. Lighting, insulating, air conditioning, or controlling technologies are now under development. The efficiency of indoor farming plants can be improved in the future using such developed technologies. Not only LCA, but also other assessment methods should be possibly incorporated into decision making. Even in the LCA-related methods, life cycle costing and social LCA (UNEP/SETAC Life Cycle Initiative, 2009) are also the elements of life cycle sustainability assessment (UNEP/SETAC Life Cycle Initiative, 2011). Making informed choices on products (UNEP/SETAC Life Cycle Initiative, 2011) and processes are strongly needed and can be supported by LCA.

REFERENCES

Bauman, H., Tillman, A.M., 2004. The Hitch Hiker's Guide to LCA. Studentlitteratur AB, Lund.

EU-JRC-IES (EU the Commission's Joint Research Centre, Institute for Environment and Sustainability), ELCD (European reference Life Cycle Database), http://eplca.jrc.ec.europa.eu/.

Haes, H.A.U., Finnveden, G., Goedkoop, M., Hauschild, M., Hetwich, E.G., Hofstetter, P., Jolliet, O., Klöepffer, W., Krewitt, W., Lindeijer, E., Müller-Wenk, R., Olsen, S.I., Pennington, D.W., Potting, J., Steen, B., 2002. Life-cycle impact assessment: striving towards best practice. SETAC Press, Pensacola.

ISO (International Organization for Standardization), 2006. ISO 14040: environmental management—life cycle assessment—principles and framework.

ISO (International Organization for Standardization), 2006. ISO 14044: environmental management—life cycle assessment—requirements and guidelines.

Itsubo, N., Inaba, A., 2010. LIME2. Maruzen, Tokyo.

JEMAI (Japan Environmental Management Association for Industry), MiLCA, http://www.milca-milca.net/.

JLCA (Life Cycle Assessment Society of Japan), JLCA-LCA database, http://lca-forum.org/database/.

Kikuchi, Y., Hirao, M., 2009. Hierarchical activity model for risk-based decision making integrating life cycle and plant-specific risk assessments. J. Ind. Ecol. 13 (6), 945–964.

Matsubae, K., Kajiyama, J., Hiraki, T., Nagasaka, T., 2011. Virtual phosphorus ore requirement of Japanese economy. Chemosphere 84 (6), 767–772.

PRé Consultants, SimaPro, http://www.pre-sustainability.com/simapro.

Reap, J., Roman, F., Duncan, S., Bras, B., 2008a. A survey of unresolved problems in life cycle assessment. Part 1: Goal & Scope Definitions and Inventory Analysis. Int. J. Life Cycle Assess. 13 (4), 290–300.

Reap, J., Roman, F., Duncan, S., Bras, B., 2008b. A survey of unresolved problems in life cycle assessment. Part 2: life cycle impact assessment and interpretation. Int. J. Life Cycle Assess. 13 (5), 374–388.

Sugiyama, H., Fischer, U., Hungerbühler, K., Hirao, M., 2008. Decision framework for chemical process design including different stages of environmental, health, and safety assessment. AIChE J. 54 (4), 1037–1053.

Swiss Center for Life Cycle Inventories, ecoinvent database, http://www.ecoinvent.org/.

UNEP/SETAC Life Cycle Initiative, 2009. Guidelines for social life cycle assessment of products, http://www.unep.fr/shared/publications/pdf/DTIx1164xPA-guidelines_sLCA.pdf.

UNEP/SETAC Life Cycle Initiative, 2011. Towards a life cycle sustainability assessment, http://www.unep.org/pdf/UNEP_LifecycleInit_Dec_FINAL.pdf.

EDUCATION, TRAINING, AND INTENSIVE BUSINESS FORUMS ON PLANT FACTORIES

25

Toshitaka Yamaguchi, Michiko Takagaki, Satoru Tsukagoshi
Center for Environment, Health and Field Sciences, Chiba University, Kashiwa, Japan

INTRODUCTION

Research and development on plant factories has made remarkable progress in recent years and is attracting much industrial interest. As plant factories have spread rapidly in Japan, it has become particularly important to train staff to ensure the success of plant factory businesses.

Five greenhouses (semi-closed system) with an environmental control system and two plant factories with artificial lighting (PFAL) with a total floor area of 10,666 m^2 were built on the campus of the Center for Environment, Health, and Field Sciences, Chiba University (hereinafter "the Center") with the support of the Ministry of Agriculture, Forestry, and Fisheries (MAFF) in 2011. An additional PFAL (floor area of 179 m^2) was built in 2014. The Center and the Japan Plant Factory Association (JPFA) have used these PFALs for research, education, and training.

The Center is conducting an education program for graduate school students on plant factories since 2011 as a part of the Plant Environment Designing Program. The Center also accepts short-stay (SS) students from overseas sister universities, who stay for 70-90 days or 10–15 days. The features of this program and its achievements are outlined in Section "Plant Environment Designing Program" of this chapter.

The Center and JPFA and/or Japan Greenhouse Horticulture Association (JGHA) have jointly held intensive business forums on plant factory information, science, business, and operation with the support of MAFF since 2010. The intensive business forums are introduced in Section "Intensive Business Forums on Plant Factories".

JPFA has offered business workshops on a wide range of issues of plant factories for engineers, researchers, and managers, by planning workshops, excursions, and instructor dispatches on a monthly basis since 2009. The business workshops are described in Section "JPFA's Business Workshops".

The education and training programs for students and business people have been conducted for two types of plant factory. One is a greenhouse with environmental control system (GECS), a semi-closed plant production system, which is often called a plant factory with solar light in Japan. The other is a PFAL. There are many common aspects in education and training between the two types, so the education and training programs do not distinguish between them on the common issues, but clearly do so when there are specific issues for either of them.

Plant Factory. http://dx.doi.org/10.1016/B978-0-12-801775-3.00025-1

PLANT FACTORIES IN THE CENTER

Five GECSs and two PFALs were built on the campus of the Center with the support of MAFF in 2011 (Figure 25.1). A new PFAL was additionally built in 2014. The plant factories are managed and operated by private companies as shown in Table 25.1, while the Center and JPFA use them for research, education, and training. Since 2011, over 5000 people have visited the campus annually for the site

FIGURE 25.1

Overview of plant factories in the Center.

Table 25.1 Plant Factories in the Center of Chiba University

Companies	Crop	Type	Feature
Seiwa Co., Ltd. (until 2014)	Tomato	GECS	Integrated environmental control
Sumitomo Electric Industries, Ltd. (from 2014)	Tomato	GECS	Sand culture system
Iwatani Agrigreen Co., Ltd.	Tomato	GECS	Multi-truss high-plant-density
JA Zen-Noh (National Federation of Agricultural Cooperation Association)	Tomato	GECS	Next generation system
Mitsubishi Plastics Agri Dream Co., Ltd.	Tomato	GECS	Single-truss-moving bed, high-plant-density
Daisen Co., Ltd.	Tomato	GECS	D-tray low-height, high-plant-density cultivation
Mirai Co., Ltd.	Leafy vegetables	PFAL	Ten-story cultivation bed
Wago Co., Ltd.	Leafy vegetables	PFAL	Cultivation of crisp head lettuce
Japan Dome House Co., Ltd. (from 2014)	Leafy vegetables	PFAL	Airtight polystyrene dome house

Table 25.2 Number of Visitors to the Center's Plant Factories

Fiscal Year (April–March)	Visitors
2011	5027
2012	6933
2013	5416
2014 (April–January)	4387

tour, with the cumulative number reaching 20,000 at the end of January 2015. These people included the general public, farmers, business professionals, students, researchers, international visitors, and so on. The number of visitors each year is shown in Table 25.2.

PLANT ENVIRONMENT DESIGNING PROGRAM

The Center, the Graduate School of Horticulture and the Department of Design Science, the Graduate School of Engineering, Chiba University, have jointly held the Plant Environment Designing Program as a part of the Campus Asia Project of the Japanese Ministry of Education, Culture, Sports, Science, and Technology since 2010. The objective of this program is to develop international human resources with the ability to plan and implement projects with technological knowledge, thus contributing to the environment through plants. The program has two categories, Urban Plant Factory Project and City Green Project. The Urban Plant Factory Project has an education program for graduate school students focusing on plant factories. This report focuses on the Urban Plant Factory Project (Onuma, 2014).

The curriculum of the Plant Environment Designing Program consists of three modules: (1) project-based learning (PBL), (2) internship, and (3) lectures.

In the PBL module, a student or group of students conducts a project on practical themes from actual issues faced by commercial plant factories and market needs. They then devise a strategy for solving the issues through workshops and carry out experiments and/or field surveys.

Each student joins one or two plant factory companies as an intern and performs various work of the company to learn practical techniques and management skills in the fields (Figure 25.2). Students can choose a type of plant factory that matches their research themes and their goals for the future.

Companies for which students worked from 2010 to 2014 as an internship in a plant factory were (1) Seiwa Co., Ltd., (2) Iwatani Agrigreen Co., Ltd., (3) JA Zen-Noh (National Federation of Agricultural Cooperatives Association), (4) Mitsubishi Plastics Agri Dream Co., Ltd., (5) Daisen Co., Ltd., (6) Mirai Co., Ltd., (7) Japan Dome House Co., Ltd., (8) Sumitomo Electric Industries, Ltd., and (9) JPFA.

Students attend lectures on protected horticulture and project management, given by faculty members of the university and lecturers from business. The themes of the protected horticulture lectures are: (1) environmental factors affecting plant growth in a greenhouse, (2) measurement of environmental factors, (3) the process of photosynthesis and the photosynthetic response, and (4) practice of cultivating vegetables. The theme of the project management lecture is the process and cycle of project management to solve issues.

The Plant Environment Designing Program accepts foreign students from overseas sister universities using the SS program of the Japan Student Services Organization (JASSO). The costs of studying are supported by JASSO via Chiba University. In this program, the Center also accepts SS (70–90 days)

FIGURE 25.2

Student's internship in a GECS in the Center.

students for education on plant factories. They attend lectures on protected horticulture as mentioned previously and have internships at plant factories in two or three locations.

This program also invites students to stay for 10 or 15 days to expose them to actual plant factories, as well as international symposiums and workshops on plant factories. Field tours to plant production sites and Japanese cultural events are also held. The SS students are mostly from Asian countries, as shown in Table 25.3.

The plant factory international symposiums were jointly held by the Center and JPFA in 2011, 2012, and 2013 (Figures 25.3–25.5), with the following themes and presentation titles each year:

2011 theme: Plant factory research and opportunities in East Asia
(1) Plant factories in Korea: research and industrialization, research approaches to solving problems confronted by the plant factory industry in Korea
(2) Plant factory research and opportunities in Taiwan
(3) Plant factory research and opportunities in China
2012 theme: Accomplishments of the Chiba University Plant Factory Project
(1) Overview and summary
(2) Productivity improvements by integrated environmental control
(3) Multi-truss high plant density cultivation of tomato
(4) Next-generation production system for tomato
(5) Reduced-chemical, single-truss, high-plant-density, moving-bench growing system
(6) Three-truss high-density tomato growing system with D-tray
(7) Low-cost Mirai type PFAL

Table 25.3 Short-Stay Students Accepted by the Center for Education on Plant Factory

Period	Fiscal Year (April-March)	Number of Students	Countries
10–15 days	2011	24	China, Indonesia, Korea, Singapore, and Thailand
	2012	7	China, Thailand, and Turkey
	2013	20	China, Germany, Indonesia, Korea, and Thailand
	2014	9	China, Indonesia, Korea, and Thailand
	Total	60	
70–90 days	2011	9	China and Thailand
	2012	15	China, Korea, and Thailand
	2013	14	Thailand, Indonesia, China, and Vietnam
	2014[a]	20	China, Indonesia, and Thailand
	Total	58	

[a]*2014: Including fourth Q's estimation.*

FIGURE 25.3

International Symposium in 2012.

(8) Stable crisp head lettuce production with artificial lighting

(9) Plant factories in downtown areas: development and popularization of small plant factories

(10) Inter-thematic areas: optimizing plant factories: multi-faceted analyses with integrated database

(11) Review and perspectives on plant factories

FIGURE 25.4

Observing the Mirai PFAL during the International Symposium in 2012.

FIGURE 25.5

International Symposium 2013.

2013 theme: Activities report on the Plant Environment Designing Program
(1) Program overview
(2) Presentation of project outcomes (Titles related to plant factory only are listed below):
    1. Development and evaluation of household plant factories for Chinese customers
    2. Evaluation of impressions on the design of a shop-style plant factory

3. Designing a plant factory as a community garden in temporary housing complexes in earthquake-stricken Natori City
4. Designing a plant factory for "within-restaurant" production and consumption
5. Use of algae in plant factories: potential and perspectives of green cell factories
6. Growth and nutrient uptake of lettuce as affected by the light quality of LED lamps
7. Tipburn in Chinese cabbage related to the cultivar and calcium

(3) Presentation of internship outcomes:
1. Trends in agriculture: Smart farmer for teenagers
2. Experiences gained from Iwatani and Mirai Company

INTENSIVE BUSINESS FORUMS ON PLANT FACTORIES

The Center and JPFA jointly held intensive business forums on plant factories covering the information, science, business, and operation of plant factories with the support of MAFF from 2010 to 2013 for people who were operating plant factories or planning to start doing so. Since May 2014, the forums have been held under the auspices of the Center, JPFA, and JGHA.

From 2010 to 2012, this forum held three courses each year: a basic course, a professional course, and a special course for field advisers in local governments. In 2013, the forum was expanded with the addition of seven specialized courses focusing on solutions to real problems in running a plant factory, in addition to the basic course and professional course. In 2014, two new courses were added at the strong request of participants in 2013: (1) methods of installing and utilizing environmental measures and (2) lectures on key points and case studies of plant factory business management. The number of participants each year are listed in Table 25.4. Figures 25.6–25.9 show hands-on practice at installing PFALs during forums.

Table 25.4 People Who Attended the Intensive Business Forum

2010		2011		2012		2013		2014	
Date	#	Date	#	Date	#	Date	#	Date	#
10/5–1/25	50	11/14–11/17	45	8/9–8/11	32	8/1–8/2	17	6/17–6/19	31
1/5–1/7	24	11/21–11/22	7	12/5–12/7	49	8/5–8/7	36	7/2	52
3/15	6	1/10–1/11	21	2/18–2/20	24	8/27–8/29	17	7/3–7/4	31
		1/23–1/26	37	3/4–3/5	17	10/1–10/3	10	9/2–9/3	32
		1/30–2/2	38			11/5–11/7		9/17–9/19	11
						12/3–12/5		10/15–10/17	
						10/17–10/18	18	11/19–11/21	
						12/10–12/12	47	10/7–10/9	32
						1/27–1/29	22	12/2–12/4	11
						2/5–2/7	15	12/16–12/18	18
						3/4–3/6		1/14–1/16	
						2/24–2/26	27	2/3–2/5	21
	80		148		122		209		240

FIGURE 25.6

Practice at installing PFAL-1.

FIGURE 25.7

Practice at installing PFAL-2.

2010 BUSINESS FORUMS

1. Basic course (16 days). Aims: (1) Deepen understanding of advanced plant factories, (2) Eliminate experience- and intuition-based agriculture by acquiring the knowledge needed for science-based agriculture, and (3) Switch to corporate agricultural management (October 5, 12, 19, 26; November 2, 9, 16, 30; December 7, 14, 21, 28, 2010, January 4, 11, 18, 25, 2011).

2. Professional course (3 days). Aims: (1) Deepen understanding of advanced plant factories, (2) Eliminate experience- and intuition-based agriculture by acquiring the knowledge needed for

FIGURE 25.8

Practice at installing PFAL-3.

FIGURE 25.9

Practice at installing PFAL-4.

science-based agriculture, (3) Switch to corporate agricultural management, and (4) Develop human resources who understand the theory, methods of environmental control and cultivation technology, and who can plan and execute cultivation by themselves (January 5–7, 2011).

3. Special course for field advisers (1 day). Aims: (1) Update plant factories and introduce Chiba University's plant factory, (2) Introduce environmental control technology in plant factories, and (3) Introduce cultivation skills in plant factories (March 15, 2011).

2011 BUSINESS FORUMS

1. Basic course (twice; 4 days). Aims: (1) Deepen understanding of advanced plant factories, (2) Eliminate experience- and intuition-based agriculture by acquiring the knowledge needed for science-based agriculture, and (3) Switch to corporate agricultural management (November 14–17, 2011) (January 23–26, 2012).
2. Professional course (4 days). Aims: (1) Deepen understanding of advanced plant factories, (2) Eliminate experience- and intuition-based agriculture by acquiring the knowledge needed for science-based agriculture, (3) Switch to corporate agricultural management, and (4) Develop human resources who understand the theory, methods of environmental control and cultivation technology, and who can plan and execute cultivation by themselves (January 30 to February 2, 2012).
3. Special course for field advisers (two times; 2 days). Aims: (1) Update plant factories and introduce Chiba University's plant factory, (2) Introduce environmental control technology in plant factories, and (3) Introduce cultivation skills in plant factories (November 21–22, 2011) (January 10–11, 2012).

2012 BUSINESS FORUMS

1. Basic course (twice; 3 days). Aims: (1) Deepen understanding of advanced plant factories, (2) Eliminate experience- and intuition-based agriculture by acquiring the knowledge needed for science-based agriculture, and (3) Switch to corporate agricultural management (August 9–11, 2012) (February 18–20, 2013).
2. Professional course (3 days). Aims: (1) Deepen understanding of advanced plant factories, (2) Eliminate experience- and intuition-based agriculture by acquiring the knowledge needed for science-based agriculture, (3) Switch to corporate agricultural management, and (4) Develop human resources who understand the theory, methods of environmental control and cultivation technology, and who can plan and execute cultivation by themselves (December 5–7, 2012).
3. Special course for field advisers (2 days). Aims: (1) Update plant factories and introduce Chiba University's plant factory and (2) Introduce the plant factory of the National Agriculture and Food Research Organization (March 4–5, 2013).

2013 BUSINESS FORUMS

1. Basics of plant factories (3 days). Aims: (1) Deepen understanding of advanced plant factories, (2) Eliminate experience- and intuition-based agriculture by acquiring the knowledge needed for science-based agriculture, and (3) Switch to corporate agricultural management. (August 5–7, 2013).
2. Basic management of culture solution (twice; 2 days). Aims: (1) Understand the importance of managing culture solution through lectures and (2) Learn the methods of chemical analysis of culture solution and data analysis (August 1–2, 2013) (October 17–18, 2013).
3. Q&A on GECS (3 days). Aims: (1) Receive all answers on any questions on GECS through workshops, (2) Deepen understanding of difficult management techniques through practice, and (3) Practice the measurement of CO_2 concentration, temperature and humidity, and estimate the ventilation rate and vapor pressure deficit (August 27–29, 2013).

4. GECS operation (3 days in October, 3 days in November, 3 days in December). Aims: (1) Practice growing tomato at the key growing timings from seeding to harvest and (2) Practice in the specially designed greenhouse for training and in commercial-size greenhouses (October 1–3, November 5–7, December 3–5, 2013). (Figures 25.10–25.12 show GECS operation practice with students.)

FIGURE 25.10

Practice—Operation of GECS-1.

FIGURE 25.11

Practice—Operation of GECS-2.

FIGURE 25.12

Practice—Operation of GECS-3.

5. Business workshop for managers (3 days). Aims: (1) Learn what to know, what to target, how to ensure profit, reduce expenses and how to manage labor, from successful managers (December 10–12, 2013).
6. Q&A on PFAL operation (3 days). Aims: (1) Gain knowledge and tips for operating a PFAL and (2) Learn the key points of PFAL operation through practice and workshops (January 27–29, 2014).
7. PFAL installation (3 days in February, 3 days in March). Aims: (1) Practice installing a PFAL and (2) Master the system structure and know-how of cultivating techniques (February 5–7, March 4–6, 2014).
8. Professional knowledge (3 days). Aims: (1) Gain expert knowledge on plant factories, (2) Gain expert knowledge to practice science-based agriculture and (3) Gain information on updated techniques to introduce them to commercial operation (February 24–26, 2014).

2014 BUSINESS FORUMS

1. Basics of plant factories (3 days). Aim: Learn the basic technical terms and knowledge of plant factories through lectures (June 17–19, 2014).
2. Basic management of culture solution (1 day). Aim: Gain basic technical knowledge of managing culture solution through lectures (July 2, 2014).
3. Advanced management of culture solution (2 days). Aim: Acquire the skill of managing culture solution through lectures and practice (July 3–4, 2014).

4. Installation and utilization of environmental measures (2 days). Aim: Understand how to visualize the status of the environment and acquire the skill through lectures and practice (September 2–3, 2014).
5. Operation of GECS (3 days in September, 3 days in October, 3 days in November) Aim: Understand the key points for the operation of GECS through practice (September 17–19, October 15–17, November 19–21, 2014).
6. Key points and case studies of plant factory business management (3 days). Aim: Gain the know-how for managing a plant factory from case studies through lectures (October 7–9, 2014).
7. Q&A on GECS (3 days). Aim: Ask questions on GECS and get answers through workshops and practice (December 2–4, 2014).
8. Operation of PFAL (3 days in December, 3 days in January). Aim: Understand the key points for PFAL through practice (December 16–18, 2014; January 14–16, 2015).
9. Q&A on PFAL (3 days). Aim: Ask questions on PFAL through lectures and experiments (February 3–5, 2015).

Business forums on plant factories are offered by other organizations in Japan too, such as the following:

1. Mie Prefecture Agricultural Research Institute
   http://www.mate.pref.mie.lg.jp/plant-factory/index.htm
2. National Agriculture and Food Research Organization—Tsukuba
   http://www.naro.affrc.go.jp/vegetea/plant_factory/
3. National Agriculture and Food Research Organization—Kyushu
   http://www.naro.affrc.go.jp/karc/plant_factory/
4. R&D Center for the Plant Factory, Research Organization for the 21st Century, Osaka Prefecture University: http://www.plant-factory.21c.osakafu-u.ac.jp/
5. Research Center for High-technology Greenhouse Plant Production, Ehime University
   http://igh.agr.ehime-u.ac.jp/

JPFA'S BUSINESS WORKSHOPS

JPFA organizes business workshops (Figure 25.13) on a wide range of issues of plant factories for engineers, researchers, and managers, and has planned workshops, held excursions, and dispatched instructors on a monthly basis since 2009.

Both PFALs and GECSs have increased in Japan recently. Many of the companies are new to this field and are eager to learn about plant factories. However, not all of their businesses are successful. Even for those who attain a good balance, there seem to be many opportunities to improve their business. There were an estimated 200 PFALs in September 2014 in Japan. It was estimated that roughly 25% were profitable, 50% break even, and 25% fall short (Kozai, 2014). There were 200 GECSs in 2014; most of the businesses were in relatively good condition, but there were still many opportunities to improve them.

The scope of opportunities for improvement or strengthening is wide, including management, production systems, workers' skills, marketing, sales, etc.

FIGURE 25.13

JPFA's business workshops.

JPFA started a business workshop on the issues both for PFAL and GECS on October 1, 2009 before it was formally founded on May 18, 2010, and has held 82 business workshops as of the end of January 2015. Each workshop had two to four presentations. The lecturers were university professors, researchers, and officers of experimental stations of the Japanese government and local governments, as well as representatives of private enterprises.

The slides of most presentations are available on DVD for business people and those who did not attend the seminars; most of them are in Japanese but a few are in English.

The following workshops were held and classified into categories by year. The number of participants each year is shown in Table 25.5.

2009 BUSINESS WORKSHOP

1. General information, status, objectives: (1) The direction of Chiba University's plant factory (October 1, 2009) and (2) Field tour to a newly developed PFAL: Sankyo Frontier Co. (January 21, 2010).
2. Overseas information: (1) The progress of development of greenhouse horticulture and environmental control technology in the Netherlands (November 19, 2009).
3. Environmental control technologies: (1) Heat pumps (October 20, 2009), (2) Ubiquitous environmental control systems (UECS) (December 3, 2009), (3) Light sources in PFAL (December 17, 2009), (4) Selection of light sources and designing lighting and air-conditioning for PFAL (February 18, 2010), and (5) Tour to heat pumps and mixing fans in plant factories and seedling production system—PFAL in Chiba University (February 18, 2010).

Table 25.5 People Who Attended JPFA's Business Workshops

2009		2010		2011		2012		2013		2014	
Date	#	Date	#	Date	#	Date	#	Date	#	Date	#
10/1	66	4/16	73	4/14	100	4/19	101	4/18	85	4/23	117
10/8	42	4/26	59	5/19	95	5/17	101	5/16	79	5/28	71
10/20	57	5/13	44	6/16	76	6/21	97	6/20	89	6/25	52
10/30	42	6/10	74	7/14	104	7/6	71	7/8	14	7/22	62
11/19	62	7/1	55	8/8	74	7/19	101	7/18	81	8/27	50
12/3	53	7/8	46	9/15	86	8/23	72	8/22	87	9/13	430
12/17	72	8/3	55	10/20	59	9/20	107	9/17	85	10/22	53
1/14	49	8/5	58	11/17	75	10/18	80	10/22	53	11/6	74
1/21	86	8/6	38	12/15	78	11/15	74	11/19	64	11/26	63
1/21	46	9/13	79	1/12	91	12/13	91	12/17	77	12/24	50
2/4	55	10/14	54	2/9	97	1/17	84	1/21	69	1/28	115
2/18	92	11/11	75	3/15	109	2/14	82	2/18	84		
2/18	43	12/9	68			3/14	85	3/18	47		
3/4	70	12/15	59								
		1/13	58								
		2/17	111								
		2/17	80								
		2/23	28								
		3/10	106								
	835		1220		1044		1146		914		1022

4. Production and production system: (1) The system of hydroponic culture (October 8, 2009; October 30, 2009) and (2) Issues of high-yield tomato production (March 4, 2010).
5. Plant breeding for plant factory cultivation: (1) Tomato and lettuce (January 14, 2010).
6. Business: (1) Vegetable processing and distribution (January 21, 2010).
7. Pest control: (1) Integrated pest management in plant factories (February 4, 2010).

2010 BUSINESS WORKSHOP

1. General information, status, objectives: (1) Outline of plant factories in five locations in Japan supported by MAFF (April 26, 2010), (2) Field tour to plant factories in Tamagawa University (July 1, 2010, August 6, 2010), (3) Ubiquitous in-town PFAL (July 8, 2010), (4) Expanded opportunities for PFAL (August 5, 2010), (5) Plant factories and a happy life (December 9, 2010), (6) Outcomes and plans of plant factories: GECS in Chiba University (February 17, 2011), and (7) Outcomes and plans of plant factories: PFAL, in-town PFAL and "Inter-thematic Areas" Consortiums (March 10, 2011).

2. Overseas information: (1) Production in large-scale greenhouse and PFAL in China, production in greenhouse in Saudi Arabia (April 16, 2010), (2) Tomato greenhouse production in the world (December 15, 2010), and (3) People training in the Netherlands and plant factories in China (February 17, 2011).

3. Environmental control technologies: (1) Integrated environmental control using multipurpose heat pump in a semi-closed production system (June 10, 2010), (2) UECS in Institute of Vegetable and Tea Science (August 3, 2010), (3) The issues of ground environmental control (November 11, 2010), (4) The issues of root zone environmental control (January 13, 2011), and (5) UECS utilization for plant factories in Chiba University and its demonstration (February 23, 2011).

4. Business: (1) The financial balance of PFAL and consultation (September 13, 2010).

5. Production and production system: (1) Management of hydroponic system (May 13, 2010) and (2) Culture mediums for seedling production and growing beds (October 14, 2010).

2011 BUSINESS WORKSHOP

1. General information, status, objectives: (1) Overall introduction of plant factories in Chiba University and their measuring and integrated environmental control (April 14, 2011), (2) JA Zen-noh's GECS in Chiba University (May 19, 2011), (3) Actual results of construction in Chiba University/Learning from failures in other cases (June 16, 2011), (4) The situation of greenhouse and plant factory production in Fukushima and Miyagi after the Great East Japan Earthquake, and the impact of blackouts and their countermeasures (July 14, 2011), (5) MKV's GECS in Chiba University (August 8, 2011), (6) Plant Factory Center in the Matsudo campus, Chiba University (September 15, 2011), (7) Daisen's GECS in Chiba University (October 20, 2011), (8) Mushroom production and issues (October 20, 2011), (9) The current situation of plant factories in Japan (October 20, 2011), (10) Nisshinbo's PFAL, (11) Iwatani-Agrigreen's GECS in Chiba University (November 17, 2011), (12) Seiwa's GECS in Chiba University (November 17, 2011), (13) Activity report of plant factories in 2011: PFAL Consortiums, "Inter-Thematic Areas" Consortiums, and "Plant Factories in Downtown Areas" Consortiums (February 9, 2012), and (14) Activity report on 2011 of "GECS" Consortiums (March 15, 2012).

2. Overseas information: (1) Updating the information of plant factories in China and Korea (May 19, 2011) and (2) The current situation of plant factories in China (December 15, 2011).

3. Environmental control technologies: (1) A household heat pump for horticultural use (November 17, 2011).

4. Production and production system: (1) Development of robots supporting agricultural work (June 16, 2011), (2) Seed coating and seedling production (August 8, 2011), (3) Spray-type hydroponic system (November 17, 2011), and (4) Effects and possibility of green lighting in plant factory production (January 12, 2012).

2012 BUSINESS WORKSHOP

1. General information, status, objectives: (1) Plant factories in downtown areas (September 20, 2012), (2) Health care and plant factories (November 15, 2012), (3) Quality of vegetables in plant factories (December 13, 2012), (4) From plant factories to vertical agriculture, urban agriculture,

and agricultural culture city (February 14, 2013), and (5) Newly established plant factories in the Tohoku district after the big earthquake (March 14, 2013).

2. Overseas information: (1) Current situation of vegetable grafting in Europe and North America (July 6, 2012) and (2) Latest status of plant factory research and development, and spread in China and Taiwan (February 14, 2013).

3. Production and production system: (1) Tomato production technique—A 50-year history (July 6, 2012) and (2) Possibility and problems of strawberry cultivation in plant factories (July 19, 2012).

4. Environmental control technologies: (1) Measures to counter high temperature in GECS (May 17, 2012), (2) Latest environmental measurements control in plant factories (August 23, 2012), and (3) Interim report on integrated environmental control development in plant factories (February 14, 2013).

5. Business: (1) How to reduce heating expenses in GECS (April 19, 2012), (2) Forefront of PFAL business (June 21, 2012), (3) The business of plant factories (October 18, 2012), (4) New development of PFAL business (January 17, 2013), and (5) Production and business strategy for low potassium lettuce (March 14, 2013).

2013 BUSINESS WORKSHOP

1. General information, status, objectives: (1) Recent trends of GECS: rebuilding project in the disaster area (April 18, 2013), (2) Current situation of MAFF subsidized projects of plant factories related to the disaster reconstruction (May 16, 2013), (3) Opinions offered concerning standard specifications by Super Hort Project Association and results of a questionnaire on plant factories in Japan and Korea (September 17, 2013), and (4) Plant factories in downtown areas (January 21, 2014).

2. Environmental control technologies: (1) Application of LED to plant cultivation (June 20, 2013), (2) Report on integrated environmental control system development in 2012 (August 22, 2013), (3) Energy-saving effect and good use of multifunction agricultural heat pump (October 22, 2013), (4) Measurement and controllers in GECS (December 17, 2013), and (5) Evaporative cooling system (March 18, 2014).

3. Business: (1) Marketability and breeding of plant factory vegetables (July 18, 2013), (2) Examples of new entrants to plant factory business (November 19, 2013), and (3) Business management in plant factories (February 18, 2014).

2014 BUSINESS WORKSHOP

1. General information, status, objectives: (1) Medical herbs (April 23, 2014), (2) Plant factory symposium: direction to attain a "platinum society" (wealthy and aging society), A method for community revolution based on plant factories, PFAL & GECS in Kashiwanoha (September 13, 2014) (Figure 25.14), and (3) MAFF's project to accelerate the expansion of next-generation GECS (November 26, 2014).

2. Overseas information: (1) Expansion of PFAL in Taiwan (July 22, 2014), (2) The status and perspectives of greenhouse horticulture in Israel (July 22, 2014), (3) Current status of the consulting business in Europe and its prospects in Japan (August 27, 2014), and (4) North American Vertical Farm (November 6, 2014).

FIGURE 25.14

Plant Factory Symposium with 430 attendees in 2014.

3. Production and production system: (1) Establishment of hydroponics for cucumbers (May 28, 2014) and (2) Cultivation of paprika and bell pepper in GECS (October 22, 2014).
4. Environmental control technologies: (1) Null balance CO_2 enrichment (June 25, 2014).

REFERENCES

Kozai, T., 2014. International Conference on Vertical Farming and Urban Agriculture 2014: 9th and 10th September 2014. The University of Nottingham, UK.

Onuma, Y., 2014. Plant environment designing program. The Center for Environment Health and Field Sciences of Chiba University, Japan.

PFALs IN OPERATION AND ITS PERSPECTIVES

SELECTED COMMERCIAL PFALs IN JAPAN AND TAIWAN

26

INTRODUCTION

As of March 2014, there were 165 plant factories with artificial lighting (PFALs) in Japan and 45 PFALs in Taiwan for commercial production. This chapter introduces selected PFALs among these 210 PFALs, and outlines their business background, location, business model, market, cost, target crops, production capacity, and cultivation systems.

In many countries, including Japan, Taiwan, Korea, USA, and China, the number of PFALs has been increasing year by year. The largest PFAL in Japan produces 23,000 lettuce heads daily. Both Taiwanese and Japanese PFAL companies export PFALs to China, Mongolia, Singapore, and/or Russia, and are planning to export to the Middle East and other Asian countries. Daily production capacity per PFAL will reach around 30,000 leaf lettuce heads soon, and more and more manual operations will be automated within the next few years. In The Netherlands, the PFAL business has recently started growing.

REPRESENTATIVE PFALs IN TAIWAN

Wei Fang

Department of Bio-Industrial Mechatronics Engineering, National Taiwan University, Taipei, Taiwan

To produce RTE (ready to eat) and RTC (ready to cook) leafy greens, a PFAL needs to be kept in a standard clean room, which is normally equipped with epoxy floor, air-tight thermally insulated roof and walls, air shower, air conditioner, and CO_2 enrichment. These are called the "essential five (E5)."

As of 2014, among the 45 organizations engaged in PFAL research and/or business, most of them use the E5 mentioned above. Among these organizations, there are three representative companies with distinctly different business models: Cal-Com Bio Corp., Glonacal Green Technology Corp., and Ting-Mao Bio-Technology Corp.

The largest PF (plant factory) with lamps, but not PFAL, in Taiwan is operated by Yasai-Lab. Corp., which uses none of the E5 due to a different design concept. This company focuses on producing EAW (eat after wash) and CAW (cook after wash) leafy greens and will be introduced in Section "Yasai-Lab Corp.: The Largest PF (Plant Factory) in Taiwan."

Plant Factory. http://dx.doi.org/10.1016/B978-0-12-801775-3.00026-3

CAL-COM BIO CORP. OF NEW KINPO GROUP

This company produces leafy greens and medicinal herbs for members only at two locations. The daily production of the first and second PFAL is 100 and 1000 plants per day, respectively (Figure 26.1a and b).

This company is unique in that it is the only PFAL company in Taiwan with ISO9001 and ISO22000 certification. They also have a seeding robot developed by themselves and a semi-automatic packaging machine.

GLONACAL GREEN TECHNOLOGY CORP.

This company is not only a PFAL construction and consulting company but also produces leafy greens which are sold to local four- and five-star restaurants. The daily production at the demonstration room is around 400 plants (Figure 26.2a). Figure 26.2b shows a window view of the demonstration room. A light-mounting-frame with adjustable hanging strings can be seen.

The company has patented a movable light-mounting fixture, which can move the lights horizontally or vertically when needed. In addition, the integrated design of the bench, production system, artificial lights, air flow movement, controls for the light cycle and nutrients, and measurement of environmental factors have been patented as a whole. Each bench is a stand-alone equipment, which not only eliminates cross-contamination but also enables the equipment to be leased. Due to the stand-alone design concept, there are no pipes and no electricity cables on the floor or between benches, so it is easy to clean the floor and to transport items using carts.

TING-MAO BIO-TECHNOLOGY CORP.

This company has a mass production center (floor area: around 3300 m^2, Figure 26.3a) and eight restaurants (brand name: Nice-Green Kitchen), of which six are equipped with a small CE bench and one with a demonstration room (132 m^2) (Figure 26.3b and c).

This company is unique in that they have multiple channels to sell their products including: (a) web site for the general public and members, (b) chain restaurants for members, (c) exchange memberships

(a)   (b)

FIGURE 26.1

(a) The first PFAL of Cal-Com Bio Corp., Taiwan. (b) The second PFAL of Cal-Com Bio Corp., Taiwan.

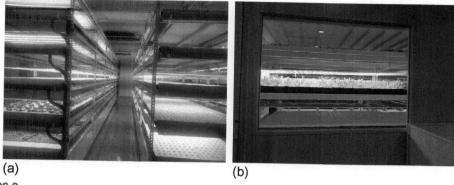

FIGURE 26.2

(a) The demonstration room of Glonacal Green Technology Corp., Taiwan. (b) Window view of the demonstration room of Glonacal Green Technology Corp., Taiwan.

with other health-related organizations such as yoga clubs, and (d) development of various lines of processed products with PF-grown vegetables as add-in ingredients.

LEE-PIN: A PFAL BUILDING INSIDE A GREENHOUSE

Most of the PFALs in Taiwan have E5 equipment as described previously, with two exceptions: Lee-Pin and Yasai-Lab. More details of Yasai-Lab will be described in the next section. Lee-Pin uses a gutter-connected greenhouse as the PF enclosure with internal fan and roof vents and retractable internal shade curtains to block excess sunlight. It is located in a paddy field in Dan-Sui (northern tip of Taiwan) as shown in Figure 26.4a. At present, the monthly production of Lee-Pin is about 5 tons, which it mainly sells to local community residents. There are 12 lines of benches, each 15 m long with six layers per bench. The upper layer relies on sunlight and the bottom five layers use artificial light. The floor of the greenhouse is epoxy as shown in Figure 26.4b. The seedling production area with air-conditioning system and the work area are in the fixed shade zone in the front end of the greenhouse as shown in Figure 26.4a.

YASAI-LAB CORP.: THE LARGEST PF IN TAIWAN

Yasai-Lab Corp. renovated an industrial factory as the PF enclosure. It is the largest PF with lamps, but not PFAL, in Taiwan and perhaps the world at present in terms of full-scale daily production. Without using any of the E5 equipment mentioned previously, it uses a pad and fan system instead of an air-conditioning system for cooling, thus limiting itself to being unable to grow off-season crops. The indoor humidity is normally high, leading to a slow growth rate compared with other PFAL companies. They sell the CAW and EAW leafy greens through supermarket chains. Thanks to the huge daily harvest, their strategy is to sell at the lowest price (0.1–0.2 NT\$/g) compared with all other PFALs (0.4–2 NT\$/g) (1 US\$ = 31.5 NT\$). At such low prices, it is almost impossible to recoup the investment

FIGURE 26.3

(a) Four views of the mass production center of Ting-Mao Bio-tech. Corp., Taiwan. (b) One of the restaurants (brand name: Nice-Green Kitchen) and CE bench style PF concept demonstration facility of Ting-Mao Bio-tech. Corp., Taiwan. (c) One of the restaurants (brand name: Nice-Green Kitchen) and PF concept demonstration room of Ting-Mao Bio-tech. Corp., Taiwan.

FIGURE 26.4

(a) Appearance of plant factory located in a paddy field, Lee-Pin Corp., Taiwan. (b) Inside view of the plant factory located in a paddy field, Lee-Pin Corp., Taiwan.

for now. Customers welcome such low-price, pesticide-free products, so this company makes other companies look bad to customers for whom price is the major concern.

The dimensions of Yasai-Lab's PF are $60 \times 30 \times 8$ m^3. All benches are 7 m high, with a total of six seedling benches with 16 layers each and 31 benches with 10 layers for mature-stage production. In full operation, the facility can grow 1.8 million plants at the same time and production is around 2.5 tons per day; 60,000 plants are harvested per day in full operation.

As shown in the upper part of Figure 26.5a, Yasai-Lab uses pink LED panels. The panels can be moved horizontally, and so at most 12 h of light per day can be provided for all crops. Several automatic devices are used, such as a seeding machine for coated seeds, and a transplanting machine as shown in the bottom part of Figure 26.5a. Figure 26.5b shows the conveyors for moving the growing panels vertically and horizontally. The company also has root cutting and root pressing machines to reduce the amount of waste. Surprisingly, no automatic or semi-automatic packaging machines are used.

SPREAD Co., Ltd.

Shinji Inada

Spread Co., Ltd., Kyoto, Japan

VISION AND MISSION

Our mission is to pass this lush and beautiful world on to the next generation. In order to achieve this, we need to utilize natural energy and develop new technologies. Innovation is essential. We aim to create a sustainable company that will allow future generations to live with peace of mind. We are striving to create a system in which high quality, nutritionally balanced food can be produced anywhere in the world, using food infrastructure that can provide for all people fairly and equally.

HISTORY AND LOCATION

Spread was founded in 2006 in Kyoto, Japan. Kyoto is a traditional producer of high-quality vegetables, and the "Kyoto Vegetable" brand is famous throughout Japan. Kyoto was the former capital of Japan for 1200 years, and has a strong culture of craftsmanship. It is a rare place where advanced production and traditional industry can still coexist. Spread takes advantage of these unique characteristics, working to develop cutting-edge technology while preserving the mindset of traditional craftsmanship. Our PFAL is in Kameoka City, Kyoto, near a flourishing agricultural area (Figure 26.6).

The reasons why we chose this area are: (1) As Kyoto vegetables are high quality and have a good brand image, they have an advantage in the marketplace; (2) The PFAL is close to a highway interchange, which reduces shipping time and delivery costs; (3) The Kameoka area has high-quality ground water, which can be used for hydroponics.

Spread is a member of the Trade Group of distribution companies. Trade was founded in 2001 as a vegetable distributor targeting the domestic wholesale market, and grew quickly in the farm industry. The Trade enterprise is made up of five companies, including distributors, wholesalers, and design companies. Its strength lies in its vegetable distribution and supply chain.

(a)

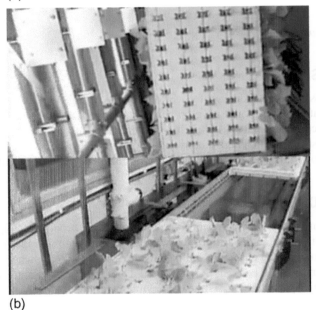

(b)

FIGURE 26.5

(a) Illuminated benches of Yasai-Lab's PF and operations of transplanting device. (b) Automatic transfer of growing panels in the vertical and horizontal directions in Yasai-Lab Corp., Taiwan.

Spread's creation was inspired by a conversation with a Japanese farmer who said he would let his farm die with his generation, rather than allow his son to continue. Upon hearing these words, we realized that the future of Japanese agriculture was in crisis.

Japan's current agricultural industry has many problems: An aging farming population, a lack of successors to inherit those farms, an increase in abandoned farming plots, and more. These issues are

FIGURE 26.6

Kameoka Plant.

structural in nature and have no simple solution. With these points in mind, we set out to create a PF where young people could find interest and secure gainful employment. We felt that to create a promising and sustainable agriculture industry, we needed to innovate and transform the business.

BUSINESS MODEL

From factory operations to sales, Spread is expanding its vegetable production and sales business while carrying out all technological development in-house. Lettuce under our Vegetus brand is delivered to approximately 2000 domestic supermarkets every day and has the largest sales area of any PF lettuce (Figure 26.7). We aim to take this brand power and technological know-how to create a packaged "factory farm system" and expand our business through partnerships and franchising at home and abroad.

FIGURE 26.7

"Vegetus" products.

FIGURE 26.8

Processing area.

Under a "partnership," the partner provides investment capital and builds the facilities. Spread will then conduct business at these facilities while paying usage fees. Spread's usage fees allow the partner to make a profit from the capital investment.

With a "franchise" the franchisee runs the actual business in addition to providing investment capital and building facilities. Spread will then support the business by offering consulting, buying produce, or offering support in developing new products. The franchisee can earn income from the sale of vegetables, and expand their current operations or start new ventures.

What makes these franchise and partnership systems possible is the strength of Spread's sales system and factory farming technology. Our sales system includes established temperature-controlled distribution, marketing attuned to our customers' needs, and branding that differentiates us from the competition. Our PF technology enables a high-quality product to be grown at low cost in a large factory through superior research and development, facility design, and administration. Our intellectual property and know-how, combined with the sales techniques and engineering cultivated at our Kameoka plant (which has the highest product output in Japan), are powerful resources in our business model (Figure 26.8).

MAIN CROPS

Assuming an average weight of 90 g per plastic container or bag, we produce 21,000 heads of frilled, pleated, sunny, and romaine lettuce every day (Figure 26.9). Supermarkets are the major buyer, accounting for 90% of sales. This lettuce is sold under the Vegetus name. The standard size of a Vegetus lettuce is 70–90 g for the regular size, and 50–60 g for the mini size. The remaining 10% of our sales comes from bulk deliveries made to restaurants, delis, and airlines. In general, factory-farmed vegetables are known for being free of agrochemicals, but Vegetus is also popular among consumers for its taste and nutritional content.

Regarding costs, our materials cost is about 50% more compared to conventional open-air farms, but the total cost including operating expenses is no more than 10% higher. When the value of

Frilly lettuce   Green and red lettuce

Romaine lettuce   Pleats lettuce

FIGURE 26.9

Main crops.

agrochemical-free production and a stable supply are also considered, the overall worth far exceeds that of produce cultivated outdoors.

OUTLINE OF PFAL

(1) Established: July 2007, (2) Land area: 4780 m², (3) Structures: Total floor area: 2868 m², Height: 15.94 m, (4) Cultivation area: Building No. 1: 10,800 m² (900 m² × 12 tiers), Building No. 2: 14,400 m² (900 m² × 16 tiers) (Figures 26.10 and 26.11), (5) Other facilities: Research and development lab, experiment lab, study area, meeting room, processing area, clean room, refrigeration room, (6) Electricity supply capacity: 1750 kW, (7) Water source: Ground water, 16 tons per day, and (8) Cultivation: Multitiered deep flow hydroponics.

COSTS BY COMPONENT AND SALES PRICE

1. Initial investment: (1) Land and structures: 612 million yen (1 US$ = 120 yen), (2) Facilities, etc.: 1046 million yen, (3) Infrastructure (electrical equipment, etc.): Japan has good electrical systems and municipal water, so the initial investment for the Kyoto PFAL was negligibly small. However, the cubicle electrical system cost about 6 million yen, and has a maximum power output of around 1700 kW. The receiving voltage is 6600 V. Water consumption is 2.5 tons per day.
2. Production costs: Current production costs are shown in Figure 26.12. Production costs as a percentage of sales (cost ratio) are approximately 90%, but new technology will reduce the percentage to 70% or less in the PFAL currently being planned.
3. Sales price: Our Vegetus brand is sold in supermarkets at around 200 yen per plastic bag, or 150 yen for the mini size.

FIGURE 26.10

Cultivation area (inside).

FIGURE 26.11

Cultivation area (emphasis is on hygiene).

MARKETS

Japan's domestic lettuce market is currently worth around 150 billion yen per year, and Spread aims to capture 10% of this. Approximately 2000 large supermarkets in the Kanto (Tokyo, Yokohama, etc.) and Kansai (Kyoto, Osaka, etc.) regions currently account for 90% of our sales. To expand our market share we will target restaurant chains, convenience stores, sandwich makers, and other businesses. This will require further reduction of the sales price and the development of lower-cost production techniques.

At Spread, we are developing next-generation technologies to achieve low-cost, sustainable PFs, and aim to complete this by the spring of 2016.

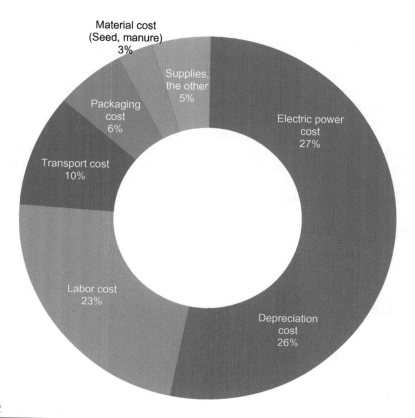

FIGURE 26.12

Cost components.

With lower costs from new cultivation techniques and increased awareness of food safety among consumers, we expect the market for our products to expand further. Demand will come from a wide area, including North America, Europe, the Middle East, and Asia. Furthermore, expansion into areas unsuitable for conventional greenhouses will be possible.

We expect to sell to a wide variety of businesses, such as supermarkets, restaurant chains, hotels, convenience stores, delis, and airlines. The stability and high-quality characteristics of factory farm produce will be increasingly sought after.

FUTURE PLANS

The history of agriculture has been a constant fight with the weather. Climate, the natural enemy of farming, has long been a problem. However, complete climate-controlled PFALs are unaffected by the weather, and are expected to become the agriculture system of the future. Spread currently holds leading technology in the worldwide PFAL industry, and has developed a low-cost, high-efficiency

PFAL method. In the future, we will strengthen our research and development to address other environmental concerns, produce new products, and expand as a highly functional PFAL business.

Spread is now working on a cutting-edge new factory project scheduled to begin operation in the spring of 2016. This new factory will be a milestone for us, combining our R&D base with a flagship factory capable of low-cost, high-volume production (30,000 heads of lettuce per day). This lower cost production will help us to expand into new domestic and foreign markets. The new facility will also raise our technological capability, which is the core of a PF's operation and new product development. We will increase our energy efficiency and achieve complete water recycling.

With our marketing techniques, we will improve supply and distribution, offering the best possible supply system for our customers' needs. We aim to create a food production system in which delicious food can be made stably at any time, and to create infrastructure that can provide that food to all the people in the world fairly and equally.

WEBSITE URL

http://www.spread.co.jp/en/

MIRAI Co., Ltd.

Shigeharu Shimamura

Mirai Co., Ltd., Tokyo, Japan

VISION AND MISSION

Current crop production in the world is estimated to be able to provide food for no more than 8.5 billion people, yet the global population is predicted to reach 9.3 billion by 2050, in which case food shortages may occur. We believe that PFALs are an important solution to this problem because a PFAL can be 100 times more effective than outside farming in terms of production efficiency based on time and space. Furthermore, the irrigation water requirement of a PFAL is only one-hundredth of that of outside farming. Seventy percent of the world's water today is used for agriculture. If PFALs spread globally, water will be saved and more land could be used for cultivating staple foods. Thus, the PFAL is essential for solving food shortages in the future.

HISTORY AND LOCATION

In 2004, Mirai Co., Ltd., an agricultural venture, was established with Shigeharu Shimamura as president. In 2008, in collaboration with the Ministry of Education, Culture, Sports, Science, and Technology, Mirai installed a small PFAL in the Showa Station at the South Pole. In 2009, as a project of the Ministry of Agriculture, Forestry, and Fisheries, the Low-cost PFAL for Future Generations Consortium was established at Chiba University in Kashiwa City, Chiba, and Mirai opened a PFAL in the campus in July 2011. In March 2013, a new PFAL with all LEDs was opened in Tagajyo City, Miyagi Prefecture, as a project subsidized by the Ministry of Economy, Trade, and Industry. Based on this experience, in June 2014, Mirai opened a new PFAL in Kashiwa City, with a production capacity of 10,000 leaf lettuce heads per day, the same capacity as the one in Tagajyo City (Figure 26.13).

FIGURE 26.13

A new PFAL in Kashiwa City.

BUSINESS MODEL

Mirai sells both the PFAL system it has developed and the vegetables produced at PFALs operated by Mirai. Mirai sells cultivation equipment to customers abroad, and PFALs are built by local contractors hired by the customers, while the market for selling vegetables is currently limited to Japan. When planning and designing a PFAL, Mirai offers a consulting service to a local design company, to ensure the quality of environmental control inside the PFAL. After the PFAL is constructed, the local customer operates the PFAL, and Mirai offers a consulting service for cultivation to the customer, if necessary. The main crops produced by PFALs currently include lettuce, green leaf, romaine lettuce, red romaine, peppermint, coriander, kale, arugula, mitsuba, watercress, mustard green, mizuna, etc.

OUTLINE OF PFAL

The main components of a PFAL are the cultivation racks, lightning system, and nutrient circulating system including fertilizer control equipment, CO_2 control equipment, and piping system. Aspects of the PFAL such as building size and floor area are fully customized and determined after the planning and designing are completed. One example is a PFAL in Miyagi, which was originally a semiconductor factory of Sony Corporation and was rebuilt as the world's largest commercial PFAL illuminated by LEDs only in July 2014. The special LED fixtures were developed by GE and emit light of the optimal wavelengths for plant growth. The PFAL is nearly half the size of a football field (25,000 square feet or 2322 m^2) and produces 10,000 heads of leaf lettuce per day. The PFAL uses 17,500 LED lamps which are spread over 18 cultivation racks rising 15 tiers high. Approximately 30 workers are employed, including both full-time employees and part-timers. Before entering the PFAL, they wash their hair and body in a hot water shower, then put on protective clothes, and finally take an air shower, to minimize the risk of contamination. Several types of leaf lettuce and herbs are cultivated, which take 35 days from seeding to harvesting. The cultivated vegetables are carried to the packaging room inside the PFAL without exposure to rain or outside air, so the vegetables can be packed and delivered to customers in a clean condition.

COSTS BY COMPONENT AND SALES PRICE

The component configuration is fully customized. In addition to the cost of the cultivation equipment, there is also the cost of constructing the PFAL. The construction cost is quoted by a local contractor and varies depending on the customer, so the exact total cost is determined after planning and designing, considering each country's situation.

MARKETS

There are potential markets for PFALs throughout the world. In particular, inquiries from the Middle East, Southeast Asia, Russia, USA, and China have been increasing. The PFAL technology of Mirai was installed in Mongolia in 2014 and in Hong Kong in 2015, and is expected to be used in Russia in 2015.

FUTURE PLANS

Mirai will expand its PFAL business to regions where vegetable cultivation is difficult such as cold regions, isolated islands, dry regions, and deserts. Mirai also aims to utilize the extra energy in urban areas and water resources in smart cities or smart communities. In the near future, Mirai plans to increase the production of herbs in addition to leaf lettuce. Production of fruit vegetables such as strawberry, root vegetables such as radish, and medicinal plants will also be increased. Through its environment-friendly and high-tech agriculture for stably producing and delivering high-quality vegetables, Mirai will contribute to a healthy life for all people in the world.

WEBSITE URL

http://miraigroup.jp/en/

JAPAN DOME HOUSE Co., Ltd.

Katsuyuki Kitagawa

Japan Dome House Co., Ltd., Kaga-shi, Ishikawa, Japan

BACKGROUND AND GEOGRAPHICAL LOCATION

Our expanded polystyrene building (registered brand name "dome house") has excellent heat-insulating and air-tightness (number of air exchanges: 0.01 per hour) properties, and can withstand severely hot and cold weather. Its structural mechanics make it resistant to natural disasters such as earthquakes, strong winds, etc. and it can therefore help resolve food production issues in case of an emergency. Using the dome house, we have developed an original PFAL.

FIGURE 26.14

The dome house being used as the PFAL (left) and mushroom factory (right).

BUSINESS MODEL

Our dome house is especially suitable for plant production in harsh environments such as in very cold areas or hot deserts. We have also developed a unique plant cultivation system which can considerably reduce the use of water and fertilizer. These advantages are particularly useful in areas where vegetable cultivation in open fields is difficult. Our PFALs provide an integral system of building, plant facilities, and plant cultivation system, etc. (Figure 26.14). We obtained approval for the PFAL from the Ministry of Economy, Trade, and Industry in Japan, and the PFAL started operation in March 2013 as a demonstration in Date city in Fukushima Prefecture.

We built a mushroom factory using the same dome house as the PFAL next to it (Figure 26.14, left). The CO_2 generated by the mushrooms is piped to the PFAL to boost plant photosynthesis, thus reducing the CO_2 enrichment cost and preventing the CO_2 generated by the mushrooms from being released into the atmosphere.

Currently, we are producing leafy vegetables including leaf lettuce, butter head lettuce, Komatsuna (Japanese mustard lettuce), potherb mustard, Korean lettuce, arugula, basil, Italian parsley, chervil, etc.

OUTLINE OF PFAL

The arch-shaped PFAL building built in Date city is 7.7 m wide and 36 m long with a floor area of 270 m^2. The building also includes all of the necessary spaces for PFAL management such as a shipping area and a rest area for employees. Eight-tier cultivation racks are placed in the middle and six-tier racks on both sides along the arch-shape (Figures 26.15 and 26.16), and have an annual production capacity of 520,000 leaf lettuce heads.

This PFAL allows the cultivation of plants with a small amount of water by circulating liquid fertilizer through a gutter. Water can be further saved by collecting drain water from air-conditioning and reusing it for watering plants. This system places less burden on the cultivation racks and so lighter rack materials can be used.

FIGURE 26.15

The arch-shaped cultivation room.

FIGURE 26.16

Eight-tier cultivation racks placed in the middle, and six-tier racks on both sides.

FIGURE 26.17

The system for automatically uploading and downloading culture panels from or to the cultivation tiers.

The automatic uploading and downloading system shown in Figure 26.17 reduces the danger of working in a high location and allows operations to be handled by a limited number of people.

Plant cultivation is divided into the following four stages depending on the size of the growing plants:

(1) Seeding and germination: In order to place the seeds in the foamed urethane mat, first remove the air using a special machine and then seed. Next, place the mat in the germination box, control the temperature and humidity, and leave the seeds to germinate for 2–3 days.

(2) Raising Seedlings A: Move the urethane mat with sprouted plants to under the light source to grow them. Grow seedlings in Rack A for 10 days. As a light source, use LEDs which are effective for promoting plant growth.

(3) Raising Seedlings B: Divide the urethane mat into separate pieces and transplant them to the float board. This is done smoothly by the transplanting robot, and then the seedlings are left to grow for 10 days on Rack B.

(4) Cultivating and Harvesting: Replant the seedlings on the float board with a wider interval between roots by using the transplanting robot. After replanting, move them smoothly to the cultivation trench by using the robot. Then cultivate the plants for 10 days while draining liquid fertilizer through the gutter. At harvest, remove the float boards from the rack and harvest each plant by the robot.

SALES PRICE OF PFAL

The PFAL has four "fixed" advantages: fixed time, fixed amount, fixed quality, and fixed price. Generally, the larger the construction size, the lower the initial cost per unit production capacity. We consider our PFAL as a complete set, from construction to cultivation system, and therefore provide the PFAL at a set price. The sales price of our PFAL in Japan as of September 2014 was approximately 250 million yen for the building size and production capacity mentioned above. This price includes the building, cultivation equipment, cultivation system, building materials, and construction work, as well as installation of the cultivation and electrical machines (the price may vary depending on the conditions of the building site and other factors).

MARKETS

(1) In Japan, the Ministry of Agriculture, Forestry, and Fisheries provides subsidies to promote the PFAL business. We consider this business will expand greatly in the future.
(2) The Ministry of Economy, Trade, and Industry of Japan is also promoting the export of the PFAL system, since there is expected to be global demand especially in the Middle East, Far East, etc.

FUTURE PLANS

We are planning to focus on overseas markets in addition to the market in Japan by accumulating advanced cultivation know-how through joint research and development with Chiba University, which is one of the leading universities in PFAL research. We have received many offers from overseas and specific plans are being pursued.

WEBSITE URL

http://www.dome-house.jp

INTERNATIONALLY LOCAL & COMPANY (InLoCo)

Yasuhito Sasaki

Internationally Local & Company, Itoman-shi, Okinawa, Japan

NO AGRICULTURAL BACKGROUND

Internationally Local & Company (InLoCo) was founded in 2008 with the aim of establishing a PFAL business in Okinawa, Japan. When the first factory went into operation in 2010, the company had no experience in agriculture or plant production at all, nor even any knowledge about growing vegetables or seedlings. The initial seedlings looked like sprouts because they had not been exposed to light at a right timing in their early growing stage, although better growth was achieved later on with the benefit of experience. Why did the company start a PFAL without any experience?

This question can be answered by considering InLoCo's company philosophy, which was first drawn up in 2001 when the present author was in graduate school. InLoCo's philosophy is to create a world where people everywhere enjoy emotional and physical good health. Whenever fundamental decisions are made, this has always been the guiding principle. If people around the world are emotionally and physically healthy, there will be fewer problems threatening lives, societies, and the environment, including war, crime, and disease. What we eat affects the way we are and has a considerable impact on our physical as well as emotional health. There are many places in the world where people do not have access to fresh and safe food. Even in developed countries, people worry

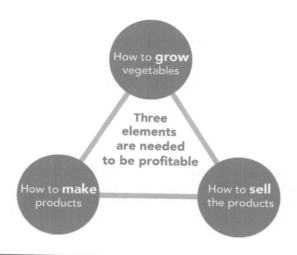

FIGURE 26.18

Three elements needed to be profitable in the PFAL business.

about whether food is safe or not. This is why InLoCo decided to start a PFAL business to supply truly good food to countries around the world. However, the initial start-up years were not easy.

THE BASIS OF PROFITABILITY

To succeed in the field of this business, companies must have fundamental capabilities. They must be able to grow, produce, and sell vegetables (Figure 26.18). There are three elements necessary for profitability in this business: (1) Ability to maintain stable growth of high-quality vegetables, (2) Ability to create attractive products from vegetables, and (3) Ability to manage product sales at the prices we set. These may sound simple steps, yet they are very difficult to accomplish.

HOW DID InLoCo ACHIEVE PROFITABILITY?

The process

InLoCo was able to develop the three elements above. The company set the selling price, added value to its products, and cut the cost of growing and producing. This process sounds simple, but it was not. It took 3 years to make a profit after the first PFAL started operation in Okinawa.

1. InLoCo's PFAL

    The first PFAL was built in the ground floor of an existing building, which had previously been used as a spice factory. The upper three floors are all residential apartments. The total area is 180 m^2, including office space. The hydroponics system has four layers and uses fluorescent lamps as light sources.

2. The three reasons for choosing Okinawa are: (1) Climatic difficulties of growing vegetables outside (summer heat and typhoons), (2) Many resort hotels as potential customers, and (3) Easy access to other Asian countries for export.

3. Step-by-step approach

   (1) 0–6 months: Over the first 6 months, in-house training focused on producing stable quantities of high-quality vegetables.

   (2) Ninth month: Three months later, a major convenience store chain signed on as a customer.

   (3) Tenth month: Around the same time, exports to Hong Kong started.

   (4) Sixteenth month: Six months later, major local supermarket chains were welcomed into InLoCo's customer base.

Products

(1) Products: Iceplant (198 yen), Babyleaf (198 yen), Wasabi-na (138 yen), and eight other leafy vegetables (138 yen) (1US$ = 120 yen) (Figure 26.19).

(2) Production costs: Initial costs: land (0%), building (0%), facilities (80%), and infrastructure (20%). Running costs: depreciation (25%), electricity (30%), labor (35%), and supplies (10%).

(3) Markets for vegetables: Mostly in Okinawa, with some sales to Hong Kong and Tokyo

(4) Clients: Supermarkets, restaurants, and wholesalers

(5) Target consumers: People seeking healthy, safe, and nutritious vegetables

Marketing has included in-store demonstrations to explain how InLoCo's vegetables are grown and why they are so clean and safe. Different products were tested, and changes were made to the quantity of vegetables, size, and package design. Sales promotions, POP design, and other efforts have been carried out; InLoCo is actively working to reach customers and currently expanding, in line (Figures 26.20–26.22) with the company's philosophy.

Nevertheless, InLoCo's expansion is limited by manpower constraints. This limitation can be overcome by marketing PFALs packaged with the company's proprietary knowledge. Recently, a new business of franchise development was commenced, in which InLoCo uses its experience to assist customers. This business is called Agrichise, a combination of the words agriculture + franchise.

FIGURE 26.19

Leafy vegetables produced in the PFAL.

FIGURE 26.20

Second PFAL Itoman, Okinawa (Renovation Type).

FIGURE 26.21

Third PFAL is used for growing strawberries hydroponically. Three other container-type units are in operation in the company's plant factory complex to pursue further potentials of plant factory; a solar-powered PFAL for off-grid use, a PFAL with aquaponics systems to grow both fish and vegetables, and a container-type unit which utilizes both sunlight and LEDs to grow a wide variety of vegetables.

In short, InLoCo possesses the know-how to produce specific quantities of vegetables within a specific period in a specific area. This allows customers using the company's PFAL system to move up the starting line. If a business does not produce sufficient quantities to become profitable, it has no chance of succeeding. In addition, InLoCo's system has been refined and streamlined, reducing the quantity of supplies necessary as well as labor and electricity while maintaining a balance of all aspects.

FIGURE 26.22

Fourth PFAL in a shopping mall in Okinawa demonstrates a perfect example of in-store production for in-store consumption, enabling shopping-mall customers to enjoy twenty-two varieties of "exceptionally fresh" vegetables grown, harvested, processed, and sold on site.

FIGURE 26.23

Container type. Housed in a 40-ft container, this model is ideal for testing cultivation prior to constructing a commercial PFAL. This plan may also be used for household planting and research purposes (price: starting at 8.5 million yen).

BUSINESS MODEL

Three phases are required in establishing a business model. Before starting a project, the needs, or potential needs, which can be satisfied are identified. Then, InLoCo does the following: (1) Tries to satisfy the needs, (2) Creates a successful and profitable model, and (3) Utilizes the model to penetrate the market.

InLoCo's current business model is profitable: the small to mid-sized PFAL profit model. InLoCo has begun marketing PFAL facilities bundled with its plant-growing and operational know-how. The prices of these PFAL facilities (Figures 26.23–26.25) are as follows.

FIGURE 26.24

New building type. A PFAL kit packaged with a new prefabricated building. The expandable units offer a flexible design to suit any location (price: starting at 10 million yen).

FIGURE 26.25

The renovation type allows customers to make efficient use of existing space, which can be reconfigured for optimal cultivation and operation (price: starting at 5 million yen).

Starting a PFAL requires a variety of knowledge and experience, ranging from advanced planning, to facility installation to cultivation techniques. A successful business also requires the development of distribution channels. InLoCo provides "Agrichise," a PFAL support program to help customers overcome these various hurdles. Proprietary experience and expertise are provided to support everything from establishing a factory to selling vegetables (Figure 26.26).

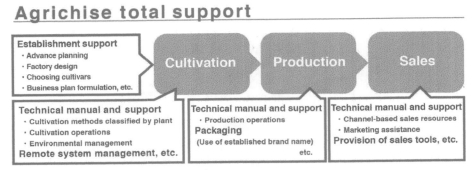

Agrichise total support

Establishment support
- Advance planning
- Factory design
- Choosing cultivars
- Business plan formulation, etc.

Cultivation → **Production** → **Sales**

Technical manual and support
- Cultivation methods classified by plant
- Cultivation operations
- Environmental management
Remote system management, etc.

Technical manual and support
- Production operations
Packaging
(Use of established brand name)
etc.

Technical manual and support
- Channel-based sales resources
- Marketing assistance
Provision of sales tools, etc.

FIGURE 26.26

Agrichise (agriculture + franchise).

FUTURE PLANS

The company will broaden its scope of business in several directions in terms of countries, locations, industries, factory sizes, plant varieties, growing techniques, and energy sources. They will be integrated to create a new style of agriculture, the New Agriculture Model:

1. Countries: Countries in Asia, Africa, and other regions
2. Locations: Deserts, urban areas, etc.
3. Industries: Supermarkets, restaurants, etc.
4. Factory sizes: Ranging in capacity from enough to feed one household to a whole city
5. Food varieties: Root vegetables, fruits, fish, etc.
6. Growing techniques: Aquaponics, mist cultivation, etc.
7. Energy sources: Solar power, wind power, etc.

There are many variables to be taken into account. However, InLoCo is focusing on small to mid-sized PFALs, the area in which it has a competitive advantage. InLoCo's businesses will be integrated into other projects which are now in the planning stage.

Success in the PFAL industry is not easy. Timing and experience matter. The PFAL industry needs to improve (or even innovate) growing technologies to raise quality and productivity as well as to cut initial and running costs. Many improvements need to be made so that people everywhere can enjoy safe and delicious vegetables. However, it is not necessary to wait for such improvements to be made as profitable business models are already established and available. Most importantly, PFALs can satisfy various needs. As we all know, people will never stop eating.

WEBSITE URL

http://www.in-lo-co.com

SCI TECH FARM Co., Ltd.

Kozo Hagiya

Sci Tech Farm Co., Ltd., Tokyo, Japan

Hiroyuki Watanabe

Tamagawa University, Tokyo, Machida-shi, Japan

VISION AND MISSION

Sci Tech Farm Co., Ltd. (STF) has set out the vision of "respecting people" and the mission of "Providing the world with food safety and security, and contributing to society, through LED farmed vegetables and production technology" for the development of PFAL, in line with the vision set out by Nishimatsu Construction Co., Ltd.

HISTORY AND LOCATION

January 2012 Tamagawa University and Nishimatsu Construction signed an agreement regarding an industry–academia joint project, based on farms of the future using the university's LED plant cultivation technology.

October 2012 Work completed on a fully artificially lit PF, based on directly cooled high-power LED technology, at Tamagawa University (LED Farm®: Tamagawa University STF). Started growing 600 plants per day, including crops such as lettuce (Figure 26.27).

February 2013 Regional supermarket chain Odakyu OX near Tokyo started selling lettuce at 18 stores.

March 2013 Obtained JGAP (Japan Good Agricultural Practice) certification.

FIGURE 26.27

Tamagawa University STF.

FIGURE 26.28

Sagamihara STF.

February 2014 Completed business model for a fully artificially lit PF using directly cooled high-power LED technology in Sagamihara (LED Farm®: Sagamihara STF). Started growing 600 plants per day, including crops such as lettuce (Figure 26.28).

October 2014 Expanded scale of cultivation at Tamagawa University STF, commencing production of 3150 plants per day, including crops such as lettuce

Tamagawa University STF: 6-1-1 Tamagawagakuen, Machida, Tokyo, Japan (Tamagawa University)

Sagamihara STF: 5-18-8 Higashi-Fuchinobe, Chuo-ku, Sagamihara, Kanagawa Prefecture, Japan.

BUSINESS MODEL

We launched the industry–academia joint project to effectively and widely use the results of research into LED plant cultivation at Tamagawa University. We also wanted to translate Nishimatsu Construction's building technology and business expertise into something that would benefit society. Figure 26.29 outlines our business model for cooperation between industry and academia. The model is based on a loop linking Tamagawa University and consumers, consisting of a down-flow from the university to consumers in one direction and an up-flow in the opposite direction. As part of the down-flow, Tamagawa University provides the operating company Sci Tech Farm Co., Ltd. (an affiliate of Nishimatsu Construction) with its accumulated expertise in LED-based plant cultivation. Using Nishimatsu Construction's building technology, Sci Tech Farm takes the relevant expertise from Tamagawa University and provides operators starting new PF businesses with resources such as cultivation systems and expertise. Each operator then grows safe, high-quality vegetables at an LED Farm®, and supplies them to consumers under the brand name *Yumesai*® (literally "dream vegetables"). In terms of the up-flow which is particularly important from the market's viewpoint, operators identify consumer needs for new products or varying levels of specific nutrients, for instance, and then work with Sci Tech Farm and Tamagawa University to meet those needs.

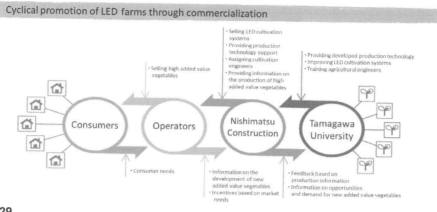

FIGURE 26.29

LED Farm® business model.

MAIN CROPS

Tamagawa University's LED Farm® system grows three varieties of leaf lettuces: Green Wave, Red Fire, and Frill Ice. Ahead of a planned expansion in the scale of production in 2014, we are also cultivating trial crops such as crown daisy, spinach, and aromatic herbs, with the aim of offering a more diverse range of crops.

In terms of quality, it is possible to control properties such as texture and levels of antioxidants, including vitamin C and polyphenols, by adjusting the light color combination of LEDs. Analysis has been carried out on leaf lettuce to determine the effects of different wavelengths on the production of ascorbic acid (vitamin C) (Figure 26.30, left) and anthocyanin (a polyphenol) (Figure 26.30, right, master's thesis by Taihei Oshima, Tamagawa Graduate School, Fiscal Year 2012). On the abscissa of these graphs, "R90B10" indicates 90 μmol m^{-2} s^{-1} PPFD of red light and 10 μmol m^{-2} s^{-1} PPFD of blue light of LEDs.

Although levels of ascorbic acid and anthocyanin both increase in proportion to blue light, the rate of increase gradually decreases. We intend to harness the properties of LED wavelengths even more effectively to maximize the potential of active ingredients in individual crops and create higher quality vegetables.

The aim of cultivating vegetables using the LED Farm® system is to produce the following qualities: (1) Completely free from pesticides, (2) Increased vitamins and other antioxidants, (3) Enhanced sweetness, flavor, aroma, etc., and (4) Improved firmness, texture, etc.

OUTLINE OF PFAL

Table 26.1 outlines the factory specifications of both the Tamagawa University STF and Sagamihara STF, the two PFs associated with the industry–academia joint project (Figures 26.31 and 26.32).

The purpose of the Tamagawa University STF is to (1) test LED farm operations and provide a technical model for PFs using LEDs, (2) contribute to undergraduate and graduate courses at university, and to science education at junior and senior high schools, (3) assist human resource development

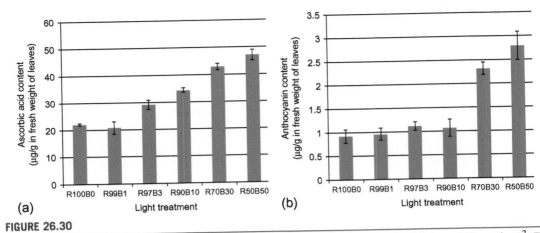

FIGURE 26.30

(a) (Left) changes in ascorbic acid. (Example) R90B10: 90 μmol m^{-2} s^{-1} PPFD of red light and 10 μmol m^{-2} s^{-1} PPFD of blue light. (b) (Right) changes in anthocyanin. (Example) R90B10: 90 μmol m^{-2} s^{-1} PPFD of red light and 10 μmol m^{-2} s^{-1} PPFD of blue light.

Table 26.1 Building and Factory Specifications for Tamagawa University STF and Sagamihara STF

	LED Farm® Tamagawa University STF	LED Farm® Sagamihara STF
Total floor area (m^2)	1000	300
Cleanliness	Class 100,000	Class 100,000
Rows/Shelves	First floor: 3 rows × 6 shelves and 3 rows × 7 shelves, second floor: 6 rows × 5 shelves (Figure 26.31)	First floor: 3 rows × 4 shelves (Figure 26.32)
Production capacity (plants/day)	3150	600
Nutrient systems	Seedling propagation: DFT Vegetables cultivation: NFT	Seedling propagation: DFT Vegetables cultivation: NFT
Light sources	Seedling propagation: LED and fluorescent lamps Vegetables cultivation: Directly cooled high-power LED	Seedling propagation: LED Vegetables cultivation: Directly cooled high-power LED
Loading/transfer of plant trays	Automatic	Manual
JGAP certification	Obtained March 28, 2013	Obtained March 19, 2014

DFT and NFT stand for "deep flow technique" and "nutrient film technique," respectively.

FIGURE 26.31

Cultivation racks in Tamagawa University STF.

FIGURE 26.32

Cultivation racks in Sagamihara STF.

in relation to PFs, (4) provide information on LED-based cultivation, and (5) sell *Yumesai*® products. The purpose of the Sagamihara STF is to (1) provide a PF business model, (2) assist human resource development in relation to PFs, and (3) sell *Yumesai*® products.

COST BREAKDOWN AND SALES PRICE

Table 26.2 shows a breakdown of production costs based on the business model at the Sagamihara STF. Using LED lights reduces utility bills compared to PFs running fluorescent lights. Another point which needs to be mentioned is high depreciation.

Table 26.3 shows a breakdown of prices for the LED Farm® PF system at the Sagamihara STF, for the specifications listed in Table 26.1. As this business model is based on using idle facilities, initial costs could be reduced and the percentages will be different if the facilities have a transformer cubicle installed or already have clean room specifications.

Table 26.2 Breakdown of Production Costs at Sagamihara STF

Item	Cost Percentage (%)
Consumables	2.6
Labor	28.7
Packaging	1.2
Waste processing	2.4
Utilities	16.4
Depreciation	36.5
Others	12.2
Total	100.0

Table 26.3 Initial Cost Breakdown for the LED Farm® PFAL at Sagamihara STF

Item	Cost Percentage (%)
Vegetable cultivation racks	28.5
Seedling propagation racks	4.3
Liquid fertilizer management system	5.1
Environmental management system	4.0
Receiver/transformer cubicle	8.3
Electrical installation	19.2
Air conditioning and water supply/drainage installation	17.6
Insulation panels	12.8
Others	0.2
Total	100.0

MARKETS

Markets for vegetables

Following a 4-month trial cultivation period at the Tamagawa University STF from October 1, 2012, three safe, high-quality *Yumesai*® products (Green Wave, Red Fire, and Frill Ice lettuce) went on sale at 18 stores operated by Odakyu OX (Odakyu Shoji Co., Ltd.) from February 1, 2013, mainly along Odakyu Electric Railway lines near Tokyo. The products were extremely popular, with monthly consumption (items sold/items delivered) remaining around 90% throughout the 20 months up until the end of September 2014. With the Sagamihara STF also operational from April 2014, cultivation was expanded to include crops such as rocket leaf, maintaining strong sales at 26 Odakyu OX stores.

We aim to expand the market in the future to include department stores, central kitchens of restaurant chains, and online sales, while continuing to focus on semi-premium supermarkets.

Markets for LED Farm® *system packages*

As a result of the industry–academia joint project, Sci Tech Farm is now selling LED Farm® system packages, containing the necessary hardware and software for a vegetable production system. Since the business model at the Sagamihara STF commenced operation in April 2014, numerous companies interested in operating PFs have visited the site.

As a business, LED Farm® is based on the concept of urban agriculture, geared towards enabling local production for local consumption, and reducing food mileage. We aim to upgrade to tailor-made vegetable production systems in the future, in line with consumer demand for more efficient production and a wider range of products.

At present, potential sales channels for LED Farm® system packages include the following: (1) Idle facilities in urban areas (empty factories, offices, etc.), (2) Facilities attached to restaurant central kitchens, (3) Areas of complex facilities, and (4) Unused land under bridges.

Maintenance and management services are also available for customers purchasing LED Farm® system packages, from installation planning to operation, as well as support for research into cultivation methods in line with new consumer demand, as outlined in Figure 26.29.

FUTURE PLANS

We intend to draw up plans for the future PFAL business, to achieve the following goals based on results to date: (1) Reduce costs, (2) Reduce initial and running costs, (3) Increase profitability, and (4) Improve yield, develop higher-quality vegetables, create a wider range of products, and improve production efficiency.

Our target for the domestic market is for vegetables produced by LED Farm® system packages to account for 10% of all produce from fully artificially lit PFs in 10 years' time, by 2024. To achieve that, we intend to improve the nutritional value and performance of *Yumesai*® products even further, and to develop a wider range of products.

In terms of overseas markets, we intend to sell LED Farm® system packages and supply fresh vegetables to countries that rely heavily on imported vegetables, such as Hong Kong and Singapore, and to dry or cold countries whose climates are not suited to growing vegetables.

We are also committed to group-wide optimization of LED Farm® system packages the world over, by efficiently exchanging technology information on cultivation, products, and distribution via ICT.

WEBSITE URL

http://www.scitechfarm.com

BERG EARTH Co., Ltd.

Kazuhiko Yamaguchi and Chan-suk Yang

Berg Earth Co., Ltd., Uwajima, Ehime, Japan

VISION AND MISSION

Japanese agriculture faces various problems including the falling number and aging of the farming population, and the increase in import of cheap agricultural products from abroad. To cope with these problems, Berg Earth Co., Ltd. has a management philosophy of "Revolutionizing Japanese agriculture" and "Creating a bright future for agriculture." It also aims to make agriculture profitable by supporting planned agricultural production.

We are mainly involved in two activities. One is the production and sale of fruit vegetable seedlings, especially grafted seedlings such as tomatoes and cucumbers. The other is the distribution business, in which we sell agricultural products, develop producing districts, and sell agricultural materials. We are also expanding into Asia by marketing safe agricultural products, making the best use of high-quality seedlings from Japan and our vegetable production know-how.

HISTORY AND LOCATION

Our company was originally established in 1986 as a one-man operation and Yamaguchi Engei Ltd. was set up in 1996. Then in 2001, Berg Earth Co., Ltd. was founded as a spin-off from the company's R&D, planning, and sales departments. Berg Earth was listed on JASDAQ in 2011.

The head farm for production of Berg Earth is located in Tsushima, Uwajima city in Ehime Prefecture, on the southwest coast of Shikoku. We have four directly managed farms, one in each of Nagano, Ibaraki, Iwate, and Ehime, and also work in partnership with about 20 companies throughout Japan, which are contracted to produce seedlings. This allows us to produce and sell seedlings for which freshness and transportation costs are strictly controlled. We also set up one of the biggest closed transplant production facilities (CTPFs) in the head farm and started selling pesticide-free and high-quality seedlings all over Japan from 2006. We founded Berg Fukushima Co., Ltd. as an affiliated company in 2014 and plan to introduce a CTPF to expand the supply of seedlings to eastern Japan.

BUSINESS MODEL

Berg Earth sells and produces made-to-order vegetable seedlings to meet customers' needs. The main product is grafted nursery plants of fruit vegetables such as tomatoes and cucumbers, with the combination of "scion" which is characterized by good taste and high yield and stress-resistant "rootstock" which spreads deep roots. In the production process, the head farm handles the raising of seedlings and grafting, and each farm then raises the seedlings after grafting. This enables each production center to sell seedlings that meet their own districts' needs. By integrating all the processes from sowing to grafting at the head farm, and by delivering grafted seedlings to each farm at low cost, each directly managed farm and outsourcing company can then sell the seedlings at competitive prices on the market. This makes it possible to produce planned seedlings and sell seedlings throughout the country. Moreover, we have developed a web-based system that shows inventory status, inventory management, and history of agrochemical usage. The system links the head farm with the directly managed farms and outsourcing companies, and enables us to offer seedlings at any time, anywhere, and in any quantity to customers, in line with our management philosophy.

Regarding branding, we focus on developing and selling new products that meet producers' needs. "Earth straight seedling" uses a biodegradable pot and "Nude make seedling" (Figure 26.33a)

(a)

(b)

(c)

FIGURE 26.33

(a) "Nude make seedling." Root pruning reduces transportation cost and sales price. (b) "e seedling" series grafting. (c) "Twin seedling."

is a low-priced grafted seedling. The "e seedling" series (Figure 26.33, which is higher quality than conventional seedlings cultured in a greenhouse, is pesticide-free yet has no blight risk, thanks to the CTPF. Moreover, "Twin seedling" (Figure 26.33 has a good reputation for high yield per seedling and decreases the workload for producers by ramification of one seedling into two. By suggesting these various new products, producers can choose and order optimal seedlings such as growing stage (use of agrochemicals, number of unfolded leaves) and shapes of containers, depending on their own cultivation methods.

MAIN CROPS

About 3 million seedlings per year have been produced in the CTPF and we mainly produce graftings and seedlings of Solanaceae (tomatoes and egg plants) and graftings of Cucurbitaceae (cucumbers, watermelons, melons). The main products are the "e seedling" and "Nude make seedling." The former is a high-quality, pesticide-free tomato seedling, and the latter is a pesticide-free grafting with low transportation cost and hence low sales price. Upon request, they can be used as seedlings before grafting, such as pot seedlings to reduce pesticides and to improve quality.

The items for each product are as follows: Most of the "e seedling" series are tomato seedlings; 47% of them are graftings and 53% are not. The "Nude make seedling" series includes tomatoes (34%), cucumbers (35%), melons (22%), and watermelons (9%). Standard products such as pot seedlings include cucumbers, melons, and watermelons (accounting for 88.8%) and tomatoes (7%).

Other than fruit vegetables, leafy vegetables such as lettuces, cabbages, Chinese cabbages, and herbs are also produced if an order is placed.

OUTLINE OF PFAL

The CTPF has a floor space of 700 m^2. In the facility, there are 21 closed nursery production devices with artificial lighting called "Nae Terrace" (Mitsubishi Plastics Agri Dream Co., Ltd.), each of which has one germination room and two nursing rooms, and has a floor space of 16.2 m^2 (Figure 26.34a). This facility alone is used for the series of tasks from seeding, seedling till grafting, grafting, and nursing acclimatization after grafting to shipment. Seedlings are separated twice from the outside by a building with air shower and a box-shaped Nae Terrace. Each Nae Terrace has a completely independent space and control, so even if germs and insects enter, they are kept separate and damage is minimized.

The Nae Terrace uses a timer-clock controlled subirrigation method and has six five-tiered shelves for seedling production with HF fluorescent lighting. Each Nae Terrace stocks 120 cell trays for raising seedlings and 2520 cell trays in total (Figure 26.34b). The supplying tank is placed outside the Nae Terrace and is sprayed either by set-up time using an outside panel or manually. The raising of seedlings after forced sprouting has its own schedule depending on the item and seedling standard. The conditions such as temperature, wind speed, photoperiod, CO_2 concentration, and spraying are set to enable the planned and stable culture of pesticide-free seedlings.

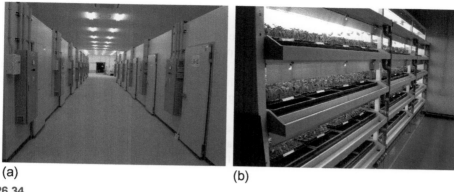

(a)   **(b)**

FIGURE 26.34

(a) Exterior of "Nae Terrace." (b) Interior of "Nae Terrace": five-tiered seedling raising shelves are placed and cell seedlings are stocked.

COSTS BY COMPONENTS AND SALES PRICE

The initial costs for the CTPF were: 40 million yen for buildings, 205 million yen for cultivation facilities, and 20 million yen for infrastructure (265 million yen in total). Operating costs per year are: 20 million yen for depreciation, 17 million yen for utilities, 25 million yen for personnel expenses, 20 million yen for materials costs, and 7 million yen for transportation costs (89 million yen in total).

The selling price of the "e seedling" series changes according to the seed price and cultivation standard: 20–68 for tomato seedlings, 118–166 for grafted tomato seedlings, and around 130 yen for cucumber or melon grafted seedlings. These prices are higher than for seedlings in a usual PF with solar power. However, because they have higher added value, sales to agricultural companies with large-scale horticulture facilities are increasing.

MARKETS

By region, 28% of the "e seedling" series are delivered to Shikoku where the head farm is located and 51% to eastern and northern Japan. Agriculture is especially popular to the north of the Kanto Region where demand for seedlings is higher than in western Japan. Seedlings in the CTPF are raised under the same environment throughout the year. For example, because the adherence of the first tomato cluster does not depend on the season or weather, the harvest time stabilizes at an early stage. Since various companies are entering the agricultural business and building large-scale facilities, the demand for the "e seedling" series is increasing year by year because productivity after planting is high and the use of less pesticide adds value.

The production of grafted seedlings of fruit vegetables reaches a peak from March to May and from July to August when producers plant seedlings. This means the low season is from November to February when seedling production decreases. We need to increase the operating rate for the whole farm throughout the year.

FUTURE PLANS

Berg Earth plans to introduce new CTPFs because of the rising demand for vegetable seedlings with less pesticide and because producers want stable production of high-value-added seedlings. Accordingly, Berg Fukushima Co., Ltd. was established as an affiliated company and we plan to increase seedling production in eastern Japan, especially for the Kanto Region, which is the biggest supplier of closed production seedlings. We are also studying the development of new products and the culture of high-value-added vegetables, utilizing the Nae Terrace during the low season.

Furthermore, we have started selling seedlings abroad to spread safe agricultural products. This includes seedling technology, technology for cultivating fruit vegetables, and building closed seedling facilities in Asia where food products are often not trusted because of environmental pollution and poor hygiene.

WEBSITE URL

http://www.bergearth.co.jp/

CHALLENGES FOR THE NEXT-GENERATION PFAL

27

Toyoki Kozai[1], Genhua Niu[2]

Japan Plant Factory Association, c/o Center for Environment, Health and Field Sciences, Chiba University, Kashiwa, Chiba, Japan[1] Texas AgriLife Research at El Paso, Texas A&M University, El Paso, Texas, USA[2]

INTRODUCTION

Plant factories with artificial lighting (PFAL) technology is multidisciplinary, and PFALs provide a unique environment for growing plants. Much of the PFAL technology differs from that used in horticulture and agriculture, although the basic science and technology are the same. Thus, new ideas for developing PFAL technology are needed.

This chapter describes examples of such ideas, including: upward lighting systems, use of green LEDs, breeding of vegetables, and medicinal plants suited to PFAL, seed propagation, a hydroponic culture system with restricted root mass, year-round production of ever-flowering berries, use of natural energy in PFAL, and data mining using big data. The ideas described below are meant to encourage the next generation of scientists to approach conventional horticultural technologies with new ideas and fresh methods.

LIGHTING SYSTEM

UPWARD LIGHTING

The photosynthetic rate of a whole plant is generally highest when the light energy is equally absorbed by the entire leaf surface assuming that the photosynthesis curve as affected by photosynthetic photon flux (PPF) is identical for all the leaves of the plant.

When solar or artificial light energy is given downward to a densely populated plant community, most light energy, especially red and blue light energy, is absorbed by the upper leaves while the PPF reaching the lower leaves is often nearly equal to or lower than the light compensation point and hence the net photosynthetic rate of the lower leaves is equal to or lower than 0. This is why the lower leaves of a densely populated plant community often become yellow-colored, shrink, or die. Under such low PPF conditions, lower leaves tend to exhibit characteristics of shade leaves with low light saturation intensity.

In order to mitigate this uneven vertical distribution of light energy over the upper and lower leaves, intercrop (sideward) lighting using LEDs is becoming popular for greenhouse crops such as tomato and rose plants in winter as well as on cloudy or rainy days in order to provide more light energy to the lower leaves. Sideward lighting was also tested on photoautotrophic plant micropropagation (Kozai et al., 1992).

Plant Factory. http://dx.doi.org/10.1016/B978-0-12-801775-3.00027-5

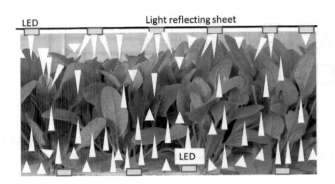

FIGURE 27.1

Diagram showing the concept of upward lighting using LEDs placed on the culture panels. In the PFAL, upward light which is not absorbed by the lower leaves is absorbed by the upper leaves or reflected back downward to the plant community by the light reflector placed above the plant canopy, to minimize the loss of upward light energy emitted from the LEDs.

In the PFAL, upward lighting using LEDs placed on the culture panels to provide more light energy to the lower leaves of leafy greens should be more beneficial than intercrop lighting (Yamori et al., unpublished). In the PFAL, upward light not absorbed by lower leaves is absorbed by upper leaves or reflected back downward to the plant community by the light reflector placed above the plant canopy, to minimize the loss of upward light energy emitted from the LEDs (Figure 27.1; Kozai, 2012). However, the upward lighting system might modify the plant morphology, physiology, and anatomy to some extent, possibly leading to new research subjects and applications. The development of an innovative lighting system specifically for PFALs is awaited.

USING GREEN LEDs

The photosynthetic rate of a leaf is directly related to PPF, not to the energy flux (Chapters 7–9). It is often said that, in a PFAL, red light (wavelength: 600–700 nm) is most effective for promoting photosynthesis of plants but some blue light (400–500 nm) is necessary mainly for photomorphological and/or phytochemical reasons for a given amount of electric energy consumption. This is because the light energy contained in one photon is inversely proportional to the wavelength. Therefore, the light energy per photon is 1.2 (=600/500) to 1.75 (=700/400) times higher in a blue photon than in a red photon.

It is well known that a thin layer of chlorophylls suspended in a liquid absorbs red and blue photons very well but absorbs little or no green photons. It is also known that a single green leaf absorbs about 50%, reflects about 20%, and transmits about 30% of green photons. These figures are for a thin layer of chlorophylls and a single leaf, respectively, not the whole plant community.

When plants are densely populated, most green photons transmitted by the upper leaves will be received by the lower leaves. In a PFAL, green photons reflected upward by the leaves are again reflected back downward by the light reflector to the plant community and are received by the leaves. As a result, most green photons emitted by the LEDs are received by the plant community vertically

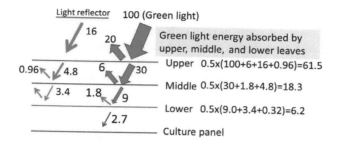

Light reflector   100 (Green light)

Green light energy absorbed by
upper, middle, and lower leaves

Upper 0.5x(100+6+16+0.96)=61.5

Middle 0.5x(30+1.8+4.8)=18.3

Lower 0.5x(9.0+3.4+0.32)=6.2

Culture panel

FIGURE 27.2

Schematic diagram showing the percentages of green light energy transmitted and reflected by upper, middle, and lower horizontal green leaves. Total green light energy absorbed by upper, middle, and lower leaves is also given. Red and blue light energy is absorbed by the upper leaf, and the middle and lower leaves receive almost no red and blue light energy. It is assumed that the percentages of reflection, transmission, and absorption of a horizontal leaf for green light energy are 20, 30, and 50%, respectively.

more evenly than red and green photons, because almost all red and blue light photons are absorbed by the upper leaves, and thus are neither transmitted to the lower leaves nor reflected upward (Figure 27.2; Kozai, 2012).

Owing to the recent widespread application of green LEDs to traffic signals, etc., the cost performance of green LEDs has improved considerably, and so green LEDs could be useful in the densely populated plant community in a PFAL. Recent studies indicate different effects of green light on plant growth and development (e.g., Kim et al., 2004; Johkan et al., 2012).

LAYOUTS OF LEDs

An LED lamp consists of many LED chips (around 0.3 mm square, 0.10–0.3 mm thick) and an LED chip consists of many LED elements. In addition to photosynthetic photon emission efficiency, spectral distribution and PPF at the plant canopy level, and the layout of LED chips in an LED lamp (a tube like a fluorescent tube or a plate of different sizes), the following are important considerations when designing LED lamps and lighting systems using LEDs: (1) optical/chemical properties and 3-D shape of the plastic material cover, (2) enclosed fluorescent substances and directivity of photon emission, and (3) heat release function. The array design of LEDs affects the distribution of 2-D light quality on a leaf on a scale of 0.1 mm. This microscale uneven distribution may cause edema of leaves (Mori and Takatsuji, 2013). Lighting system technology using LEDs for PFALs is still at the initial stage and much progress is expected to be made.

BREEDING AND SEED PROPAGATION
VEGETABLES SUITED TO PFAL

The most important genetic characteristic of field-grown and greenhouse-grown crops is their tolerance to diseases caused by pathogens, insects, and environmental stresses such as high/low temperatures, dry/wet soil/air, salt, toxic heavy metal accumulation, macro/micronutrient deficiency/excess, etc.

Growers will not want to commercially grow any crop that does not have tolerance to diseases or environmental stresses, even if the crop provides a high yield and high quality under narrowly specified conditions. Such high-yield, high-quality crops with little tolerance are suitable for growing in a PFAL, because plants grown in a PFAL need little tolerance to any disease caused by airborne pathogens or to environmental stresses (although it may be difficult to eliminate all pathogens in culture beds with nutrient solution).

Most breeders and breeding scientists may not imagine that they could breed a new cultivar showing a high growth rate and high quality just by removing the genes related to tolerance from a crop. However, such an "impractical" breeding approach has never been tried, and could lead to a new field of breeding and genetics. On the other hand, in the long history of crop breeding many breeding trials focusing on high-yield, high-quality crops have failed because of low tolerance to disease and/or environmental stress. Such breeding approaches could be used for production in PFALs.

SEED PROPAGATION AND BREEDING USING PFAL

Commercial seed propagation is currently conducted once a year mainly in highlands in developing countries where the climate is often favorable for seed propagation, labor cost is relatively low, and pollination with unexpected plants can be avoided. However, the climate and weather in such areas are becoming increasingly changeable, labor costs are rising, and social infrastructure is sometimes unstable.

Many annual plant species can flower and pollinate within a few months after flower bud development. The success of flower bud formation, flowering, and pollination is largely influenced by the environment, especially light (photoperiod, quality, and intensity) and temperature. Therefore, the seeds of many annual plant species could be obtained three or four times a year in a PFAL by controlling the environment optimally for seed propagation. This method could be used for seed propagation and breeding of annual herbaceous plants. Or, seeding could be conducted every day or every week and plants with seeds could be harvested daily or weekly starting around 3 months later.

MEDICINAL PLANTS

In recent years, more people are demanding natural medicines, cosmetics, supplements, food additives, etc. In general, existing efforts to breed medicinal plants are not as advanced as those for horticultural and agricultural crops. Genetic traits of a certain species or variety are often diverse in a plant community, with wide variations in concentrations of medicinal components (secondary metabolites). In fact, many commercially used medicinal plants are still wild and harvested in their natural habitat, although some are selected for better genetic traits by visual observation by growers and then cultivated. In other words, most medicinal plants are not bred based on modern science such as molecular biology. This is one of the reasons why vegetative propagation of an elite mother plant, which is costly, is predominant over its reproductive seed propagation.

Concentrations and yields of medicinal components are largely affected by the environment. It is also difficult to obtain high-quality medicinal plants with high yields every year. On the other hand, the commercial value of medicinal plants per unit dry weight is much higher than that of leaf vegetables. Thus, if we can breed a high-quality, high-yield medicinal plant, register it as a new cultivar, and grow it under a controlled environment, it may be a good business opportunity.

In the case of medicinal plants, it is essential to grow them at a fast growth rate until near the end of the cultivation cycle, and then provide environmental stresses several days before harvest to sharply enhance the production of medicinal components. Note that it is possible to use both the aerial and root parts of traditional root medicinal plants such as wasabi and *Angelica sinensis*, when grown in PFALs.

CULTIVATION
CULTURE SYSTEM WITH RESTRICTED ROOT MASS

The major function of plant roots is absorption of water, inorganic fertilizer, and dissolved oxygen. Other minor functions of roots include: (1) supporting the aerial part of the plant, (2) symbiosis with microorganisms, (3) barrier to infection by pathogens, (4) production of phytohormones such as cytokinin, and (5) leaching of organic acids to the root zone.

In a PFAL, electric energy is consumed to produce the roots. Thus, if the roots are not salable, the root mass must be minimized provided that doing so does not affect the growth of the aerial part.

In culture beds, plant roots require increasing amounts of water, fertilizer, and dissolved oxygen as the aerial parts of the plants grow. Thus, the flow rate and/or strength of nutrient solution must be increased as the plants grow. On the other hand, the flow rate of nutrient solution around the roots tends to decrease as the roots grow, due to the increase in resistance to the flow of nutrient solution in the culture beds. Considering this trade-off, a new culture bed and nutrient solution circulation system which does not slow down the flow rate needs to be developed to promote the growth rate of the aerial parts of plants with minimum root mass.

EVER-FLOWERING BERRY PRODUCTION IN PFALs

Year-round production of strawberry fruit using ever-flowering cultivars in a PFAL is attracting much research interest and could be a global business opportunity. This business has already been started by a Japanese company in 2013, and more groups have been conducting experimental research on this subject.

Propagation and transplant production of blueberry plants have already been commercialized in North America (Chapter 19, Section 4). Aung et al. (2014) confirmed that continuous blueberry (*Vaccinium* spp.) production is possible throughout the year, including the off-season, through a combination of an open field, plastic houses/glasshouses, and controlled environment rooms with artificial lighting. Blueberry plants could be flowered year round and thus year-round production of blueberry fruits made possible (Ogiwara and Arie, 2010). Likewise, plants of raspberry, blackberry, etc. may be flowered and their fruits may be produced year round in PFALs.

PFAL WITH SOLAR CELLS

Why are PFALs needed in addition to greenhouses with free solar energy to grow leafy vegetables? To answer this frequently asked question, we must consider three points: (1) there are large differences in light energy use efficiency (LUE) among PFAL, greenhouse, and open field, (2) the availability of solar energy largely depends on the time of day, weather, season, and location, and (3) the efficient use of free solar energy in a greenhouse is costly because other environmental factors need to be controlled. Answers to points 1 and 2 have already been described in Chapters 3 and 2, respectively. The answer to point 3, which is related to the answer to point 2, is given here.

In an open field, it is difficult to use solar energy efficiently to grow plants in winter because the temperature is often too low, photosynthetically active radiation (PAR) flux is also often too low, and/ or day length is much shorter than in summer. In order to use free solar energy efficiently during the cold winter in a greenhouse, the greenhouse must be heated, which is costly. To reduce the heating cost, a thermal screen needs to be installed, which incurs its own cost. To improve the yield and quality of crops in winter, the additional cost of installing a supplemental lighting system may be needed, while in the hot summer, costs are incurred for ventilation, shading and/or evaporative cooling. If the greenhouse is simply left empty in winter and/or summer, the annual depreciation cost is high. In short, the insufficiency, excess and uncontrollability of external solar energy are major factors limiting the economic profitability of plant production. This limitation is one of the reasons why leafy greens can be produced commercially in PFALs in very cold and very hot regions, even though PFALs consume much electricity.

Is it feasible to use solar energy for PFAL operation by generating electricity using photovoltaic (PV) cells? Intuitively, it seems unreasonable to use solar energy to generate electrical energy by PV cells, store the surplus in a battery, and then convert it to artificial light energy using LEDs to grow plants in the PFAL.

The percent conversions from solar radiation energy to chemical energy fixed in plants in a PFAL, a greenhouse heated in winter, and an open field can be estimated as follows. The percent conversion of PV cells from solar to electric energy is 20–40% at most, while that of LEDs from electric to PAR is 30–40% at most, and the percent conversion of PAR to chemical energy fixed in plants is 2–4%. Multiplying all these factors together, the light use efficiency or percent conversion from solar energy to chemical energy is 0.12% ($0.2 \times 0.3 \times 0.02 \times 100$) to 0.64% ($0.4 \times 0.4 \times 0.04 \times 100$) at most.

On the other hand, the percent PAR to solar radiation energy is no more than 50%, and the annual average percent conversion from PAR to chemical energy in usable parts of plants is roughly estimated to be in the order of 0.10% in an open field and 1.0–1.7% in a heated greenhouse (Kozai, 2013). Thus, the overall percent conversion from solar energy to chemical energy in usable plants is roughly 0.05% ($=0.5 \times 0.001$) in the open fields and 0.5–0.85% (0.5×0.01–0.5×0.017) in a greenhouse. So, these percentages in PFAL and the greenhouse are not significantly different. Thus, it is too early to conclude that the use of solar energy using PV cells for PFAL is not feasible at all even in arid and/or hot regions. The use of wind and geothermal energy in cold regions for PFAL with PV cells is also an interesting challenge. Note that the yield per unit land area is around 10 times and 100 times higher in a PFAL than in a greenhouse and open field, respectively.

Other challenging issues in PFAL R&D include: (1) the design and construction of an innovative building structure with facilities, (2) data mining using big data, (3) developing self-taught educational programs, (4) PFAL for farming on the moon, (5) international guidelines for the design, measurement, control, and management of PFAL, (6) developing an optimal environmental control system using minimum resources, (7) introducing molecular biology for breeding and solving physiological disorders of plants such as tip burn, (8) flexible automation of handling plants and supplies, and of cleaning the culture room, (9) creating new markets for PFAL-grown plants such as head-forming leafy vegetables harvested before heading, mini-carrots, mini-radishes, etc. with tasty leaves/shoots, and mineral-rich leafy greens, (10) service design of PFAL, (11) PFAL connected with other biological and/or abiotic systems, (12) improving concentration and composition of functional components by environmental control, and (13) breeding of dwarf plants suited to PFAL.

REFERENCES

Aung, T., Muramatsu, Y., Horiuchi, N., Che, J., Mochizuki, Y., Ogiwara, I., 2014. Plant growth and fruit quality of blueberry in a controlled room under artificial light. J. Jpn. Soc. Hortic. Sci. 83 (4), 273–281.

Johkan, M., Shoji, K., Goto, F., Hashida, S., Yoshida, Y., 2012. Effect of green light wavelength and intensity on photomorphogeneis and photosynthesis in *Lactuca sativa* L. Environ. Exp. Bot. 75, 128–133.

Kim, H.H., Golins, G.D., Wheeler, R.M., Sager, J., 2004. Green light supplementation for enhanced lettuce growth under red- and blue-light-emitting diodes. HortScience 39 (7), 1617–1622.

Kozai, T., 2012. Plant factory with artificial light (written in Japanese: Jinkoko-gata shokubutsu kojo). Ohmsya Pub, p. 227.

Kozai, T., 2013. Resource use efficiency of closed plant production system with artificial light: concept, estimation and application to plant factory. Proc. Jpn. Acad. Ser. B 89, 447–461.

Kozai, T., Kino, S., Jeong, B.R., Hayashi, M., Kinowaki, M., Ochiai, M., Mori, K., 1992. A sideward lighting system using diffusive optical fibers for production of vigorous micropropagated plantlets. Acta Hortic. 319, 237–242.

Mori, Y., Takatsuji, M., 2013. How to design and build plant factories with LEDs (written in Japanese: LED shokubutsu kojo no tachiagekata susumekata). Nikkan Kogyo Pub, p. 171.

Ogiwara, I., Arie, T., 2010. Development on year round production method of blueberry fruits in plant factory with artificial four seasons. In: Nikkei, B.P., Institute, Cleantech, Monozukuri, Nikkei (Eds.), Plant Factory Encyclopedia (Written in Japanese: Shokubutsu kojo taizen). Nikkei Business Publications, Inc., Tokyo, pp. 40–46

CONCLUSIONS: RESOURCE-SAVING AND RESOURCE-CONSUMING CHARACTERISTICS OF PFALs

28

Toyoki Kozai[1], Genhua Niu[2]

Japan Plant Factory Association, c/o Center for Environment, Health and Field Sciences, Chiba University, Kashiwa, Chiba, Japan[1] Texas AgriLife Research at El Paso, Texas A&M University, El Paso, Texas, USA[2]

This chapter summarizes the resource-saving and resource-consuming characteristics of plant factories with artificial lighting (PFAL) as conclusions, since resources are the most important factor that determines the opportunities and challenges of PFALs, which consist of a thermally well-insulated and highly airtight structure containing multiple tiers with lighting and hydroponic units, CO_2 supply unit, air conditioners (or heat pumps), and control unit.

ROLES OF PFALs IN URBAN AREAS

In addition to food production, PFALs as one form of agriculture play environmental, social, and cultural roles, as described in Chapter 2. PFALs are expected to become a key component in urban areas to help alleviate the following local and global issues:

(1) Climate change (global warming), pest insects, and natural disasters (strong wind, rain, flood, drought, etc.), and resultant vulnerabilities of crop yield and quality.
(2) Increasing urban population.
(3) Increasing demands for local production for local consumption to improve food security and safety.
(4) Increasing demands for fresh food, nutritious food, functional food for health care and/or higher quality of life.
(5) Increasing shortages and/or prices of water for irrigation and fossil fuels.
(6) Decreasing population involved in agriculture due to aging of farmers.
(7) Decreasing arable land area due to urbanization, soil contamination, and salt accumulation at soil surface, etc.
(8) Decreasing opportunities for growing plants or producing foods by people living in urban areas.

BENEFITS OF PRODUCING FRESH VEGETABLES USING PFALs IN URBAN AREAS

The benefits of producing fresh vegetables using PFALs in urban areas include:

1) Compared with field agriculture, over 100 times higher annual productivity per unit land area regardless of weather and season, by the use of multiple tiers, optimal environmental control, minimum loss of plants to pest insects, etc. A large land area is not required for producing fresh foods.
2) PFALs can be built in shaded and/or unused areas such as an empty space in a building, and even the basement, because neither soil nor sunlight is needed.
3) The distance between production and consumption sites is minimized, thus reducing fuel consumption, CO_2 emissions, and traffic for food transportation.
4) Citizens can enjoy fresh food at lower prices.
5) Losses of fresh produce and/or fuel for cooling the produce during transportation are minimized.
6) Safe, enjoyable, and light-work job opportunities under comfortable environments are created for a wide range of people including elderly and handicapped persons.
7) Waste water, vegetable waste, and CO_2 produced in urban areas can be reused, after proper processing, as essential resources (water, CO_2, and fertilizer) for growing plants in PFALs. Waste heat energy can be used for heating greenhouses in winter and for other purposes.
8) The PFAL can be combined with aquaculture, mushroom culture, and fermentation systems for mutual efficient use of their respective wastes as resources.
9) Surplus electricity supply at nighttime compared with daytime can be used.

RESOURCE-SAVING CHARACTERISTICS OF PFALs

Compared with a greenhouse, PFALs can greatly reduce the consumption of resources other than electricity for high-quality plant production. The percentages of resources saved and improvement in produce quality and yield in a PFAL compared with a greenhouse are as follows:

1) 100% reduction in pesticide application by keeping the culture room clean and free from pest insects.
2) 95% reduction in water consumption by recycling water vapor that is transpired by plant leaves and condensed into liquid water at the cooling panels of air conditioners.
3) 90% reduction in land area thanks to more than 10 times increase in annual productivity per unit land area.
4) 90% reduction in variation of yield and quality by accurate and optimal environmental control regardless of weather and season.
5) 50% reduction in fertilizer consumption by recycling with little drainage of circulating nutrient solution.
6) 50–70% reduction in labor hours per unit yield by optimal layout of culture systems, organized operations, automation, and fewer physiological disorders of plants.
7) 10–30% reduction in plant residue by less loss caused by physical, chemical, and biological damage of produce.

The experimental evidence and theoretical explanations for these percentages are given in relevant chapters of this book. However, more rigid experimental evidence and deeper theoretical considerations are needed for more widespread use of PFALs.

POSSIBLE REDUCTIONS IN ELECTRICITY CONSUMPTION AND INITIAL INVESTMENT

The key tasks for PFALs are: (1) to reduce electricity consumption for lighting and air conditioning, (2) to reduce the initial resources and financial investment, and (3) to generate electrical energy from solar power, wind power, biomass energy, and geothermal energy.

ELECTRICITY CONSUMPTION

At present, roughly 10 kWh (36 MJ) of electrical energy are consumed to produce 1 kg of salable fresh lettuce heads. This electricity consumption can be reduced by more than half (to less than 5 kWh/kg fresh weight) in the near future by:

1) Using LEDs with optimal light quality, lighting cycle, and lighting schedule,
2) Using a well-designed lighting system with reflectors to maximize the ratio of light energy received by plant leaves to the light energy emitted from light sources and providing uniform lighting to all leaf surfaces,
3) Optimally controlling temperature, CO_2 concentration, and nutrient supply to maximize plant photosynthesis, growth, and functional components for a given consumption of electrical energy,
4) Optimizing the transplanting schedule or automatic spacing to maximize the planting density on the culture panels, and
5) Maximizing the percent salable parts of plants (minimizing plant residue).

These statements are based on converted electrical energy. The economic value per kilogram of produce or per kilowatt-hour (joule) of electrical energy can be further improved by environmental control.

INITIAL RESOURCE INVESTMENT

Cement and iron are the two major materials whose consumption should be greatly reduced in PFAL construction, because a large amount of energy is required and a large amount of CO_2 is emitted during their production and the construction of a PFAL. Wood may be a possible alternative material for parts of the PFAL.

INCREASING THE PRODUCTIVITY AND QUALITY

Annual productivity of leafy vegetables per unit land area with existing PFALs is over 100 times that in open fields. Next-generation PFALs will offer 200 times greater annual productivity, as well as higher economic value thanks to higher quality.

DEALING WITH POWER CUTS

A power cut for about 10 h in a PFAL does not significantly affect plant growth. In general, plants in a PFAL receive 15–16 h of light daily. If a light period is only 10 h due to a power cut, the following light period can be extended to 20 h or so.

Also, since PFALs are thermally well insulated and highly airtight, the air temperature in the culture room changes very slowly with the air temperature outside if all the lamps and air conditioners are off. Therefore, plant growth is hardly affected for a half day or so, even if the outside temperature is extremely high or low. Nevertheless, a back-up system should be installed in the PFAL to operate the pump that circulates the nutrient solution.

CHALLENGES

Challenging issues regarding the efficiency of resource use in a PFAL, as described in Chapter 27, include:

(1) Improving the lighting system: (1) Upward lighting, (2) Use of green LEDs, and (3) Layout of LED chips in an LED lamp.
(2) Breeding and seed propagation: (1) Selection and breeding of vegetables suited to PFAL, (2) Seed propagation using the PFAL, and (3) Breeding using the PFAL.
(3) Selection, breeding, and environmental control for efficient secondary metabolite production of medicinal and/or functional plants.
(4) Production of pharmaceuticals by transient expression methods.
(5) Development of a new culture system with restricted root mass to increase the weight ratio of salable part to total plant weight.
(6) Ever-flowering strawberry, blueberry, and other berry production in PFALs.
(7) PFAL with solar cells with or without batteries networked with other natural energy systems.
(8) Production of raw materials for processed foods such as paste, sauce, pickled vegetables, juice, cosmetics, aromas, etc.
(9) Integrating PFALs into ecological urban planning and management.
(10) Micro-, mini-, and small PFALs for education, self-learning, and as a hobby.
(11) Biochemical and molecular biological analyses of the environmental and genetic effects on functional components and physiological phenomena, and their application for breeding and environmental control.
(12) Data mining using big data obtained from the PFAL system.
(13) Introducing information and communication technology into the PFAL.

There may be many other issues to tackle for conserving resources in a PFAL.

In order to improve the resilience and sustainability of society, more diverse plant production systems including open fields and greenhouses are required. The PFAL is one such plant production system.

This book has emphasized the usefulness of PFALs in urban areas. However, PFALs can also be useful in rural and local areas where electricity is supplied from renewable energy. PFALs can also be used for various purposes where plants cannot be grown due to conditions such as harsh weather, shortage of water, and soil degradation, as long as electricity is available.

Finally, we must distinguish the technologies and costs of current PFALs from those of PFALs 5, 10, and 20 years later. Research, technology, business, and applications of PFALs have only just started and will progress rapidly, offering many opportunities as well as challenges.

Index

Note: Page numbers followed by *f* indicate figures and *t* indicate tables.

A

Abiotic environmental factors, 153–162
Absolute humidity (AH), 132
Airborne microorganisms, 288–289
Air current speed, 83, 138, 157–158
Algae control, 285–286
Aquaponics, 16, 103, 103*f*
Aseptic culture room, 279–280
Automated technology, plant factories, 313–319
Automatic seeding machine, 226, 228*f*

B

Beijing Kingpeng International Hi-Tech Corporation, 55
Beneficial elements, 169
Berg Earth Co., Ltd., 382–386
Biological factor management, 285–292
Blueberry, CTPS, 257–260
Breeding and seed propagation, next-generation PFAL, 389–391
Building integrated agriculture, 107–110

C

Ca^{2+}, 173–175
Cal-Com Bio Corp., 352
Candidate crops, PMPs, 194–195
Canopy photosynthesis, 148
Carbon fixation and metabolism, 143–144
Carotenoids, 141–143
China, PFAL, 53–58
Chinese cabbage, 23, 24*f*
Chlorine, 286
Chlorophylls, 141–143
Chung-hwa Plant Factory Association (CPFA), 41–42
Citizen-centered horticulture science, 102
Closed hydroponic systems, 214–216
Closed plant production system (CPPS), 72*f*
Closed transplant production system (CTPS)
  blueberry, 257–260
  ecophysiology, 243–250
  electricity costs, 240–242
  light source, air conditioners, and small fans, 239–240
  main components, 238
  nutrient solution supply, 242
  photosynthetic characteristics, light environment, 251–257
  strawberry transplants, 260–266
CO_2 concentration, 137–138, 156–157, 188
Coefficient of performance (COP), 79, 87–88, 304–306
Colony forming units (CFUs), 288

Color rendering index (CRI), 104
Commercial PFALs
  Berg Earth Co., Ltd., 382–386
  Internationally Local & Company, 368–374
  Japan Dome House Co., Ltd., 364–368
  Mirai Co., Ltd., 362–364
  Sci Tech Farm Co., Ltd., 375–382
  Spread Co., Ltd., 355–362
  Taiwan, 352–355
Commercial UV sterilizer, 220, 220*f*
Continuous row farming, 106–107
Controlled environment agriculture (CEA), 51
Coriander plants, 181–183, 182*f*
CO_2 use efficiency (CUE), 75–76
C_3 plants, 144
C_4 plants, 144
Crassulacean acid metabolism (CAM), 144
Crisp head lettuce plant, 23, 24*f*
CTPS. *See* Closed transplant production system (CTPS)
Cultivation panel washer, 318–319
Culture panel cleaning machine, 209*t*, 210*f*
Culture room, PFAL, 208–211

D

Daily light integral (DLI), 117–118, 154–155
Deep flow technique (DFT), 152–153, 213, 214*f*
Development of major components for IT-LED based plant factories, 51
Dewpoint temperature, 135
Dry bulb temperature, 134

E

Ecophysiology, 243–250
Electrical conductivity (EC)-based hydroponic system, 216–217
Electrical energy use efficiency, 78–79
  Electricity consumption and reduction, 303–308
Electron transport and bioenergetics, 143
Energy balance, 129–130
Energy use reductions, 109–110
Environmental factors, medicinal components, 188–191
Environmental-Symbiotic City, 101
Environmental testing, 288–289
Environment conditions, PMPs, 197–200
Essential elements, 165–168
Europe, PFAL, 61–66
Ever-flowering berry production, 391

F

Floor plan, PFAL, 205–206
Fluorescent lamps (FLs), 3, 126–127
Flux type LEDs, 121, 121*f*
Food deserts, 9
Food safety, 212
Forced ventilation system, 276–277

G

Genetically modified (GM) plant factories construction, 195–197
Global warming potential (GWP), 324–325
Glonacal Green Technology Corp., 352
Glycyrrhizin, 189–190
GM rice plants, 197, 197*f*
GM strawberry plants, 196, 196*f*
Good food revolution, 9
Green furniture m-PFAL, 92–94, 93*f*
Greenhouse with environmental control system (GECS), 331
Green LEDs, 388–389
Green roofs. *See* Rooftop plant production (RPP)
Growing degree days concept, 153–154
Growth respiration, 145

H

Hazard analysis and critical control point (HACCP), 212
Heat conduction and convection, 130–131
Herbs, 27, 28*f*, 181–183
Heterotrophs, 11–12
High-intensity discharge (HID) lamps, 118
Household m-PFAL, 92–94, 93*f*
Humidity, 132, 156
HydroGarden, 65–66
Hydrogen peroxide, 285–286
Hydroponic butter head lettuce, 173, 174*f*
Hydroponic greenhouse growing, 107
Hydroponic systems, 213–220, 214*f*
Hypericum perforatum plants, 188

I

Incandescent lamp, 119
Indoor vertical farms, 3
Inorganic fertilizer use efficiency (FUE$_I$), 79
Intensive business forums, 337–343, 337*t*
Internationally Local & Company (InLo&Co.), 368–374
International Vertical Farming and Urban Agriculture Conference (VFUA), 66
Interplant lighting and upward lighting, 81–82
Inventory data collection/impact assessment, 326–327
In vitro medicinal plants, 278–280

Ion-selective sensors, 43
Ion-specific nutrient management, 217–220
Ion uptake rate, 87
ISO22000, 212

J

Japan Dome House Co., Ltd., 364–368
Japan Greenhouse Horticulture Association (JGHA), 331
Japan, PFAL, 35–39
Japan Plant Factory Association (JPFA), 39, 331
JPFA's business workshops, 344–348

K

Korea, PFAL, 50–53

L

Lamp type LED, 121, 121–122*f*
Latent heat, 131
Layout, PFAL, 205–211
LCA. *See* Life cycle assessment (LCA)
Leaf area index (LAI) and light penetration, 147–148
Leafy vegetables, 27, 28*f*, 180–181
LED4CROPS facility, 65
LED Farm® system packages, 381
Lee-Pin Corp., 353
Life cycle assessment (LCA), 321–327
Life cycle impact assessment (LCIA), 324–325
Life cycle inventory (LCI) analysis, 323–324
Light
    discharge light emission, 119
    electroluminescence, 119
    fluorescent lamps, 126–127
    LEDs, 120–125
    measurement, 117–118
    physical properties, 115–116
Light-emitting diodes (LEDs), 3, 83, 120–126, 388–389
Light energy use efficiencies, 77–78, 81–84
Lighting system (LS), 298, 300–303, 387–389
Light quality control, 180–183
Light quality effects, 243–245, 249–250
Low-carbon, green growth campaign, 51
Low-nitrate vegetables, 178–179
Low-potassium spinach, 177
Low-potassium vegetables, 177–178

M

Macronutrients, 165, 166*t*
Maintenance respiration, 145
Medicinal components, 187–191
Medicinal *Dendrobium officinale*, 254–257

Medicinal plants, 37, 38*f*, 390–391
Melatonin, 190–191
Micro- and mini-PFALs (m-PFALs)
  applications, 92–98
  biosystems, 103
  challenges, 102
  characteristics and types, 91–92
  design concept, 98
  ecosystem, 102
  Internet, 99–101
  light source and lighting system design, 104
  plant growth visualization, 102
  productivity and benefits, 102
  virtual, 102
Microbiological testing, 287
Micronutrients, 165, 167*t*
Microorganism management, 287–292
Miraculin, 198
Mirai Co., Ltd., 362–364
Moist air properties, 133–136
m-PFALs. *See* Micro- and mini-PFALs (m-PFALs)

N

National Agriculture and Food Research Organization
  (NARO), 39
Natural ventilation method, 272–276
Net photosynthetic rate, 85
Next-generation PFAL
  breeding and seed propagation, 389–391
  cultivation, 391
  lighting system, 387–389
  solar cells, 391–393
Nitrate reductase (NR) activity, 178
Nitrosoamines, 178
Nondestructive plant growth measurement system, 43–47
North America, PFAL, 59–61
Number of air exchanges per hour, 139
Nutrient and root zone, 158–159
Nutrient film technique (NFT), 152–153, 213, 214*f*
Nutrient management systems, 214–217
Nutrient solution, 170
Nutrient solution supply, 242
Nutrient uptake and movement, 169–170

O

Open hydroponic systems, 214–216
Operation room, PFAL, 207
Organic fertilizers, 15–16
Ozonated water, 286
Ozone (O_3) systems, 220

P

PAM. *See* Photoautotrophic micropropagation (PAM)
Perillaldehyde, 181–183, 182*f*
Perilla plants, 181–183, 182*f*
PFAL. *See* Plant factory with artificial lighting (PFAL)
PFAL-design and management (D&M) system
  electricity consumption and reduction, 303–308
  lighting system, 298, 300–303
  logical structure of equations, 300
  plant growth measurement, analysis, and control, 309–311
  structure and function, 295–297
  structure of design system, 297, 297*f*
  structure of software, 298–299
Photoautotrophic micropropagation (PAM)
  advantages and disadvantages, 272
  development, 271–272
  forced ventilation system, 276–277
  natural ventilation method, 272–276
  photosynthesis, 271
  secondary metabolite production, 278–280
Photoautotrophs, 11–12
Photonics Industry & Technology Development Association
  (PIDA), 41–42
Photoperiodism, 159–160
Photorespiration, 146–147
Photosynthesis
  carbon fixation and metabolism, 143–144
  C_3, C_4 and CAM, 144
  electron transport and bioenergetics, 143
  estimation of rates, 85–87
  LAI and light penetration, 147–148
  light absorption, 141–143
  single leaf and canopy, 148
Photosynthetic photon flux density (PPFD), 118
Physical environmental factors and properties
  air current speed, 138
  CO_2 concentration, 137–138
  energy balance, 129–130
  heat conduction and convection, 130–131
  latent heat, 131
  moist air properties, 133–136
  number of air exchanges per hour, 139
  radiation, 130
  temperature measurement, 131–132
  water vapor, 132–133
Physiological response
  dark respiration, 308
  net photosynthesis, 308
  translocation, 162
  transpiration, 160–162
  water uptake, 308
Phytomass, 8–9

Plan-do-check-act (PDCA) cycle, 204
Plan for promotion of agri-food and ICT convergence, 50
Plant Environment Designing Program, 333–337
Plant factories
    Center of Chiba University, 332–333, 332t
    greenhouse with environmental control system, 331
    intensive business forums, 337–343, 337t
    Japan Greenhouse Horticulture Association, 331
    JPFA's business workshops, 344–348
    number of visitors, 332–333, 333t
    Plant Environment Designing Program, 333–337
Plant factory of Zhejiang University, 55–57
Plant factory with artificial lighting (PFAL)
    advantages, 4
    application, 87
    China, 53–58
    coefficient of performance of heat pump, 87–88
    CO_2 use efficiency, 75–76
    definition, 17–18, 71–72
    description, 3
    disadvantages, 4–5
    earth, space farms, autonomous cities, 31
    electrical energy use efficiency, 78–79
    electricity consumption, 80
    electricity cost, 23–24, 80
    Europe, 61–66
    factors, 30–31
    FLs, 3
    grafted and nongrafted seedlings, 4
    indoor vertical farms, 3
    initial cost, 22
    inorganic fertilizer use efficiency, 79
    ion uptake rate, 87
    Japan, 35–39
    Korea, 50–53
    labor cost, 25
    land price, 27
    layout, 205–211
    leafy greens, 4, 27–29
    light energy use efficiencies, 77–78, 81–84
    motion economy, 203–204
    net photosynthetic rate, 85
    North America, 59–61
    plan-do-check-act cycle, 204
    plant production process, 204–205
    plants suited and unsuited, 19
    principal components, 71–72
    production cost, 22–23
    profit, 26–27
    resource use efficiency, 72, 79–80
    sanitation control, 211–212
    scientific benefits, 18–19
        social needs and interest, 19–21
        sustainable, 29–30
        Taiwan, 39–49
        taste and nutrition, 25–26
        transpiration rate, 86
        water consumption, 27
        water uptake rate, 86–87
        water use efficiency, 73–75
Plant growth and development
    environmental factors, 153–159
    photoperiodism, 159–160
    shoot and root, 151–153
    translocation, 162
    transpiration, 160–162
Plant growth measurement, analysis, and control, 309–311
PlantLab, 63–64
Plant-made pharmaceuticals (PMPs)
    advantages, 193
    aerial environmental factors, 200
    candidate crops, 194–195
    environment conditions, 197–200
    functional protein-encoding genes, 193, 194t
    GM plant factories construction, 195–197
Plant-mushroom culture system, 103, 103f
Plant production process, PFAL, 204–205
PMPs. *See* Plant-made pharmaceuticals (PMPs)
PPFD distribution, 300–302

Q

Quality testing, bacteria and fungi, 290
Quantitative management of nutrients, 171–172

R

Radiation, 130
Raised-bed production, 106
Red leaf lettuce plants, 181
Relative humidity (RH), 132
Resource saving and resource consuming characteristics,
    PFALs, 395–399
Resource use efficiency (RUE), 17, 72, 79–80
Respiration, 144–146
Ribulose 1,5-bisphosphate carboxylase/oxygenase
    (RuBisCO), 144
Rice plants, 198–200
Rooftop Agriculture, 105–106
Rooftop Farming, 105–106
Rooftop Gardening, 105–106
Rooftop greenhouses (RTG), 64
Rooftop plant production (RPP), 105–110
Root-zone environmental factors, 214
RPP. *See* Rooftop plant production (RPP)

S

Sanitation control, PFAL, 211–212
Sci Tech Farm Co., Ltd., 375–382
Seeding device, 314–315
Seedling production, PFAL
    preparation, 223–226
    seeding, 226–229
    transplanting, 229–235
Seedling selection robot system, 315–317
Shuttle-type transfer robot, 317–318
Single leaf and canopy, 148
Sink–source relationship, 162
Small root vegetables and medicinal plants, 27–28, 28f
Soil fertility, 15–16
Soilless culture systems, 165–172
Solar cells, 391–393
Solution pH and nutrient uptake, 171
Spectral quality and UV radiation, 189–191
Spread Co., Ltd., 355–362
Sterilization system, 220
Stormwater management, 107–108
Strawberry plants, 197–198
Strawberry transplants, CTPS, 260–266
Surface mount device (SMD) type LED, 121, 121f
Sustainable PFAL, 29–31
Sweet potato transplants, 37, 38f

T

Tabletop m-PFAL, 92–94, 92f
Taiwan, PFAL, 39–49, 352–355
Taiwan Plant Factory Industrial Development Association
    (TPFIDA), 41–42
Temperature measurement, 131–132
Temperature stress, 188–189
Thermocouple, 131–132
Ting-mao Bio-Technology Corp., 352–353
Tipburn, 173–175, 174f
Tomato plants, 198
Transgenic lettuce circadian rhythm germination
    period, 315, 316f

Translocation, 162
Transpiration, 86, 131, 160–162
Transplantation of seedling, 229–235

U

Upward lighting, 387–388
Urban agriculture, 61–62
Urban fresh food production
    biological systems, 13–15
    interrelated global issues, 7–9
    key indices, 17
    organic fertilizers and microorganisms, 15–16
    PFAL, 17–31
    photoautotrophs and heterotrophs, 11–12
    resource inflow and waste outflow, 9–10
    stability and controllability of environment, 16–17
    waste recycling, 12–13
Urban Plant Factory Project, 333

V

Vapor pressure deficit (VPD), 132–133, 156
Vegetable transplant production, 251–254
Vertical farming, 9
VertiCrop technology, 65, 65f
Virtual m-PFAL, 102

W

Waste heat, 109–110, 110f
Waste recycle/reuse-bioproduction closed system, 12–13, 13f
Water diffusion process, 160–161
Water stress, 189
Water uptake rate, 86–87
Water use efficiency (WUE), 73–75
Water vapor, 132–133
Wet bulb temperature, 134–135
Wireless sensor network (WSN), 43

Y

Yasai-Lab Corp., 353–355

Edwards Brothers Malloy
Ann Arbor MI. USA
July 12, 2016